The Subnuclear Series • Volume 48

Proceedings of the International School of Subnuclear Physics

WHAT IS KNOWN AND UNEXPECTED AT LHC

THE SUBNUCLEAR SERIES*

Series Editor: ANTONINO ZICHICHI, *European Physical Society, Geneva, Switzerland*

Volume 1 was published by W. A. Benjamin, Inc., New York; 2–8 and 11–12 by Academic Press, New York and London; 9–10 by Editrice Compositori, Bologna; 13–29 by Plenum Press, New York and London; 30–48 by World Scientific, Singapore.

*For the complete list of titles in this series, please go to

http://www.worldscientific.com/series/ss

The Subnuclear Series • Volume 48

Proceedings of the International School of Subnuclear Physics

WHAT IS KNOWN AND UNEXPECTED AT LHC

Edited by

Antonino Zichichi
European Physical Society
Geneva, Switzerland

 **World Scientific**

NEW JERSEY · LONDON · SINGAPORE · BEIJING · SHANGHAI · HONG KONG · TAIPEI · CHENNAI

Published by

World Scientific Publishing Co. Pte. Ltd.

5 Toh Tuck Link, Singapore 596224

USA office: 27 Warren Street, Suite 401-402, Hackensack, NJ 07601

UK office: 57 Shelton Street, Covent Garden, London WC2H 9HE

The Subnuclear Series — Vol. 48
WHAT IS KNOWN AND UNEXPECTED AT LHC
Proceedings of the Internaitonal School of Subnuclear Physics 2010

ISBN 978-981-4522-47-2

In-house Editor: Rhaimie Wahap

Printed in Singapore by World Scientific Printers.

PREFACE

In August/September 2010, a group of 66 physicists from 42 laboratories in 19 countries met in Erice to participate in the 48th Course of the International School of Subnuclear Physics. The countries represented by the participants were: Canada, China, France, Georgia, Germany, Hungary, India, Iran, Italy, Japan, Mexico, Netherlands, Poland, Russia, Switzerland, Taiwan, Ukraine, United Kingdom and the United States of America.

The School was sponsored by the Academies of Sciences of Estonia, Georgia, Lithuania, Russia and Ukraine; the Chinese Academy of Sciences; the Commission of the European Communities; the European Physical Society; the Italian Ministry of Education, University and Scientific Research; the Sicilian Regional Government; the Weizmann Institute of Science; the World Federation of Scientists and the World Laboratory.

The purpose of the School was to focus the attention on all the sectors of Subnuclear Physics. The most recent theoretical developments were discussed, particularly in the field of Quantum Gravity and in Cosmology, and experimental highlights from the most relevant sources of new data were presented.

An original feature of the School, introduced in 1996, is a series of special sessions devoted to "New Talents", with the aim of encouraging and promoting young physicists to achieve recognition at an international level. This is a serious problem in Experimental Physics, where collaborations count several hundreds of participants and it is almost impossible for young fellows to be known. Even if with much less emphasis, the problem exists also in Theoretical Physics. So it was decided to offer 29 young fellows the possibility to give an open presentation of the results of their studies, followed by a discussion. Among them 6 best "New Talents" have been selected for publication in this volume, and 4 were given an award. In particular, one award for an original presentation in theoretical physics, one for an original presentation in experimental physics, one for an original presentation in detector physics and one for an original presentation in mathematical physics were assigned.

During the organization and the running of the Course, I enjoyed the collaboration of my colleague and friend, Gerardus 't Hooft, who shared with me the Directorship of the School. I would like to thank him, together with the group of invited scientists and all the people who have contributed to the success of the Course.

I hope the reader will enjoy the book as much as the students attending the lectures and the discussion sessions, which, as every year, have been the focal point of this School activity. Thanks to the work of the Scientific Secretaries, the discussions have been reproduced as faithfully as possible. At various stages of my work I have enjoyed the collaboration of many friends whose contributions have been extremely important for the School and are highly appreciated. I thank them most warmly. A final acknowledgement to all those in Erice, Bologna and Geneva, who have helped me in so many occasions and to whom I feel very indebted.

Antonino Zichichi

Geneva, October 2010

CONTENTS

SPECIAL SESSIONS FOR NEW TALENTS

CLOSING CEREMONY

Harmony of Scattering Amplitudes:
From QCD to $N = 8$ Supergravity

Z. Bern[1] and H. Ita

Department of Physics and Astronomy
UCLA, Los Angeles, CA
90095-1547, USA

Abstract

In these lectures we discuss some remarkable properties of scattering amplitudes in gravity and gauge theories. We start by summarizing the unitarity method and explain how it reveals surprising structures in loop amplitudes. We describe various recent state-of-the-art applications of the unitarity method to QCD and maximally supersymmetric gauge and gravity theories. We also review a recently discovered duality between color and kinematics in gauge theories, and an associated double-copy property of gravity amplitudes. These properties have recently been conjectured to hold to all loop orders.

1 Overview

Recent years have taught us that there is remarkable harmony and simplicity in gauge and gravity scattering amplitudes. A key idea behind this understanding has been that new amplitudes can be fully constructed using only on-shell information [1, 2]. This avoids unphysical gauge-dependent quantities in intermediate steps, which can obscure important properties of the amplitudes. Studies have uncovered a number of novel unexpected structures in gauge and gravity amplitudes, revealing that they are much simpler than anyone had anticipated. Planar amplitude in maximally supersymmetric gauge theories have an even richer structure. A rather striking example is Witten's observation that amplitudes in massless gauge theories are associated to curves in twistors space [3]. Other examples are the iterative structure leading to an all-loop-order solution of the simplest of the planar amplitudes of $\mathcal{N} = 4$ super-Yang-Mills theory, valid at both weak and strong coupling [4, 5], a new symmetry explaining this structure [6, 7] and a Grassmannian structure [8]. In these lectures we will not describe these interesting results, which are well discussed in other recent reviews [9].

Instead, here we begin with a brief summary of a basic general-purpose approach for carrying out loop-level calculations of scattering amplitudes: the unitarity method. To illustrate its utility we describe a few applications. This includes state-of-the-art predictions of LHC physics (for example, see refs. [10, 11, 12]). In particular, the first next-to-leading-order QCD calculation with five final state objects (including jets) [12] was done this way. Other examples which summarize include studies of supersymmetric

---

[1]Presenter

1

gauge-theory amplitudes [4, 13, 14, 15], and their ultraviolet properties in $D > 4$ dimensions [16, 17, 18]. It has been used in calculations of $\mathcal{N} = 8$ supergravity amplitudes, demonstrating that they are better behaved in the ultraviolet than had previously been appreciated [19, 17, 20]. (For other recent discussions of this issue see refs. [21, 22, 23].) More generally, the on-shell methods have allowed ever more complex calculations to be carried out. Such calculations in turn reveal an ever richer structure in the scattering amplitudes which can then be used to find improvements in the calculational methods.

An important aspect of the harmony between gauge and gravity theories is that many of the same basic techniques used to carry out calculations for one theory can then be applied to the other theory. But as we shall discuss the relationship goes much deeper, with the detailed multiloop structure of perturbative gauge and gravity theories intimately intertwined.

Gravity has many similarities with gauge theories. Both are based on the idea of local symmetries and therefore share a number of formal properties. Nevertheless, their structural and dynamical behavior appears to be quite different. Non-abelian gauge theories have a rather different dynamical behavior than gravity. Quantum chromodynamics, for example, exhibits confinement of particles while gravity does not. Moreover, consistent quantum gauge theories have existed for more than a half century, but as yet no satisfactory point-like quantum field theory of gravity has been constructed. The structures of the Lagrangians of the two theories are also rather different: the non-abelian Yang-Mills Lagrangian contains only up to four-point interactions while the Einstein-Hilbert Lagrangian contains infinitely many interactions.

Nevertheless, recent developments [25, 26] show that in a precise diagrammatic sense, perturbative gravity is a double copy of gauge theories,

$$\text{gravity} \sim (\text{gauge theory}) \times (\text{gauge theory}). \tag{1}$$

An early version of this property was understood at tree level more than 25 years ago using string theory, via the Kawai-Lewellen-Tye (KLT) relations [27]. These relations hold as well in field theory, as the low-energy limit of string theory. In this limit, the KLT relations for four-, and five-point amplitudes are

$$
\begin{aligned}
M_4^{\text{tree}}(1,2,3,4) &= -i s_{12} A_4^{\text{tree}}(1,2,3,4)\widetilde{A}_4^{\text{tree}}(1,2,4,3), \\
M_5^{\text{tree}}(1,2,3,4,5) &= i s_{12} s_{34} A_5^{\text{tree}}(1,2,3,4,5)\widetilde{A}_5^{\text{tree}}(2,1,4,3,5) + \\
&\quad + i s_{13} s_{24} A_5^{\text{tree}}(1,3,2,4,5)\,\widetilde{A}_5^{\text{tree}}(3,1,4,2,5).
\end{aligned}
\tag{2}
$$

Here the M_n^{tree}'s are amplitudes in a gravity theory stripped of couplings, the A_n^{tree}'s and $\widetilde{A}_n^{\text{tree}}$'s are two distinct color-ordered gauge-theory amplitudes and the Mandelstam invariants are $s_{ij} \equiv (k_i + k_j)^2$, with k_i being the outgoing momentum of leg i. The color-ordered gauge-theory amplitudes [28] correspond to the coefficient of color traces with a given ordering of matrices. The gravity states are direct products of gauge-theory states for each external leg. (Explicit formulæ for n-point amplitudes may be found in refs. [29].)

The KLT relations show that gravity and gauge theory are intertwined, but as we shall explain, the heuristic relation (1) is now understood much more simply. Today

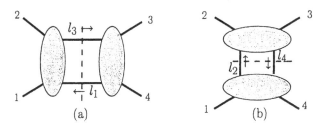

Figure 1: The two-particle cuts of a one-loop four-point amplitude. The exposed intermediate states are placed on shell.

we have a simple and direct manifestation of the relation (1), conjectured to extend to all loop orders: the numerators of gravity diagrams are two copies of appropriately arranged gauge-theory diagram numerators. Besides offering a potent tool for carrying out loop calculations in gravity, this remarkable harmony between gravity and gauge theory strongly suggests that they belong together in a unified quantum theory, perhaps along the lines of string theory.

The lecture is organized as follows: In section 2 we give a brief overview of the unitarity method. Then in section 3, we summarize a few nontrivial applications to perturbative QCD, super-Yang-Mills and supergravity. In section 4 we then show a remarkable harmony between gauge and gravity theories, focusing on the double-copy property of gravity scattering amplitudes in terms of gauge-theory ones. Then in section 5 we explain how to arrange the Lagrangians so that they exhibit the double-copy property. Finally in section 6, we give our conclusions and outlook for the future.

2 Unitarity Method

The unitarity method was originally developed in the context of one-loop supersymmetric amplitudes [1], but with further refinements [34, 35, 36, 14, 37, 17, 18, 7], it offers a powerful formalism for any massless theory at any loop order, including non-planar contributions. This method has been reviewed numerous times [38, 39, 40], so here we give only a brief summary.

Unitarity has been a basic principle in quantum field theory since its inception. For a description of unitarity during the 1960's, see ref. [41]. However, a variety of difficulties prevented its widespread use as a means of constructing amplitudes, especially after the rise of gauge theories in the 1970's. These difficulties include nonconvergence of dispersion relations and the inapplicability to massless particles. It was also unclear how one could fully reconstruct loop amplitudes beyond four points from their unitarity cuts. The unitarity method overcomes these basic difficulties, allowing for the complete construction of loop amplitudes at any loop order. It does so by avoiding dispersion relations, and instead using the existence of an underlying covariant Feynman diagram representations to fully reconstruct amplitudes. By construction the obtained Feynman-like integrands have the correct unitarity cuts in all channels.

4

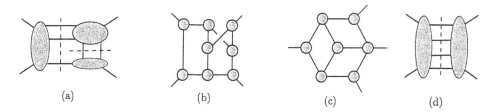

(a)   (b)   (c)   (d)

Figure 2: Examples of cuts used to determine the a three-loop four-point amplitude. The cut (a) decomposes the amplitude into a product of tree amplitudes. The cut (b) is a maximal cut and require the loop momenta to be complex momenta. Diagram (c) shows a near-maximal cut and diagram (d) is a four-particle cut. In this figure all exposed intermediate lines are placed on shell.

Over the years there have been a number of important refinements to the unitarity method. Generalized unitarity [41] (where multiple internal lines are placed on shell, subdividing a loop amplitude into more than two pieces) was successfully applied in ref. [35] as a means for greatly simplifying loop calculations. An important more recent development is the use of complex momenta [32] by Britto, Cachazo and Feng [36], leading to the realization that at one-loop in four dimensions, quadruple cuts directly determine the coefficients of all box integrals by freezing the loop integration. Powerful new methods for dealing with triangle and bubble integrals at one-loop, as well as rational terms have also been developed [37, 42]. (These have been described in other recent reviews [43].) At higher loops, efficient means of constructing the integrands of amplitudes, including non-planar contributions, have also been devised [14, 17, 18, 44].

A key feature of the unitarity method is that it can construct amplitudes at any loop order directly from on-shell tree amplitudes. It does so from cuts of the form

$$C = \sum_{\text{states}} \prod_j A^{\text{tree}}_{(j)} \tag{3}$$

where the sum over states runs over all physical states that can propagate via the cut lines, which are all placed on shell. The product runs over all tree amplitudes composing the cut. For example, for a one-loop four-point amplitude, the two unitarity cuts shown in fig. 1 are sufficient to reconstruct the complete amplitude. This generalizes to more complicated cases; for example, in fig. 2 various generalized cuts are shown for a three-loop four-point amplitude. This ability to completely reconstruct loop amplitudes from tree amplitudes, allows tree-level properties to be carried directly into loop calculations. If the on-shell tree amplitudes have a particular symmetry or property, then the unitarity cuts (3) automatically carry the property, since they are simply sums of products of tree amplitudes. This feature makes the unitarity method a convenient tool for identifying and proving properties of loop level amplitudes. (This was highlighted in a recent review using dual conformal symmetry and the duality between color and kinematics as examples [39].)

Although the unitarity method applies just as well to supersymmetric and non-supersymmetric theories, it is usually much simpler to deal with the supersymmetric cases because they have a simpler analytic structure. Indeed, the original application of the unitarity method was to construct one-loop supersymmetric amplitudes with arbitrary numbers of external legs [1]. The better power counting of supersymmetric theories allows all terms in one-loop scattering amplitudes to be readily constructed using cuts evaluated in four dimensions. Naively, this is not allowed because the amplitudes are divergent and need to be regularized. Nevertheless, a more careful study shows that for one-loop supersymmetric amplitudes it gives complete results [1].

For QCD or supersymmetric theories at higher loops, the situation is more complex. For one-loop QCD by using four-dimensional cuts, we can drop certain rational terms in the amplitudes [34]. These terms arise from $\mathcal{O}(\epsilon)$ terms in the integrand hitting ultraviolet poles in ϵ. Such terms can be recaptured by using loop-level on-shell recursion relations allowing the entire calculation to be performed using four-dimensional massless momenta [42]. More generally, by computing in $D = 4 - 2\epsilon$ dimensions, all rational terms are automatically captured and one can fully reconstruct amplitudes at any loop order [34]. A convenient way to implement this [7, 45, 39], is via six-dimensional helicity [46, 47]. This offers many of the advantages of four-dimensional helicity, but meshes well with regularization of the divergences, in a manner consistent with unitarity.

3   Applications of the Unitarity Method

An important question is whether the new techniques and understanding has led to any new calculations or applications. In this section we describe two rather different nontrivial applications of the unitarity method. The first of these is an example from collider physics. The second are calculations which illuminate multiloop ultraviolet properties in gauge and gravity theories.

3.1   Applications to QCD

Today the Large Hadron Collider (LHC) is operational and people are eagerly awaiting new results. In many scenarios of new physics, a sharp resonance appears, but in others the first signatures would be slight alterations of various distributions. For many new physics searches the signals are excesses in broader distributions of jets, along with missing energy. In particular, for supersymmetry searches, such multi-parton final states and missing energy are typical. The physics program at the Large Hadron Collider relies, to some extent, on having ever improved theoretical control over modeling of high-energy collisions. A detailed theoretical understanding not only increases the reach in new physics and particle searches, but also provides studies of the fundamental dynamics and properties of particles.

A first step towards a theoretical understanding of QCD is the evaluations of cross sections at leading order (LO) in the strong coupling $\alpha_S(\mu^2)$. Many tools [48] are available to generate predictions at leading order. Some of the methods applied incorporate higher-multiplicity leading-order matrix elements into parton-showering programs [49, 50], using matching (or merging) procedures [51].

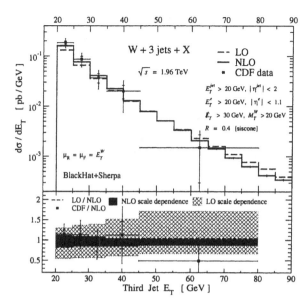

Figure 3: The cross section as a function of the transverse energy of the third jet, for inclusive $W + 3$-jet production, compared to the NLO prediction. In the upper panel the NLO distribution is the solid (black) histogram, and CDF data points are the (red) points, whose error bars denote the statistical and total uncertainties (excluding the luminosity error). The LO QCD predictions are shown as dashed (blue) lines. The lower panel shows the distribution normalized to the NLO prediction, using the CDF experimental bins (that is, averaging over bins in the upper panel). The scale-dependence bands are shaded (gray) for NLO and cross-hatched (brown) for LO.

However, cross sections in QCD can have strong sensitivity to higher-order corrections, motivating the challenging quest for calculating them. Next-to-leading (NLO) order predictions significantly reduce renormalization- and factorization-scale dependence. This becomes more important with increasing jet multiplicity (see e.g. [52]). Fixed-order results at NLO can also be matched to parton showers [53] with the prospect of complete event generation at next to leading order in the strong coupling. Here we will discuss a few parton-level NLO QCD calculations where the new unitarity techniques have had an impact.

NLO cross sections are built from several ingredients: virtual corrections, computed from the interference of tree-level and one-loop amplitudes; real-emission corrections; and a mechanism for isolating and integrating the infrared singularities in the latter. In the past, a bottleneck to these studies was posed by one-loop amplitudes involving six or more partons [52]. On-shell methods, have successfully broken past this bottleneck, by avoiding gauge-noninvariant intermediate steps, and reducing the problem to much smaller elements analogous to tree-level amplitudes. Approaches based on Feynman diagrams have also led to new results with six external partons, exemplified by the NLO cross section for producing $t\bar{t}b\bar{b}$ at hadron colliders [54]. On-shell methods should be

especially advantageous for processes involving many external gluons, which often domi-
nate multi-jet final states. Various new results [10, 11, 12] already indicate the suitability
of these methods for a general-purpose numerical approach to high-multiplicity one-loop
amplitudes.

Here we will give an example to illustrate some of the advances of recent years. In
particular, we will focus on an important class of processes in the search for new physics;
vector-boson production in association with jets. Processes yielding Z and W bosons in
association with jets have a particularly rich phenomenology at the electroweak symmetry
breaking scale, being important backgrounds to many searches of new physics, for Higgs
physics, and will be important to precision top-quark measurements.

In LO or NLO QCD the jets correspond to either single partons or pairs of partons
(which then hadronize). The state-of-the-art in perturbative QCD for hadron colliders is
now reaching parton level next-to-leading order computations with up to five final-state
objects. In fact, the only such process to be computed so far is $W + 4$-jet production [12],
using unitarity and on-shell methods.

A rather nontrivial example of a new result where the unitarity methods has lead to
useful phenomenology is $W + 3$-jet production at the Tevatron. Fig. 3 gives the cross
section as a function of the transverse energy of the third most energetic jet, compared
against experimental results from the CDF collaboration [55]. Besides showing very
good agreement between theory and experiment, within the experimental uncertainties,
it also illustrates the genetic feature that NLO results have a much smaller dependence
on renormalization and factorization scales than LO results. The NLO scale dependence
is shown as the gray band in the lower panel of the figure, while the LO dependence
is shown in the cross-hatched region. In each case, the scale dependence is determined
by varying the renormalization and factorization scales up and down by a factor of 2,
following the traditional prescription. Further details may be found in refs. [10, 12].

3.2 Applications to Maximally Supersymmetric Gauge and Gravity Amplitudes

Another arena where the new ideas and methods have had a significant impact are studies
of ultraviolet properties gauge and gravity theories. The theories of choice for studying
this at high loop orders are maximally supersymmetric theories, both for their technical simplicity and because supersymmetry tends to mitigate ultraviolet divergences.
$\mathcal{N} = 4$ super-Yang-Mills theory was proven to be ultraviolet finite in four dimensions
long ago [56]. The ultraviolet behavior of $\mathcal{N} = 8$ supergravity [57] in four dimension
is, however, still under study. Below we summarize concrete calculations in maximal
super-Yang-Mills theory and supergravity that shed light on this fundamental question.
Recent reviews discussing the ultraviolet properties of $\mathcal{N} = 8$ supergravity in more detail
are given in refs. [21].

Conventional wisdom holds that it is impossible to construct point-like ultraviolet
finite theories of gravity (see *e.g.* refs. [58]). Indeed, simple power-counting arguments
point to the difficulty of doing so. In a classic paper, 't Hooft and Veltman showed that
gravity coupled to matter generically diverges at one loop in four dimensions [30, 31].

Due to the dimensionful nature of the coupling, the divergences cannot be absorbed by a redefinition of the original parameters of the Lagrangian, rendering the theory non-renormalizable. Pure Einstein gravity does not possess a viable counterterm at one loop, delaying the divergence to at least two loops [30, 59]. The two-loop divergence of pure Einstein gravity was established by Goroff and Sagnotti and by van de Ven through direct computation [32, 33].

Supersymmetry offers a mechanism for delaying the onset of divergences in gravity theories. No supergravity theory can diverge until at least three loops, because the potential three loop counterterm built out of three Riemann tensors cannot be made supersymmetric [58]. This potential counterterm is an R^4 term corresponding to four powers of the Riemann tensor with the indices appropriately contracted, to allow for a supersymmetric completion. Supersymmetry alone cannot delay the ultraviolet divergences in gravity theories because of the increasingly worse divergences at each loop order in gravity theories. This leads to the general question for supergravity theories of whether a given potential counterterms identified by power counting and symmetry arguments is actually present.

Power counting arguments assume that all symmetries and relevant properties are known and accounted for. For the case of $\mathcal{N} = 8$ supergravity we do know that there are unexpected ultraviolet cancellations in the theory to all loop orders [60, 61]), though it is still not clear if these are powerful enough to render the theory finite to all loop orders. (These cancellations are related to a well studied property of one-loop $\mathcal{N} = 8$ amplitudes: in four dimensions triangle and bubble integrals drop out of the amplitudes, when expressed in a basis of scalar integral in four dimensions [62].) Some hints also follow from string-theory dualities [63]. We also know that gravity loop amplitudes are much more closely tied to better behaved gauge-theory amplitudes than had been believed [26]. While these arguments do not offer a proof of finiteness, they do suggest that it would be wise to reexamine the ultraviolet properties of gravity theories. (For other approaches to trying to make quantum field theories of gravity sensible in the ultraviolet see refs. [64].)

Motivated by the hint of high-loop cancellations, explicit cancellations were then carried out in refs. [19, 17, 20] to directly investigate the ultraviolet properties of $\mathcal{N} = 8$ supergravity. The result of this was to definitively rule out the potential three loop R^4 counterterm in $D = 4$. Although no potential counterterm exists at four loops (because of an "accidental" cancellation similar to the one preventing a pure gravity counterterm at one loop), direct calculation establishes that the four-loop four-point amplitude of $\mathcal{N} = 8$ supergravity has the same power counting in D dimensions as $\mathcal{N} = 4$ super-Yang-Mills theory (which is known to be finite in $D = 4$).

The result of the four loop calculation is that the four-loop four-point amplitude is of the form,

$$M_4^{4\text{-loop}} \sim D^8 R^4 \times \text{loop integrals} \tag{4}$$

where the $D^8 R^4$ factor corresponds to 20 powers of momentum in the numerators of the integrals coming out as external momentum. This factors correspond to covariant derivatives contracted to four Riemann tensors in an appropriate way. If we assume that no further ultraviolet cancellations exist, and that no further powers of loop momenta can come out of the integrals as external momenta as the loop order increases, simple power counting shows that in four dimensions the first divergence would occur at seven loops.

Indeed, this is in line with recent comprehensive studies of the potential counterterms of $\mathcal{N} = 8$ supergravity [22], proving that no counterterm compatible with the $E_{7(7)}$ symmetry is available until seven loops. Based on these and other results [23, 65, 66] a consensus has formed that supersymmetry and the $E_{7(7)}$ symmetry alone cannot prevent divergences in $D = 4$ starting at seven loops and that the theory will likely diverge. There is, however, a more optimistic view [67]. (The previously claimed delay from supersymmetry of potential ultraviolet divergences in $\mathcal{N} = 8$ supergravity until nine loops [68] has been retracted [23].)

Is it possible that there are further symmetries or structures that prevent the widely expected seven loop divergences? Powercounting arguments using symmetries to rule out potential counterterms can, of course, never prove the existence of divergences, only that protection against divergences holds to a certain level; if a symmetry or structure is missed then it may turn out the bound is too loose. More generally, the only way we can be certain that the coefficient of a potential counterterm respecting the known symmetries is non-zero is to carry out the explicit calculation to show that the numerical value is nonzero.

Today, it is not yet practical to carry out a seven-loop computation. However, a simple way to lower the loop order in which a given potential counterterm can correspond to a divergence is to work in higher dimensions. By raising the dimension $\mathcal{N} = 4$ super-Yang-Mills is no longer ultraviolet finite, allowing this theory to be used as a playground for sharpening our understanding of counterterms in maximally supersymmetric theories [17, 69, 18]. Explicit calculations [16, 14, 19, 17, 20, 18] show that at least for four-point amplitudes through four loops, both $\mathcal{N} = 8$ supergravity and $\mathcal{N} = 4$ super-Yang-Mills theory are ultraviolet finite for

$$D < \frac{6}{L} + 4 \qquad (L > 1) , \tag{5}$$

where D is the dimension of space-time and L the loop order. (The case of one loop, $L = 1$, is special, with the amplitudes finite for $D < 8$, not $D < 10$.) For $\mathcal{N} = 4$ super-Yang-Mills this bound was proposed in ref. [16] and has been confirmed in ref. [70] using superspace techniques. Explicit computations summarized below demonstrate this bound is saturated in $\mathcal{N} = 4$ super-Yang-Mills theory through at least four loops [71, 16, 60]. For $\mathcal{N} = 8$ supergravity we know that the bound holds through four loops and that it is saturated at three loops [16, 19, 17, 20].

The first step for carrying out a loop calculation of a gravity scattering amplitude is to compute the corresponding amplitude in super-Yang-Mills theory. We may obtain $\mathcal{N} = 8$ supergravity amplitudes by considering the amplitudes of $\mathcal{N} = 4$ super-Yang-Mills theory. Following the strategy of refs. [16], by taking unitarity cuts of the gauge-theory amplitude which decompose it into a sum of products of tree amplitudes, we can use the KLT relations to assemble a spanning set[2] of unitarity cuts for the gravity amplitudes

---

[2]By a spanning set of unitarity cuts, we mean a set of cuts sufficient to construct the complete amplitudes; examples of spanning sets may be found in ref. [18, 39].

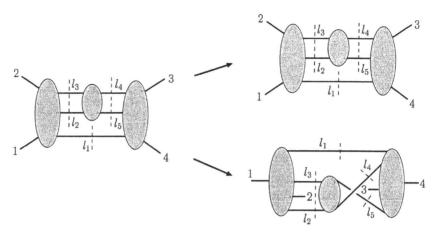

Figure 4: An example of showing how the KLT relations allow us to map gravity cuts into sums over pairs of gauge-theory cuts. Here we display a pair of gauge-theory cuts needed to evaluate the gravity cut on the left side. The remaining pairs are obtained by permuting the external legs $1 \leftrightarrow 2$ and $3 \leftrightarrow 4$.

using cuts of gauge-theory amplitudes. As an example of how this works, consider the cut for an $\mathcal{N} = 8$ supergravity four-point amplitude shown on the left side of fig. 4,

$$C_{\mathcal{N}=8} = \sum_{\mathcal{N}=8 \text{ states}} M_5^{\text{tree}}(1, 2, l_3, l_2, l_1) \, M_4^{\text{tree}}(-l_2, -l_3, l_4, l_5) \, M_5^{\text{tree}}(3, 4, -l_1, -l_5, -l_4) \, . \quad (6)$$

The sum runs over all physical states that cross any cut line. The required KLT relations are relabelings of the basic relations in eq. (2). Inserting the relabeled KLT relations into the cut (6), we obtain [17],

$$i \, (l_4 + l_5)^2 (l_1 + k_1)^2 (l_3 + k_2)^2 (l_4 - k_3)^2 (l_1 - k_4)^2$$
$$\times \left(\sum_{\mathcal{N}=4 \text{ states}} A_5^{\text{tree}}(1, 2, l_3, l_2, l_1) \, A_4^{\text{tree}}(-l_2, -l_3, l_4, l_5) \, A_5^{\text{tree}}(3, 4, -l_1, -l_5, -l_4) \right)$$
$$\times \left(\sum_{\mathcal{N}=4 \text{ states}} A_5^{\text{tree}}(1, l_1, l_3, 2, l_2) \, A_4^{\text{tree}}(-l_2, -l_3, l_5, l_4) A_5^{\text{tree}}(3, -l_4, -l_1, 4, -l_5) \right)$$
$$+ \{1 \leftrightarrow 2\} + \{3 \leftrightarrow 4\} + \{1 \leftrightarrow 2, \ 3 \leftrightarrow 4\} \, . \quad (7)$$

The relation is depicted in fig. 4, for one of the four terms in the sum over external-leg permutations. One of the gauge-theory cuts is planar, while the second is nonplanar. Remarkably, this equation gives the $\mathcal{N} = 8$ supergravity cut directly in terms of products of two $\mathcal{N} = 4$ super-Yang-Mills cuts and allows us to build the supergravity amplitude using only gauge-theory results. Using this strategy, together with method of maximal cuts [14], the three- and four-loop four-point amplitudes were calculated in refs. [19, 17, 20].

Figure 5: The Feynman rules of gauge theories have three- and four-point vertices.

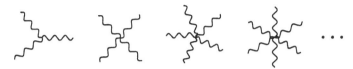

Figure 6: Gravity theories have an infinite number of higher-point contact interactions when using Feynman diagrams.

To determine whether the finiteness bound (5) might hold for $L \geq 5$ will require new information and better tools. The duality between color and kinematics and the double copy structure of gravity [26, 40], summarized below in section 4.2, is a promising direction. Using these results we should be able to immediately extract gravity amplitudes from gauge-theory ones, simply by rearranging corresponding gauge-theory amplitudes into a form where they satisfy a duality between color and kinematics [25]. This reorganization is well suited for addressing the issue of the ultraviolet properties of gravity because it converts complicated calculations in quantum gravity directly into much simpler gauge-theory calculations. This may make it possible to go to even higher loop orders than had been previously possible, and finally settle the question of the ultraviolet properties of $\mathcal{N} = 8$ supergravity.

4 Harmoneous Examples

In the previous section we noted that there is a strong similarity between gravity and gauge-theory scattering amplitudes. We now make this more precise.

4.1 A Comparison of Gravity To Gauge Theory

It is useful to start by comparing gravity to gauge theory using off-shell Feynman diagrammatic methods. The Feynman rules are generated starting from the Einstein-Hilbert and Yang-Mills Lagrangians,

$$\mathcal{L}_{\text{YM}} = -\frac{1}{4} F^a_{\mu\nu} F^{a\,\mu\nu}\,, \qquad\qquad \mathcal{L}_{\text{EH}} = \frac{2}{\kappa^2} \sqrt{-g} R\,. \qquad (8)$$

From the point of view of Feynman rules these two Lagrangians have some rather different properties. As illustrated in figs. 5 and 6, with standard gauge choices gauge theories have three- and four-point interactions, while gravity has an infinite number of contact interactions. Perhaps more striking than the infinite number of interactions is the remarkably complexity of these interactions.

To be more concrete consider the three-gluon vertex in Feynman gauge,

$$V_{3\mu,\nu,\sigma}^{abc}(k_1, k_2, k_3) = g f^{abc}\Big[(k_1 - k_2)_\sigma \eta_{\mu\nu} + \text{cyclic}\Big], \tag{9}$$

where g is the coupling, f^{abc} the usual group theory structure constants, the $\eta_{\mu\nu}$ the flat metric and the k_i the momenta of the vertex. This vertex is relatively simple. We may compare this to the three-graviton interaction in, for example, de Donder gauge,

$$G_{3\mu\alpha,\nu\beta,\sigma\gamma}(k_1, k_2, k_3) = i\frac{\kappa}{2}\Big[-\frac{1}{2}k_1 \cdot k_2 \eta_{\mu\alpha}\eta_{\nu\beta}\eta_{\sigma\gamma} - \frac{1}{2}P_6(k_{1\nu}k_{1\beta}\eta_{\mu\alpha}\eta_{\sigma\gamma})$$
$$+ \frac{1}{2}k_1 \cdot k_2 \eta_{\mu\nu}\eta_{\alpha\beta}\eta_{\sigma\gamma} + \cdots\Big], \tag{10}$$

where we have displayed the first three terms out of about 100. Here the coupling κ is related to Newton's constant by $\kappa^2 = 32\pi^2 G_N$. In total the vertex has on the order of 100 terms. The precise form of the vertex depends on the gauge, but in general the three vertex is a rather involved and unenlightening object. The complete expression can be found in refs. [72, 30].

Comparing the vertex in eq. (9) to the one in eq. (10), it is clear that gravity is much more complicated than gauge theory. Moreover, there does not appear to be any simplicity or obvious relation between the gauge and gravity vertices. The former leads to complicated diagrams but the latter is a hopeless mess. One can do somewhat better with special gauge choices and appropriate field redefinitions [33, 73], considerably simplifying the Feynman rules. Still, multiloop Feynman diagram calculations in (super) gravity extremely difficult, and in many cases out of reach even using the most powerful computers.

Now let us reconsider the same process but using on-shell methods. If we take the three-graviton vertex in eq. (10) and dot the three legs with physical polarizations tensors satisfying the physical state conditions, $k_i^2 = 0$, $\varepsilon_i^{\mu\nu}k_{i\mu} = \varepsilon_i^{\mu\nu}k_{i\nu} = \varepsilon^\mu{}_\mu = 0$, we obtain a greatly simplified vertex,

$$G_3(k_1, k_2, k_3) = -i\kappa\varepsilon_1^{\mu\alpha}\varepsilon_2^{\nu\beta}\varepsilon_3^{\sigma\gamma}\Big[(k_1)_\sigma\eta_{\mu\nu} + \text{cyclic}\Big]\Big[(k_1)_\gamma\eta_{\alpha\beta} + \text{cyclic}\Big]. \tag{11}$$

Remarkably, up to overall factors, this is just a double copy of the kinematic part of the on-shell Yang-Mills vertex,

$$V_3^{abc}(k_1, k_2, k_3) = 2\varepsilon_1^\mu\varepsilon_2^\nu\varepsilon_3^\sigma g f^{abc}\Big[(k_1)_\sigma\eta_{\mu\nu} + \text{cyclic}\Big], \tag{12}$$

where the polarization vector satisfies $\varepsilon_i^\mu k_{i\mu} = 0$. To make the comparison, we identify the graviton polarization tensor as a product of gluon polarization vectors, $\varepsilon_i^{\mu\nu} = \varepsilon_i^\mu \times \varepsilon_i^\nu$. Similar considerations allow us to express all three-point vertices in supergravity as products of super-Yang-Mills vertices. Using BCFW recursion [2, 74], these three vertices are sufficient to construct any tree-level gauge or gravity amplitude. The unitarity method

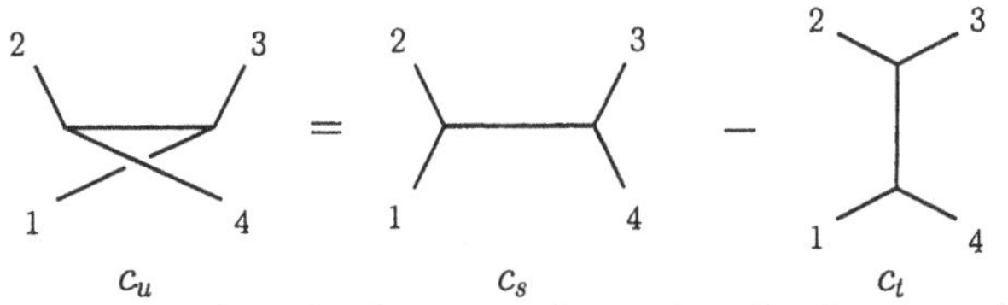

Figure 7: The Jacobi relation for color factors at four points for the three channels labeled by s, t and u. Diagram numerators satisfy the same relations.

then allows us to construct any loop amplitude (though expressions valid in $D > 4$ are needed to ensure that no terms are dropped because of regularization issues). In this way the entire gravitational S-matrix is encoded in the three-point vertex (11).

Clearly, there is a rather striking relationship between gravity and gauge theory, but to make it visible we need to keep external states on shell. As we shall see below the double-copy structure in eq. (11) is not accidental, but appears likely to extend to *all* loop orders. As such, it reflects a profound and important property of quantum gravity.

4.2 A Duality Between Color and Kinematics

As a rather striking example of a powerful structure in scattering amplitudes, we consider a rather surprising duality between color and kinematics [25, 26]. As we discuss below this duality is intimately connected to the double copy relationship between gravity and gauge theory mentioned earlier. In generally, we can write any n-point tree level amplitude with all particles in the adjoint representation as,

$$\mathcal{A}_n^{\text{tree}}(1, 2, 3, \ldots, n) = \sum_i \frac{n_i \, c_i}{\prod_{\alpha_i} p_{\alpha_i}^2}, \qquad (13)$$

where the sum runs over the set of n-point L-loop diagrams with only cubic vertices and we suppressed factors of the coupling constant. These include distinct permutations of external legs. We suppress factors of the coupling constant for convenience. The product in the denominator runs over all propagators of each cubic diagram. The c_i are the color factors obtained by dressing every three vertex with an $\tilde{f}^{abc} = i\sqrt{2}f^{abc}$ structure constant, and the n_i are kinematic numerator factors depending on momenta, polarizations and spinors. The form (13) can be obtained straightforwardly, for example, from Feynman diagrams, by representing all contact terms as inverse propagators in the kinematic numerators that cancel propagators. For supersymmetric amplitudes expressed in superspace, there will also be Grassmann parameters in the numerators.

In general, the n_i may be deformed under any shifts, $n_i \to n_i + \Delta_i$, where the Δ_i are arbitrary functions independent of color satisfying the constraint [25, 26, 75],

$$\sum_i \frac{\Delta_i c_i}{\prod_{\alpha_i} p_{\alpha_i}^2} = 0. \qquad (14)$$

We may think of these transformations as generalized gauge transformations. Some of

14

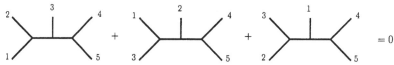

Figure 8: A Jacobi relation at five points. The color factors associated with each diagram automatically satisfy this relation. The duality states that numerators can be rearranged so that they satisfy exactly the same relation.

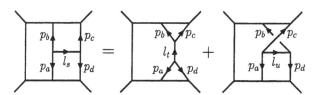

Figure 9: An example of a duality relation satisfied by numerators for diagrams of the three-loop four-point amplitude. Both color factors and numerator factors satisfy these relations.

this gauge freedom corresponds to gauge transformations in the traditional sense but most does not.

The duality conjectured in ref. [25] requires there to exist such a transformation from any valid representation to one where the numerators satisfy equations in one-to-one correspondence with the Jacobi identity of the color factors,

$$c_i = c_j - c_k \implies n_i = n_j - n_k \,. \tag{15}$$

This duality is conjectured to hold to all multiplicity at tree level in a large variety of theories, including supersymmetric extensions of Yang-Mills theory.

At tree level a consequence of this duality is non-trivial relations between the color-ordered partial tree amplitudes of gauge theory [25, 76, 77]. The duality has also been studied in string theory [78, 79] and from a Lagrangian vantage point [75]. An alternative trace-based representation of the duality (15) was recently given in ref. [80], emphasizing the underlying group theoretic structure of the duality.

Perhaps more remarkable than the duality itself is a related conjecture that once the gauge-theory amplitudes are arranged into a form satisfying the duality (15), corresponding gravity amplitudes can be obtained simply by taking a double copy of gauge-theory numerator factors [25, 26],

$$-i\mathcal{M}_n^{\text{tree}}(1, 2, \ldots, n) = \sum_i \frac{n_i \tilde{n}_i}{\prod_{\alpha_i} p_{\alpha_i}^2} \,, \tag{16}$$

where the $\tilde{n}_i$ represent numerator factors of a second gauge-theory amplitude, the sum runs over the same set of diagrams as in eq. (13). This is expected to hold in a large class of gravity theories, including theories that are the low-energy limits of string theories. (As for the gauge-theory case, we suppress factors of the coupling constants.) At tree level, this double-copy property encodes the KLT relations between gravity and gauge theory [27]. The double-copy formula (16) has been proven via on-shell recursion [2] for pure gravity and for $\mathcal{N} = 8$ supergravity tree amplitudes, whenever the duality (15) holds in the corresponding gauge theories [75].

The above conjectures have been extended to loop level [26], so that at any loop order L,

$$\mathcal{A}_m^{\text{loop}} = \sum_j \int \prod_{l=1}^L \frac{d^D p_l}{(2\pi)^D} \frac{1}{S_j} \frac{n_j c_j}{\prod_{\alpha_j} p_{\alpha_j}^2},$$

$$\mathcal{M}_m^{\text{loop}} = \sum_j \int \prod_{l=1}^L \frac{d^D p_l}{(2\pi)^D} \frac{1}{S_j} \frac{n_j \tilde{n}_j}{\prod_{\alpha_j} p_{\alpha_j}^2}, \tag{17}$$

where $\mathcal{A}_n^{\text{loop}}$ and $\mathcal{M}_n^{\text{loop}}$ are L-loop gauge and gravity amplitudes. As before we removed factors of the coupling constants. The sums now run over all distinct m-point L-loop diagrams with cubic vertices. These include distinct permutations of external legs, and the S_j are the symmetry factors of each diagram. As at tree level, at least one family of numerators (n_j or $\tilde{n}_j$) for gravity must be constrained to satisfy the duality (15). (For pure gravity, extra projectors are needed to obtain loop-level amplitudes from the direct product of two pure Yang-Mills theories.)

This loop-level extension has been tested in the rather nontrivial case of three-loop four-point amplitude of $\mathcal{N} = 4$ super-Yang-Mills theory and $\mathcal{N} = 8$ supergravity [26]. This amplitude has already been studied in some detail in refs. [19, 17]. To impose the duality (15) on the amplitude, the duality relation for every propagator in each diagram must be enforced. On any diagram, we can describe any internal line, carrying some momentum l_s, in terms of formal graph vertices $V(p_a, p_b, l_s)$, and $V(-l_s, p_c, p_d)$ where the p_i are the momenta of the other legs attached to l_s, as illustrated on the left side of fig. 9. The duality (15) requires that,

$$n(\{V(p_a, p_b, l_s), V(-l_s, p_c, p_d), \cdots\}) = n(\{V(p_d, p_a, l_t), V(-l_t, p_b, p_c), \cdots\}) \tag{18}$$
$$+ n(\{V(p_a, p_c, l_u), V(-l_u, p_b, p_d), \cdots\}),$$

where the ns are the numerators associated with the diagrams specified by the vertices. The omitted vertices are identical in all three diagrams, and $l_s \equiv (p_c + p_d)$, $l_t \equiv (p_b + p_c)$ and $l_u \equiv (p_b + p_d)$ in the numerator expressions. There is one such equation for every propagator in every diagram. Solving the system of distinct equations enforces the duality conditions (15).

It turns out that imposing the duality completely fixes the form of the amplitude, if all diagrams have the proper power counting of $\mathcal{N} = 4$ super-Yang-Mills, and that all numerators are polynomials in the kinematic invariants once we extract an overall factor the tree amplitude. Moreover, only the 12 diagrams shown in fig. 10 contribute, with the numerator factors given in table 1. These numerator factors give the correct amplitude, as checked against the unitarity cuts in ref. [26].

The fact that this arrangement can be done for the three loop four-point amplitude of $\mathcal{N} = 4$ super-Yang-Mills theory is rather striking, but it is much more surprising that we

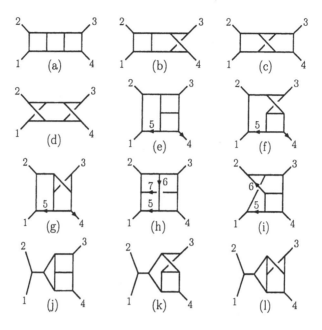

Figure 10: Loop diagrams contributing to the three-loop four-point amplitudes of both $\mathcal{N} = 4$ super-Yang-Mills theory and $\mathcal{N} = 8$ supergravity. Integrals are specified by combining their propagators with numerator factors given in table 1. The symmetry factor for diagram (d) is 2, and the rest are unity.

Table 1: The numerator factors of the integrals $I^{(x)}$ in fig. 10. The first column labels the integral, the second column the relative numerator factor for $\mathcal{N} = 4$ super-Yang-Mills theory. The square of this is the relative numerator factor for $\mathcal{N} = 8$ supergravity. An overall factor of $st A_4^{\text{tree}}$ has been removed, s, t, u are Mandelstam invariants corresponding to $(k_1 + k_2)^2, (k_2 + k_3)^2, (k_1 + k_3)^2$ and $\tau_{ij} = 2k_i \cdot l_j$, where k_i and l_j are momenta as labeled in fig. 10.

Integral $I^{(x)}$	$\mathcal{N} = 4$ Super-Yang-Mills ($\sqrt{\mathcal{N}} = 8$ supergravity) numerator
(a)–(d)	s^2
(e)–(g)	$\left(s\left(-\tau_{35} + \tau_{45} + t\right) - t\left(\tau_{25} + \tau_{45}\right) + u\left(\tau_{25} + \tau_{35}\right) - s^2 \right)/3$
(h)	$\left(s\left(2\tau_{15} - \tau_{16} + 2\tau_{26} - \tau_{27} + 2\tau_{35} + \tau_{36} + \tau_{37} - u\right) \right.$ $\left. +t\left(\tau_{16} + \tau_{26} - \tau_{37} + 2\tau_{36} - 2\tau_{15} - 2\tau_{27} - 2\tau_{35} - 3\tau_{17}\right) + s^2 \right)/3$
(i)	$\left(s\left(-\tau_{25} - \tau_{26} - \tau_{35} + \tau_{36} + \tau_{45} + 2t\right) \right.$ $\left. +t\left(\tau_{26} + \tau_{35} + 2\tau_{36} + 2\tau_{45} + 3\tau_{46}\right) + u\,\tau_{25} + s^2 \right)/3$
(j)–(l)	$s(t - u)/3$

can immediately obtain $\mathcal{N} = 8$ supergravity results simply by squaring the numerators of each diagram. Indeed, this was confirmed in ref. [26] by comparing against a spanning set of unitarity cuts of the known result for $\mathcal{N} = 8$ supergravity [19, 17]. The ability to organize the amplitude into such a form has also been confirmed for the simplest of the two-loop four-point amplitude of QCD [26].

5   Towards a Gravity Lagrangian

We now turn to the question of finding a Lagrangian which generates amplitudes with numerators that manifestly satisfy the BCJ duality (15). If a local Lagrangian of this type could be found, it would enable us to construct a corresponding gravity Lagrangian whose squaring relations with Yang-Mills theory are manifest.

The Yang-Mills Lagrangian of the type we seek can differ from the conventional Yang-Mills Lagrangian only by terms that do not affect on the amplitudes. Such a Lagrangian with manifest BCJ duality exists, and differs from the conventional Lagrangian by terms whose sum is identically zero by the color Jacobi identity [75]. Although the added terms sum to zero, they cause the necessary rearrangements in the diagrams so that the duality holds manifestly.

We write the Yang-Mills Lagrangian as

$$\mathcal{L}_{YM} = \mathcal{L} + \mathcal{L}_5' + \mathcal{L}_6' + \dots \tag{19}$$

where $\mathcal{L}$ is the conventional Yang-Mills Lagrangian and $\mathcal{L}_n'$, $n > 4$ are the additional terms required so that the duality is satisfied. At four points, the duality is trivial and holds in any gauge [81, 25], so $\mathcal{L}$ by itself will generate diagrams whose numerators satisfy eq. (15). For simplicity we choose Feynman gauge for $\mathcal{L}$. The $\mathcal{L}_n'$ are required to leave scattering amplitudes unchanged, and they must rearrange the numerators of diagrams in a way so that the BCJ duality is satisfied.

By imposing the constraint that the generated five-point diagrams satisfy the BCJ duality (15), we find the Lagrangian,

$$\mathcal{L} = \frac{1}{2} A_\mu^a \Box A^{a\mu} - g f^{a_1 a_2 a_3} d\partial_\mu A_\nu^{a_1} A^{a_2 \mu} A^{a_3 \nu} - \frac{1}{4} g^2 f^{a_1 a_2 b} f^{b a_3 a_4} A_\mu^{a_1} A_\nu^{a_2} A^{a_3 \mu} A^{a_4 \nu} , \tag{20}$$

and

$$\mathcal{L}_5' = -\frac{1}{2} g^3 f^{a_1 a_2 b} f^{b a_3 c} f^{c a_4 a_5} \left(\partial_{[\mu} A_{\nu]}^{a_1} A_\rho^{a_2} A^{a_3 \mu} + \partial_{[\mu} A_{\nu]}^{a_2} A_\rho^{a_3} A^{a_1 \mu} \right.$$
$$\left. + \partial_{[\mu} A_{\nu]}^{a_3} A_\rho^{a_1} A^{a_2 \mu} \right) \frac{1}{\Box} (A^{a_4 \nu} A^{a_5 \rho}) . \tag{21}$$

The additional terms in $\mathcal{L}_5'$ are necessarily nonlocal, at least if we want a covariant Lagrangian without auxiliary fields. The numerators n_i are derived from this action by first computing the contribution from the three-point vertices, which gives a set of three-vertex diagrams with unique numerators. Then the contributions from the four- and five-point interaction terms are assigned to the various diagrams with only three-point vertices according to their color factors. Since these terms will contain fewer propagators than those obtained by using only three-point vertices, if we put back all the propagators their contributions to the numerators contain inverse propagators, which cancel some propagators.

As previously mentioned, $\mathcal{L}_5'$ is identically zero by the color Jacobi identity. This can be easily seen after relabeling color indices to obtain,

$$
\begin{aligned}
\mathcal{L}_5' = -\frac{1}{2} g^3 (f^{a_1 a_2 b} f^{b a_3 c} + f^{a_2 a_3 b} f^{b a_1 c} + f^{a_3 a_1 b} f^{b a_2 c}) f^{c a_4 a_5} \\
\times \partial_{[\mu} A_{\nu]}^{a_1} A_\rho^{a_2} A^{a_3 \mu} \frac{1}{\Box} (A^{a_4 \nu} A^{a_5 \rho}) .
\end{aligned}
\tag{22}
$$

These terms cause a non-trivial rearrangement amongst the diagrams, although all the added terms must add up to zero by the color Jacobi identity. Nevertheless, they alter the numerators of the individual diagrams such that the duality (15) is satisfied.

It turns out that the set of terms with the desired properties is not unique. Indeed, terms that satisfy the BCJ duality by themselves can also be added. This ambiguity is due to the residual generalized gauge invariance that remains after enforcing the duality identities. The additional terms that can be added to Yang-Mills at five points which preserve the duality (15) are

$$
\begin{aligned}
\mathcal{D}_5 = \frac{-\beta}{2} g^3 f^{a_1 a_2 b} f^{b a_3 c} f^{c a_4 a_5} \Big(\partial_{(\mu} A_{\nu)}^{a_1} A_\rho^{a_2} A^{a_3 \mu} + \partial_{(\mu} A_{\nu)}^{a_2} A_\rho^{a_3} A^{a_1 \mu} \\
+ \partial_{(\mu} A_{\nu)}^{a_3} A_\rho^{a_1} A^{a_2 \mu} \Big) \frac{1}{\Box} (A^{a_4 \nu} A^{a_5 \rho}) ,
\end{aligned}
\tag{23}
$$

where β is an arbitrary parameter. $\mathcal{D}_5$ also vanishes identically by the color Jacobi identity. Since $\mathcal{D}_5$ does not serve to correct lower-point contributions to make the BCJ duality relations hold through five points, we do not need to include it. It does however show that there are multiple Lagrangians with the desired properties.

5.1 A Gravity Lagrangian from Gauge Theory

Now that we have a Lagrangian that gives the desired numerators n_i for gauge theory, we can use it to construct the tree-level gravity Lagrangian by demanding that it gives diagrams whose numerators are a double copy of the gauge-theory numerators, as in eq. (16). However, we need to first bring the Yang-Mills Lagrangian into a cubic form to achieve this. We can do so by introducing an auxiliary field $B_{\mu\nu\rho}^a$ into the Lagrangian. To give the auxiliary fields the same dimensions as the dynamical fields we need to make them propagate. (The propagation is trivial because we recover the original Lagrangian by integrating them out.) At four points this leads to the Lagrangian [75],

$$
\mathcal{L}_{YM} = \frac{1}{2} A^{a\mu} \Box A_\mu^a - B^{a\mu\nu\rho} \Box B_{\mu\nu\rho}^a - g f^{abc} (\partial_\mu A_\nu^a + \partial^\rho B_{\rho\mu\nu}^a) A^{b\mu} A^{c\nu} ,
\tag{24}
$$

where the equation of motion for the auxiliary field $B_{\mu\nu\rho}^a$ becomes

$$
\Box B_{\mu\nu\rho}^a = \frac{g}{2} f^{abc} \partial_\mu (A_\nu^b A_\rho^c) .
\tag{25}
$$

Five points becomes more complicated and a new set of auxiliary fields must be introduced to re-express the nonlocal terms into a local and cubic form. The result is [75],

$$
\begin{aligned}
\mathcal{L}_5' \rightarrow\ & Y^{a\mu\nu} \Box X_{\mu\nu}^a + D_{(3)}^{a\mu\nu\rho} \Box C_{(3)\ \mu\nu\rho}^a + D_{(4)}^{a\mu\nu\rho\sigma} \Box C_{(4)\ \mu\nu\rho\sigma}^a \\
& + g f^{abc} \left(Y^{a\mu\nu} A_\mu^b A_\nu^c + \partial_\mu D_{(3)}^{a\mu\nu\rho} A_\nu^b A_\rho^c - \tfrac{1}{2} \partial_\mu D_{(4)}^{a\mu\nu\rho\sigma} \partial_{[\nu} A_{\rho]}^b A_\sigma^c \right) \\
& + g f^{abc} X^{a\mu\nu} \left(\tfrac{1}{2} \partial_\rho C_{(3)\ \ \mu}^{b\rho\sigma} \partial_{[\sigma} A_{\nu]}^c + \partial_\rho C_{(4)\ \ \nu[\mu}^{b\rho\sigma} A_{\sigma]}^c \right).
\end{aligned}
\tag{26}
$$

From the above Lagrangian we can directly construct a gravity Lagrangian valid through five-points which is manifestly a double copy of the gauge-theory one. At four points we do so by taking two copies of the Lagrangian (24) and identifying the fields as,

$$
\begin{aligned}
A^\mu \tilde{A}^\nu &\rightarrow\ h^{\mu\nu}, \\
A^\mu \tilde{B}^{\nu\rho\sigma} &\rightarrow\ g^{\mu\nu\rho\sigma}, \\
B^{\mu\rho\sigma} \tilde{A}^\nu 1 &\rightarrow\ \tilde{g}^{\mu\rho\sigma\nu}, \\
B^{\mu\rho\sigma} \tilde{B}^{\nu\tau\lambda} &\rightarrow\ f^{\mu\rho\sigma\nu\tau\lambda},
\end{aligned}
\tag{27}
$$

Here $h_{\mu\nu}$ is the graviton field A^μ and $\tilde{A}^\nu$ gauge fields and all the remaining ones are auxiliary fields. With a similar procedure we can obtain five-point terms in the gravity Lagrangian. Beyond this the Lagrangians become more complicated. Such a procedure has not yet been carried out for supergravity, though it should work as well.

We note that above Lagrangians were constructed order by order in perturbation theory. Of course, it would be much better to have a principle for finding all order forms of such Lagrangians.

6   Conclusions and Outlook

In these lectures we summarized the unitarity method and some of its applications. We discussed applications to collider physics, and also to the question of ultraviolet properties of maximally supersymmetric gauge and gravity theories in various dimensions. We also illustrated a remarkable harmony between gravity and gauge theories, stemming from a gauge-theory duality between color and kinematics [25, 26]. Although the duality remains a conjecture, we can exploit it to guide loop computations, simply by enforcing the duality and verifying the consistency with the unitarity cuts.

There are a number of interesting future directions. The unitarity method offers a general purpose tool for carrying out amplitude calculations in both supersymmetric and non-supersymmetric theories, including their non-planar contributions. In QCD there are many more NLO calculations that need to be carried out, especially those involving vector bosons, Higgs bosons, top quarks, in associations with many jets. In particular, a high-priority calculation is Z-boson production in association with four jets. Further improvements would allow the unitarity method to tackle ever more complicated important amplitudes. Beyond this, a major step would be to merge these types of calculations with parton showering [53].

In $\mathcal{N} = 8$ supergravity, a consensus has formed that the standard symmetries of $\mathcal{N} = 8$ supergravity cannot protect the theory against divergences, starting at seven loops [22] (though there is at least one opinion to the contrary [67]). It would be very interesting to directly determine the ultraviolet properties as a function of dimension at five and higher loops. If this can be done it should greatly clarify the ultraviolet behavior of $\mathcal{N} = 8$ supergravity in four dimensions, checking the hypothesis that it may be a an ultraviolet finite theory. The duality between color and kinematic numerators and the associated double-copy property of gravity, offer a promising avenue for attacking this issue.

There are also a number of interesting open problems related to the duality between color and kinematic numerators. In particular, it would be helpful to carry out further checks of the duality for multiloop processes. More generally an all-orders proof of the duality would be important, especially if it leads to new insight into the group theoretic origins of the duality. We would also like to have Lagrangians whose diagrams satisfy the duality to all orders, and which give gravity Lagrangians as double copies, along the lines described in ref. [75], but constructed in a more systematic fashion. The duality and gravity double-copy structure likely have important non-perturbative implications. In particular, these properties suggest that all classical solutions in gravity theories may be expressible as double copies of classical solutions in gauge theories. The fact that the same kinematic building blocks appear in gravity and in gauge theories, apparently to all loop orders, is rather striking and may be taken as theoretical evidence of an underlying unification of gauge and gravity theories, perhaps along the lines of string theory.

In summary, the unitarity method is by now a mature formalism for carrying out state-of-the-art loop calculations for phenomenological and theoretical purposes. It has also played an important role in uncovering remarkable structures in scattering amplitudes, including a double-copy relationship between gravity and gauge theory. We look forward to many new exciting developments in the coming years.

Acknowledgments

We thank L. J. Dixon, J. J. M. Carrasco, F. Febres Cordero, D. Forde, H. Johansson D. A. Kosower, and R. Roiban for many enlightening discussions and collaboration on work described in this lecture. ZB's work was supported by the US Department of Energy under contract DE-FG03-91ER40662. HI's work is supported by a grant from the US LHC Theory Initiative through NSF contract PHY–0705682.

References

[1] Z. Bern, L. J. Dixon, D. C. Dunbar and D. A. Kosower, Nucl. Phys. B **425**, 217 (1994) [hep-ph/9403226]; Nucl. Phys. B **435**, 59 (1995) [hep-ph/9409265].

[2] R. Britto, F. Cachazo, B. Feng and E. Witten, Phys. Rev. Lett. **94**, 181602 (2005) [hep-th/0501052].

[3] E. Witten, Commun. Math. Phys. **252**, 189 (2004) [arXiv:hep-th/0312171]; R. Roiban, M. Spradlin and A. Volovich, Phys. Rev. D **70**, 026009 (2004) [hep-th/0403190].

[4] C. Anastasiou, Z. Bern, L. J. Dixon and D. A. Kosower, Phys. Rev. Lett. **91**, 251602 (2003) [hep-th/0309040]; Z. Bern, L. J. Dixon and V. A. Smirnov, Phys. Rev. D **72**, 085001 (2005) [hep-th/0505205].

[5] L. F. Alday and J. Maldacena, JHEP **0706**, 064 (2007) [0705.0303 [hep-th]].

[6] J. M. Drummond, J. Henn, V. A. Smirnov and E. Sokatchev, JHEP **0701**, 064 (2007) [hep-th/0607160]; Z. Bern, M. Czakon, L. J. Dixon, D. A. Kosower and V. A. Smirnov, Phys. Rev. D **75**, 085010 (2007) [hep-th/0610248]; J. M. Drummond, G. P. Korchemsky and E. Sokatchev, Nucl. Phys. B **795**, 385 (2008) [0707.0243 [hep-th]]; A. Brandhuber, P. Heslop and G. Travaglini, Nucl. Phys. B **794**, 231 (2008) [0707.1153 [hep-th]]; J. M. Drummond, J. Henn, G. P. Korchemsky and E. Sokatchev, 0808.0491 [hep-th]; J. M. Drummond, J. Henn, G. P. Korchemsky and E. Sokatchev, Nucl. Phys. B **828**, 317 (2010) [0807.1095 [hep-th]]; A. Brandhuber, P. Heslop and G. Travaglini, JHEP **0910**, 063 (2009) [0906.3552 [hep-th]]; A. Brandhuber, P. Heslop and G. Travaglini, Phys. Rev. D **78**, 125005 (2008) [0807.4097 [hep-th]]; S. Caron-Huot and D. O'Connell, 1010.5487 [hep-th]; T. Dennen and Y. t. Huang, 1010.5874 [hep-th].

[7] Z. Bern, J. J. Carrasco, T. Dennen, Y. t. Huang and H. Ita, 1010.0494 [hep-th], to appear in Phys. Rev. D.

[8] N. Arkani-Hamed, F. Cachazo, C. Cheung and J. Kaplan, JHEP **1003**, 020 (2010) [0907.5418 [hep-th]]; M. Bullimore, L. Mason and D. Skinner, JHEP **1003**, 070 (2010) [0912.0539 [hep-th]].

[9] F. Cachazo and P. Svrcek, PoS **RTN2005**, 004 (2005) [hep-th/0504194]. L. F. Alday and R. Roiban, Phys. Rept. **468**, 153 (2008) [0807.1889 [hep-th]]. L. F. Alday and R. Roiban, Acta Phys. Polon. B **39**, 2979 (2008). J. M. Henn, 1103.1016 [hep-th]. A. Brandhuber, B. Spence and G. Travaglini, arXiv:1103.3477 [hep-th].

[10] C. F. Berger *et al.*, Phys. Rev. Lett. **102**, 222001 (2009) [0902.2760 [hep-ph]]; Phys. Rev. D **80**, 074036 (2009) [0907.1984 [hep-ph]].

[11] R. K. Ellis, K. Melnikov and G. Zanderighi, Phys. Rev. D **80**, 094002 (2009) [0906.1445 [hep-ph]]; C. F. Berger *et al.* [BlackHat Collaboration], Phys. Rev. D **80**, 074036 (2009) [0907.1984 [hep-ph]]; G. Bevilacqua, M. Czakon, C. G. Papadopoulos, R. Pittau and M. Worek, JHEP **0909**, 109 (2009). [0907.4723 [hep-ph]]; K. Melnikov and G. Zanderighi, Phys. Rev. D **81**, 074025 (2010) [0910.3671 [hep-ph]]; G. Bevilacqua, M. Czakon, C. G. Papadopoulos *et al.*, Phys. Rev. Lett. **104**, 162002 (2010). [1002.4009 [hep-ph]]. C. F. Berger *et al.*, Phys. Rev. D **82**, 074002 (2010) [1004.1659 [hep-ph]]; T. Melia, K. Melnikov, R. Rontsch and G. Zanderighi, JHEP **1012**, 053 (2010) [1007.5313 [hep-ph]]; Z. Bern *et al.*, 1103.5445 [hep-ph].

[12] C. F. Berger *et al.*, arXiv:1009.2338 [hep-ph].

[13] Z.Bern, M. Czakon, L. J. Dixon, D. A. Kosower and V. A. Smirnov, Phys. Rev. D **75**, 085010 (2007) [hep-th/0610248].

[14] Z. Bern, J. J. M. Carrasco, H. Johansson and D. A. Kosower, Phys. Rev. D **76**, 125020 (2007) [0705.1864 [hep-th]].

[15] F. Cachazo, M. Spradlin and A. Volovich, Phys. Rev. D **74**, 045020 (2006) [hep-th/0602228]; Z. Bern, M. Czakon, D. A. Kosower, R. Roiban and V. A. Smirnov, Phys. Rev. Lett. **97**, 181601 (2006) [hep-th/0604074]; Z. Bern, L. J. Dixon, D. A. Kosower, R. Roiban, M. Spradlin, C. Vergu and A. Volovich, Phys. Rev. D **78**, 045007 (2008) [0803.1465 [hep-th]]; F. Cachazo, M. Spradlin and A. Volovich, Phys. Rev. D **78**, 105022 (2008) [0805.4832 [hep-th]]; M. Spradlin, A. Volovich and C. Wen, Phys. Rev. D **78**, 085025 (2008) [0808.1054 [hep-th]]; C. Vergu, 0908.2394 [hep-th]; D. A. Kosower, R. Roiban and C. Vergu, Phys. Rev. D **83**, 065018 (2011) [1009.1376 [hep-th]].

[16] Z. Bern, L. J. Dixon, D. C. Dunbar, M. Perelstein and J. S. Rozowsky, Nucl. Phys. B **530**, 401 (1998) [hep-th/9802162].

[17] Z. Bern, J. J. M. Carrasco, L. J. Dixon, H. Johansson and R. Roiban, Phys. Rev. D **78**, 105019 (2008) [0808.4112 [hep-th]].

[18] Z. Bern, J. J. M. Carrasco, L. J. Dixon, H. Johansson and R. Roiban, Phys. Rev. D **82**, 125040 (2010) [1008.3327 [hep-th]].

[19] Z. Bern, J. J. Carrasco, L. J. Dixon, H. Johansson, D. A. Kosower and R. Roiban, Phys. Rev. Lett. **98**, 161303 (2007) [hep-th/0702112].

[20] Z. Bern, J. J. Carrasco, L. J. Dixon, H. Johansson and R. Roiban, Phys. Rev. Lett. **103**, 081301 (2009) [0905.2326 [hep-th]].

[21] Z. Bern, J. J. M. Carrasco and H. Johansson, 0902.3765 [hep-th]; H. Nicolai, Physics, **2**, 70, (2009); R. P. Woodard, Rept. Prog. Phys. **72**, 126002 (2009) 0907.4238 [gr-qc]; L. J. Dixon, 1005.2703 [hep-th]; Z. Bern, J. J. Carrasco, L. Dixon, H. Johansson and R. Roiban, 1103.1848 [hep-th]. H. Elvang, D. Z. Freedman and M. Kiermaier, 1012.3401 [hep-th].

[22] G. Bossard, P. S. Howe and K. S. Stelle, 1009.0743 [hep-th]; H. Elvang, D. Z. Freedman and M. Kiermaier, 1003.5018 [hep-th]; N. Beisert, H. Elvang, D. Z. Freedman, M. Kiermaier, A. Morales and S. Stieberger, Phys. Lett. B **694**, 265 (2010) [1009.1643 [hep-th]].

[23] M. B. Green, J. G. Russo and P. Vanhove, JHEP **1006**, 075 (2010) [1002.3805 [hep-th]]; P. Vanhove, 1004.1392 [hep-th]; J. Bjornsson and M. B. Green, JHEP **1008**, 132 (2010) [1004.2692 [hep-th]].

[24] J. M. Maldacena, Adv. Theor. Math. Phys. **2**, 231 (1998) [Int. J. Theor. Phys. **38**, 1113 (1999)] [hep-th/9711200]; S. S. Gubser, I. R. Klebanov and A. M. Polyakov, Phys. Lett. B **428**, 105 (1998) [hep-th/9802109].

[25] Z. Bern, J. J. M. Carrasco and H. Johansson, Phys. Rev. D **78**, 085011 (2008) [arXiv:0805.3993 [hep-ph]].

[26] Z. Bern, J. J. M. Carrasco and H. Johansson, Phys. Rev. Lett. **105**, 061602 (2010) [arXiv:1004.0476 [hep-th]].

[27] H. Kawai, D. C. Lewellen and S. H. H. Tye, Nucl. Phys. B **269**, 1 (1986); Z. Bern, Living Rev. Rel. **5**, 5 (2002) [gr-qc/0206071].

[28] M. L. Mangano and S. J. Parke, Phys. Rept. **200**, 301 (1991); L. J. Dixon, in *QCD & Beyond: Proceedings of TASI '95*, ed. D. E. Soper (World Scientific, 1996) [hep-ph/9601359].

[29] Z. Bern, L. J. Dixon, M. Perelstein and J. S. Rozowsky, Nucl. Phys. B **546**, 423 (1999) [hep-th/9811140].

[30] G. 't Hooft and M. J. Veltman, *Annales Poincare Phys. Theor.* A **20**, 69 (1974);

[31] S. Deser and P. van Nieuwenhuizen, *Phys. Rev.* D **10**, 411 (1974); S. Deser, H. S. Tsao and P. van Nieuwenhuizen, *Phys. Rev.* D **10**, 3337 (1974).

[32] M. H. Goroff and A. Sagnotti, Phys. Lett. B **160**, 81 (1985); Nucl. Phys. B **266**, 709 (1986).

[33] A. E. M. van de Ven, *Nucl. Phys.* B **378**, 309 (1992).

[34] Z. Bern and A. G. Morgan, Nucl. Phys. B **467**, 479 (1996) [hep-ph/9511336]; Z. Bern, L. J. Dixon, D. C. Dunbar and D. A. Kosower, Phys. Lett. B **394**, 105 (1997) [hep-th/9611127]; Z. Bern, L. J. Dixon and D. A. Kosower, JHEP **0001**, 027 (2000) [hep-ph/0001001].

[35] Z. Bern, L. J. Dixon and D. A. Kosower, Nucl. Phys. B **513**, 3 (1998) [hep-ph/9708239]; Z. Bern, L. J. Dixon and D. A. Kosower, JHEP **0408**, 012 (2004) [hep-ph/0404293]; Z. Bern, V. Del Duca, L. J. Dixon and D. A. Kosower, Phys. Rev. D **71**, 045006 (2005) [hep-th/0410224].

[36] R. Britto, F. Cachazo and B. Feng, Nucl. Phys. B **725**, 275 (2005) [hep-th/0412103].

[37] C. Anastasiou, R. Britto, B. Feng, Z. Kunszt and P. Mastrolia, Phys. Lett. B **645**, 213 (2007) [hep-ph/0609191]; R. Britto and B. Feng, JHEP **0802**, 095 (2008) [0711.4284 [hep-ph]]; R. Britto and B. Feng, Phys. Rev. D **75**, 105006 (2007) [hep-ph/0612089]; G. Ossola, C. G. Papadopoulos and R. Pittau, Nucl. Phys. B **763**, 147 (2007) [hep-ph/0609007]; R. Britto, B. Feng and P. Mastrolia, Phys. Rev. D **78**, 025031 (2008) [0803.1989 [hep-ph]]; D. Forde, Phys. Rev. D **75**, 125019 (2007) [0704.1835 [hep-ph]]; S. D. Badger, JHEP **0901**, 049 (2009) [0806.4600 [hep-ph]].

[38] Z. Bern, L. J. Dixon and D. A. Kosower, Ann. Rev. Nucl. Part. Sci. **46**, 109 (1996) [hep-ph/9602280]; Z. Bern, L. J. Dixon and D. A. Kosower, Annals Phys. **322**, 1587 (2007) [0704.2798 [hep-ph]].

[39] Z. Bern and Y. t. Huang, 1103.1869 [hep-th].

[40] J. J. M. Carrasco and H. Johansson, 1103.3298 [hep-th].

[41] R. J. Eden, P. V. Landshoff, D. I. Olive, J. C. Polkinghorne, *The Analytic S Matrix* (Cambridge University Press, 1966).

[42] Z. Bern, L. J. Dixon and D. A. Kosower, Phys. Rev. D **71**, 105013 (2005) [hep-th/0501240]; Phys. Rev. D **72**, 125003 (2005) [hep-ph/0505055]; Phys. Rev. D **73**, 065013 (2006) [hep-ph/0507005]; D. Forde and D. A. Kosower, Phys. Rev. D **73**, 065007 (2006) [hep-th/0507292]; Phys. Rev. D **73**, 061701 (2006) [hep-ph/0509358]; C. F. Berger, Z. Bern, L. J. Dixon, D. Forde and D. A. Kosower, Phys. Rev. D **75**, 016006 (2007) [hep-ph/0607014].

[43] Z. Bern, L. J. Dixon and D. A. Kosower, Annals Phys. **322**, 1587 (2007) [0704.2798 [hep-ph]]; C. F. Berger and D. Forde, 0912.3534 [hep-ph]; R. Britto, 1012.4493 [hep-th]; H. Ita, to appear in "Scattering Amplitudes in Gauge Theories", special issue of Journal of Physics A, R. Roiban(ed), M. Spradlin(ed), A. Volovich(ed).

[44] N. Arkani-Hamed, J. L. Bourjaily, F. Cachazo, S. Caron-Huot and J. Trnka, 1008.2958 [hep-th]; N. Arkani-Hamed, J. L. Bourjaily, F. Cachazo and J. Trnka, 1012.6032 [hep-th].

[45] A. Brandhuber, D. Korres, D. Koschade and G. Travaglini, 1010.1515 [hep-th].

[46] C. Cheung and D. O'Connell, JHEP **0907**, 075 (2009) [0902.0981 [hep-th]].

[47] T. Dennen, Y. t. Huang and W. Siegel, JHEP **1004**, 127 (2010) [0910.2688 [hep-th]].

[48] T. Stelzer and W. F. Long, Comput. Phys. Commun. **81**, 357 (1994) [hep-ph/9401258]; J. Alwall *et al.*, JHEP **0709**, 028 (2007) [arXiv:0706.2334 [hep-ph]]; A. Pukhov *et al.*, hep-ph/9908288; M. L. Mangano, M. Moretti, F. Piccinini, R. Pittau and A. D. Polosa, JHEP **0307**, 001 (2003) [hep-ph/0206293]; A. Kanaki and C. G. Papadopoulos, Comput. Phys. Commun. **132**, 306 (2000) [hep-ph/0002082]; A. Cafarella, C. G. Papadopoulos and M. Worek, 0710.2427 [hep-ph].

[49] H. U. Bengtsson and T. Sjöstrand, "The Lund Comput. Phys. Commun. **46**, 43 (1987); T. Sjöstrand, P. Eden, C. Friberg, L. Lönnblad, G. Miu, S. Mrenna and E. Norrbin, Comput. Phys. Commun. **135**, 238 (2001) [hep-ph/0010017]; G. Marchesini and B. R. Webber, Cavendish-HEP-87/9; G. Marchesini,*et al.*, Comput. Phys. Commun. **67**, 465 (1992); G. Corcella *et al.*, hep-ph/0210213.

[50] T. Gleisberg *et al.*, JHEP **0902**, 007 (2009) [0811.4622 [hep-ph]].

[51] S. Catani, F. Krauss, R. Kuhn and B. R. Webber, JHEP **0111**, 063 (2001) [hep-ph/0109231]; M. L. Mangano, M. Moretti, F. Piccinini and M. Treccani, JHEP **0701**, 013 (2007) [hep-ph/0611129]; S. Mrenna and P. Richardson, JHEP **0405**, 040 (2004) [hep-ph/0312274].

[52] Z. Bern *et al.*, 0803.0494 [hep-ph].

[53] S. Frixione and B.R. Webber, JHEP **0206**, 029 (2002) [hep-ph/0204244]; S. Frixione, P. Nason and B.R. Webber, JHEP **0308**, 007 (2003) [hep-ph/0305252]; S. Frixione, P. Nason and C. Oleari, JHEP **0711**, 070 (2007) [0709.2092 [hep-ph]]; S. Alioli, P. Nason, C. Oleari and E. Re, JHEP **0807**, 060 (2008) [0805.4802 [hep-ph]].

[54] A. Bredenstein, A. Denner, S. Dittmaier and S. Pozzorini, JHEP **0808**, 108 (2008) [0807.1248 [hep-ph]]; Phys. Rev. Lett. **103**, 012002 (2009) [0905.0110 [hep-ph]].

[55] T. Aaltonen *et al.* [CDF Collaboration], Phys. Rev. D **77**, 011108 (2008) [0711.4044 [hep-ex]].

[56] S. Mandelstam, Nucl. Phys. B **213**, 149 (1983); P. S. Howe, K. S. Stelle and P. K. Townsend, Nucl. Phys. B **214**, 519 (1983); L. Brink, O. Lindgren and B. E. W. Nilsson, Phys. Lett. B **123**, 323 (1983).

[57] E. Cremmer, B. Julia and J. Scherk, Phys. Lett. B **76**, 409 (1978); E. Cremmer and B. Julia, Phys. Lett. B **80**, 48 (1978); Nucl. Phys. B **159**, 141 (1979).

[58] M. T. Grisaru, Phys. Lett. B **66**, 75 (1977); E. T. Tomboulis, Phys. Lett. B **67**, 417 (1977); S. Deser, J. H. Kay and K. S. Stelle, Phys. Rev. Lett. **38**, 527 (1977); S. Ferrara and B. Zumino, Nucl. Phys. B **134**, 301 (1978); P. S. Howe and K. S. Stelle, Phys. Lett. B **137**, 175 (1984). P. S. Howe and K. S. Stelle, Int. J. Mod. Phys. A **4**, 1871 (1989); N. Marcus and A. Sagnotti, Nucl. Phys. B **256**, 77 (1985).

[59] R. E. Kallosh, Nucl. Phys. B **78**, 293 (1974); P. van Nieuwenhuizen and C. C. Wu, J. Math. Phys. **18**, 182 (1977).

[60] Z. Bern, L. J. Dixon and R. Roiban, Phys. Lett. B **644**, 265 (2007) [hep-th/0611086].

[61] Z. Bern, J. J. Carrasco, D. Forde, H. Ita and H. Johansson, Phys. Rev. D **77**, 025010 (2008) [arXiv:0707.1035 [hep-th]].

[62] Z. Bern, L. J. Dixon, M. Perelstein and J. S. Rozowsky, Nucl. Phys. B **546**, 423 (1999) [hep-th/9811140]; Z. Bern, N. E. J. Bjerrum-Bohr and D. C. Dunbar, JHEP **0505**, 056 (2005) [hep-th/0501137]; N. E. J. Bjerrum-Bohr, D. C. Dunbar and H. Ita, Phys. Lett. B **621**, 183 (2005) [hep-th/0503102]; N. E. J. Bjerrum-Bohr, D. C. Dunbar, H. Ita, W. B. Perkins and K. Risager, JHEP **0612**, 072 (2006) [hep-th/0610043]; N. E. J. Bjerrum-Bohr and P. Vanhove, JHEP **0804**, 065 (2008) [0802.0868 [hep-th]]; JHEP **0810**, 006 (2008) [0805.3682 [hep-th]]; N. Arkani-Hamed, F. Cachazo and J. Kaplan, JHEP **1009**, 016 (2010) [0808.1446 [hep-th]].

[63] G. Chalmers, hep-th/0008162; M. B. Green, J. G. Russo and P. Vanhove, JHEP **0702**, 099 (2007) [hep-th/0610299].

[64] S. Weinberg, in *Understanding the Fundamental Constituents of Matter,* ed. A. Zichichi (Plenum Press, New York, 1977); S. Weinberg, in General Relativity, S. W. Hawking and W. Israel (Cambridge University Press, 1979) p. 700; M. Niedermaier and M. Reuter, Living Rev. Rel. **9**, 5 (2006); P. Horava, Phys. Rev. D **79**, 084008 (2009) [0901.3775 [hep-th]].

[65] R. Kallosh and P. Ramond, 1006.4684 [hep-th];

[66] H. Elvang and M. Kiermaier, JHEP **1010**, 108 (2010) [1007.4813 [hep-th]].

[67] R. Kallosh, JHEP **1012**, 009 (2010) [1009.1135 [hep-th]]; R. Kallosh, 1103.4115 [hep-th].

[68] M. B. Green, J. G. Russo and P. Vanhove, Phys. Rev. Lett. **98**, 131602 (2007) [hep-th/0611273].

[69] G. Bossard, P. S. Howe and K. S. Stelle, Phys. Lett. B **682**, 137 (2009) [0908.3883 [hep-th]]; N. Berkovits, M. B. Green, J. G. Russo and P. Vanhove, JHEP **0911**, 063 (2009) [0908.1923 [hep-th]]; G. Bossard, P. S. Howe, U. Lindstrom, K. S. Stelle and L. Wulff, 1012.3142 [hep-th].

[70] P. S. Howe and K. S. Stelle, Phys. Lett. B **554**, 190 (2003) [hep-th/0211279].

[71] Z. Bern, J. S. Rozowsky and B. Yan, Phys. Lett. B **401**, 273 (1997) [hep-ph/9702424].

[72] B.S. DeWitt, Phys. Rev. 162:1239 (1967); M. Veltman, in *Les Houches 1975, Proceedings, Methods In Field Theory*, eds. R. Balian and J. Zinn-Justin (North-Holland, Amsterdam, 1976); S. Sannan, Phys. Rev. D34:1749 (1986).

[73] Z. Bern and A. K. Grant, Phys. Lett. B **457**, 23 (1999) [hep-th/9904026];

[74] J. Bedford, A. Brandhuber, B. J. Spence and G. Travaglini, Nucl. Phys. B **721**, 98 (2005) [hep-th/0502146]; F. Cachazo and P. Svrcek, hep-th/0502160; A. Brandhuber, S. McNamara, B. Spence and G. Travaglini, JHEP **0703**, 029 (2007) [hep-th/0701187]; P. Benincasa, C. Boucher-Veronneau and F. Cachazo, JHEP **0711**, 057 (2007) [hep-th/0702032]; P. Benincasa and F. Cachazo, 0705.4305 [hep-th]; N. Arkani-Hamed and J. Kaplan, JHEP **0804**, 076 (2008) [0801.2385 [hep-th]]; A. Hall, Phys. Rev. D **77**, 124004 (2008) [0803.0215 [hep-th]]; C. Cheung, 0808.0504 [hep-th].

[75] Z. Bern, T. Dennen, Y. t. Huang and M. Kiermaier, Phys. Rev. D **82**, 065003 (2010) [arXiv:1004.0693 [hep-th]].

[76] N. E. J. Bjerrum-Bohr, P. H. Damgaard and P. Vanhove, Phys. Rev. Lett. **103**, 161602 (2009) [0907.1425 [hep-th]]; S. Stieberger, 0907.2211 [hep-th].

[77] B. Feng, R. Huang and Y. Jia, 1004.3417 [hep-th].

[78] N. E. J. Bjerrum-Bohr, P. H. Damgaard, T. Sondergaard and P. Vanhove, 1003.2403 [hep-th]; C. R. Mafra, 1007.3639 [hep-th].

[79] S. H. Henry Tye and Y. Zhang, JHEP **1006**, 071 (2010) [1003.1732 [hep-th]].

[80] Z. Bern and T. Dennen, 1103.0312 [hep-th].

[81] D. Zhu, Phys. Rev. D **22**, 2266 (1980); C. J. Goebel, F. Halzen and J. P. Leveille, Phys. Rev. D **23**, 2682 (1981).

CHAIRMAN: Z. BERN

Scientific Secretaries: M. D. Azmi, T. Dennen

DISCUSSION I

- *Schmidt-Sommerfeld:*

There is no S-matrix in N=4 super Yang-Mills, and that has to do with the conformality. So there are two questions: Why is there no S-matrix in N=4 super Yang-Mills, and what do you compute when you say you compute the S-matrix?

- *Bern:*

That's a very good question. The problem of the non-existence of the S-matrix turns out to be *almost* identical to the non-existence of the S-matrix in QCD. Strictly speaking, there is no S-matrix for a fixed number of particles, which has to do with the infrared singularities. When we say "the S-matrix," we're talking fast and loose, but it's really no worse than what you do in QCD. The issues are actually very similar to what they are in QCD, or even in QED or in gravity. It's one of those statements that people repeated over and over and over again: "There is no S-matrix in any conformal field theory." And strictly speaking, that statement is true. But actually for practical purposes, we know how to deal with that issue all the time anyway, so it's really not a real issue. If you really want to define something sensible, you have to have a finite resolution, so then you can get a well-defined inclusive quantity. There's another way of actually saying this: if you look at the N=4 super-Yang-Mills amplitudes, they are a piece of a QCD amplitude in a very well-defined sense, that you can do a decomposition of QCD into N=4 plus other pieces. After doing this it is clear that the key issues of defining a sensible cross section is essentially identical to the issues that we face in perturbative QCD. So in conclusion the statements are true of the nonexistence, but they're irrelevant.

- *Ortiz-Velasquez:*

I'm wondering if you have compared your predictions, for example the W+3jets production, with the available data from Fermilab?

- *Bern:*

Yes in fact we have. We do a lot better than leading order programs, because next-to-leading order is in general superior. Of course, when you have a NLO program, the very first thing you do is to try to compare against existing data, and since there's no LHC data you do the comparison of course against Tevatron data.

- *Gerbaudo:*

So when you say "comparison," you mean comparison for example of p_T spectra or jet quantities?

- *Bern:*

Yes, indeed, p_T spectra. That's basically the only available data . We have a recent paper – six months ago – on Z+3jets, that's compared against the D_0 data on p_T spectra.

- *Gerbaudo:*

The other programs for comparison are not an option?

- Bern:

We don't do those comparisons, but in their paper they have their comparisons against ALPGEN and other programs.

- Burda:

I want to to ask something about the spinor and twistor tricks that you use. Are they applicable only in the case of massless theories?

- Bern:

Yes. The MHV rules of Cachazo, Svrcek and Witten, are limited to the massless case. So you could say these rules are not as useful as we would like. We would like a formalism that you could apply in all cases, like for top or W bosons or what you have. That formalism is the on-shell recursion formalism for trees and the unitarity method for loops. This is much better because it allows for masses, and basically all cases are covered.

- Burda:

This on-shell approach is similar to the approach used in the Landau/Lifshitz textbook on QED, which also uses cuts and dispersion relations and optical theorems. Then we restore the full amplitude using just the analyticity.

- Bern:

That's right, there is some connection. However, that formalism was limited in the sense that the only thing people were able to do was two-to-two scattering. That was the Mandelstam representation or the double-dispersion relation. Today we can go beyond that in the sense that we can do any number of legs or loops. Also, there are no subtractions which were a problem due to the non-convergence of dispersion relations. In the way modern formalism works, these aren't issues.

- Burda:

Could you then give some precise reference where this approach is explained in detail?

- Bern:

Here's maybe the first place you should look: hep-ph/9602280 and arXiv:0704.2798. Also the 2nd edition of Tony Zee's book, *Quantum Field Theory in a Nutshell,* contains four chapters on the subject.

- Arnowitt:

How much of this depends on being N=8?

- Bern:

Well, the construction of the S-matrix, or the technology of how we do it, depends very little on it. But for practical purposes, we like N=8 supergravity a lot because we can go to very high loop orders and the expressions are not out of control. If you're asking about the ultraviolet properties, there it does depend on there being supersymmetry. I don't know if we need the full maximal supersymmetry for finiteness, assuming we can prove it's finite, but we definitely need enough supersymmetry, at least N=5, to control the divergences. Pure gravity, as you know, is not finite, as has been decisively demonstrated by explicit calculation, so you need some extra help to bring the power counting down. You certainly need *enough* supersymmetry, but whether you need the full N=8 or not, I don't know. We've stuck mainly to doing N=8 supergravity because we can go up to very high loop order since that is the most interesting question.

- Schmidt-Sommerfeld:
When you use the unitarity method to construct amplitudes, it wasn't quite clear to me what set of diagrams you have to write down that you cut.

- Bern:
These are a set of cuts, not really diagrams, they're the set of cuts that you have to do to fully determine the amplitude.

- Schmidt-Sommerfeld:
Why is it *this* set of cuts?

- Bern:
Basically the way you find is you write down all possible terms that can be in the amplitude, and then you make sure that somewhere in this set of cuts you can see every single term that could possibly be there. Whether they're actually there or not, never mind, but you write down an ansatz with arbitrary coefficients for every single possible term that could be in the amplitude, and you make sure that you can see each term in at least one of these cuts.

- Schmidt-Sommerfeld:
And how do you get the outcome?

- Bern:
One way we like to work is to construct an ansatz that basically has every possible term with arbitrary coefficients. Then you solve for the coefficients by demanding that the ansatz has the correct cuts, where the cuts are products of the tree amplitudes. This allows you can solve for the coefficients.

- Borsten:
What's your feeling on gravity as a double copy of gauge theory? Is it something really deep, or is it some trick for the amplitudes? Could it be that you could write gravitational solutions using solutions to gauge theory as a double copy?

- Bern:
We don't have any proof that goes beyond perturbation theory, but I'm willing to wager a lot of money that in fact it's a very general property – that any solution of gravity can be expressed as a double copy. I'm not going to tell you exactly how to do that, because I don't know, but I'm pretty sure such a mapping exists. But why don't we hold off on that until tomorrow, because first I need to show you how it works in perturbation theory, and once you see how it works in perturbation theory, I think you'll be convinced that this is not an accident; that this is of fundamental significance for quantum gravity.

- 't Hooft:
I have two unrelated questions. One is really planar Yang-Mills, first when you look at the theory non-perturbatively, you expect meson resonances, so the theory rearranges to make mesons.

- Bern:
Yes, but this is N=4 super Yang-Mills, which is unfortunately not like QCD at all; there's a finite radius of convergence in this theory, so the perturbation theory is everything. This is very special to N=4. In some sense, I wish it wasn't like that, because then we'd be closer to QCD, but then it would make it much harder to solve planar N=4 theory.

- 't Hooft:

The other question may obviously be related to what you want to say tomorrow. I was thinking if gravity is a double copy of gauge theory, maybe there's a way to rewrite it in terms functional integrals. Of course if we have a very compact way of writing everything once you see it as a functional integral. Say for instance, the vierbein formalism comes sort of halfway. If you write the metric as a product of two vierbeins, and if you could interpret every single vierbein as a gauge field, then you're nearly there.

- Bern:

As you noted, there's a problem with that, so you are only halfway there. When you write the metric as a product of vierbeins, there's an index which contracts across the vierbeins. That index is evil, because the way the double copy works is that there should be no index that contracts across the two copies. So it's not really like the vierbein, except maybe if you can do something in some special gauge. You have to stop that index from talking between the two vierbein.

- 't Hooft:

Turn the vierbein into a one-bein?

- Bern:

Yes, that's right, we would want something like this.

- Giddings:

In the context of gravity, the non-perturbative regime is obviously very interesting. Coming back to this doubling business, do you think that there's some way of representing, say, Schwartschild as a double of some gauge-theory function?

- Bern:

I was just talking to Michael Duff about exactly that question. I believe the answer is yes, but I have no idea how to do it. I really can't say more than that. Why don't we hold that question until tomorrow, because maybe after I explain how it works in perturbation theory then we can discuss exactly this point. The question of how to do this is definitely something on my mind.

- Dennen:

Supposing there's an all-orders solution in the next year or two to scattering amplitudes in planar N=4 super Yang-Mills, what would that say about topologically non-trivial configurations?

- Bern:

Well, the thing is this is in the planar limit. It's also what I remarked to Gerard 't Hooft, that the planar Yang-Mills almost certainly has a finite radius of convergence. So there's nothing outside perturbation theory. Perturbation is the entire planar theory in this case because of the finite radius of convergence. Even when you go to strong coupling, it's just by analytic continuation outside the circle of convergence. It would be really fantastic if we solved this theory in the planar limit, but unfortunately it does not mean that all of a sudden all sorts of non-perturbative issues are under great control.

DISCUSSION II

- *Schmidt-Sommerfeld:*
Does the double copy property relate gravity to the planar limit of Yang-Mills?

- *Bern:*
The duality relates planar and non-planar Yang-Mills diagrams, so the non-planar diagrams are mostly determined by this relation from planar diagrams. To get gravity out of it, you take all of the numerators and square them, regardless of whether they are planar or nonplanar. You may notice in the list of diagrams contributing to the three-loop amplitude, described in my talk, there are nonplanar diagrams listed. If you square these nonplanar numerators, they give you exactly the corresponding numerators in supergravity.

- *Schmidt-Sommerfeld:*
What is the meaning of the rank of the gauge group in gravity?

- *Bern:*
In gravity, there is no number of colors. To get gravity you replace a color factor with a kinematic numerator factor so the color no longer appears. Unfortunately, this makes gravity much more complicated than gauge theory. For gauge theory we will have big celebration if we can solve even the planar limit. But in gravity there is no way to separate planar and non-planar using anything like color; it is all mixed together.

- *Burda:*
I have a question about this duality between color and kinematics. As I understand, Yang-Mills theory is simply related to the usual Einstein gravity. Could we find an even simpler example of this duality? Do we need the theory to be non-abelian? For example, if we take a very simple abelian scalar theory with self-interactions, which gravity theory will be dual to this scalar theory?

- *Bern:*
The Yang-Mills is already so simple in how it works. It's difficult to come up with some example where we can learn more than what we have here. It turns out that there are some constructions that are built on scalar field theories. The basic idea is that instead of having a kinematic numerator in the amplitude, you can construct a scalar field theory where this numerator is replaced with second copy of the color. There, you can make some analogies to what we have with Yang-Mills. But at the end, it's not really simpler than what we already have.

- *Burda:*
About this statement that gravity is double gauge theory, that the metric tensor is a product of two gauge fields. It looks more or less analogous to the vierbein formulation of Einstein gravity. Is there any connection?

- *Bern:*
You're right that it is an obvious thing to think about. In fact, many years ago, when I was first thinking about this, I thought that it might be where the explanation lies. But on the first line

you have a problem with the flat-space index which contracts between the two vierbein. That's not what we want, because we want something that truly factorizes as it does in the double copy.

- Burda:

Aren't the color indices on the gauge fields here already traced out?

- Bern:

The color index is actually irrelevant and basically get thrown away. In our double copy, there are no indices which contract across the two copies, not even color. A lot of the problem of understanding the double-copy is not on the gravity side; it is on the Yang-Mills side. You have to modify Yang-Mills in just the right way; you have to add something to the Lagrangian which happens to be zero yet puts Yang-Mills in a form related to gravity via the double copy. The color indices don't play a role in this.

- Bettini:

In this correspondence between gravity and Yang-Mills, what is the role of the cosmological constant? I mean, gauge theories have one fundamental constant – the coupling constant – while gravity has two – the coupling and cosmological constants.

- Bern:

Unfortunately, we work in perturbation theory around flat space so we don't see this. That's not to say that your question isn't an important one. Your question is connected to the question of constructing an entire Lagrangian to all orders, and then solving equations of motion away from flat space. So at the moment, we have nothing to say about the cosmological constant; we simply don't see it in flat-space perturbation theory.

- Bettini:

Is this because the cosmological constant is a very long-range thing?

- Bern:

Yes. This is flat-space perturbation theory, so by definition we set our cosmological constant to zero, when we set up a perturbative expansion. You might say, "Oh look, you're not capturing important physics!" But, our motivation for doing this is we are interested in the ultraviolet properties of gravity which is a short distance question. So the reason why your question isn't relevant is that at present it's just not part of the framework. Now, that's not to say that we wouldn't want to make it part of the framework, but in order to do that, we would really need to understand equations of motion beyond perturbation theory. We would need to figure out the double-copy Lagrangian to all orders in perturbation theory, and then we can start thinking about putting a cosmological constant into that Lagrangian and to see how it works. For the moment, it's bit too far out of reach in order to address your question.

- Giddings:

In the usual perturbative story of gravity, the problem is that we have in fact many theories of gravity, because there are all these counter terms that we could add with arbitrary coefficients. In such a theory where you add all these terms, do you believe you would have this factorization property, or would the factorization property be a selection principle which would eliminate some of the counter terms?

- Bern:

That's a very good question. If you go back to the history of how we knew about double copy and so on, it comes from string theory, via the Kawai-Lewellen-Tye relations. In string theory such higher order operators appear. I am pretty sure that everything I'm saying here about the double copy and these types of formulas, will be apparent in the higher order terms of string theory which will have all these properties.

- Giddings:

I'm not sure if that's answering the question; let me rephrase it. If we start from the Einstein action as opposed to this factorization property, let's suppose I add to that R^{18} and R^{20}, will that respect the factorization property, or does the statement that you want the factorization property restrict the counter terms?

- Bern:

I'd be surprised if any set of coefficients would work. There's a paper by Bjerrum-Bohr, who looked into the first few orders of string theory, and he found that if you fiddle with the coefficients of the operators, you can make something work, but you can't fiddle too much or put in extra operators that didn't already appear or things won't work. So there are definitely some constraints, but if I remember the conclusion of this paper, more things are allowed than just string theory, at least to the first few orders in alpha-prime. I think that maybe the first place people should address this question systematically is by looking at string theory, at the infinite tower of operators that you could add, and then seeing what happens as the coefficients of the operators are modified.

- Galakhov:

Is it possible to construct doubled color factors? Will it have some useful meaning?

- Bern:

It does, actually. Double color factors turn out to be a scalar field theory, just phi^3 theory, but it carries two group indices. That's actually just ordinary phi^3 field theory, but with two f^{abc}s contracted into the vertex. It's a very simple field theory, and it does have a physical meaning. That's what I was alluding to earlier, that you can do something supposedly simpler, but it's actually not terribly enlightening compared to the real thing, which is gravity and Yang-Mills.

- Alba:

What is the gauge group of this sYM and what will we obtain in the case of another symmetry group?

- Bern:

Here we have SU(N) non-abelian theory. But, in fact, all these properties are independent on the specific group, all properties just rely on the Jacobi identity.

- Alba:

What does N stand for? What does N mean in gravity?

- Bern:

If you look at two formulas here, you notice I took this color factor, and I just got rid of it; it's not there any more. I just replaced it with the numerator factor, so the color factor is gone in the transition from gauge theory to gravity. So whatever you want it to be, SO(N), SU(N) and so on it's not relevant for making this transition.

- Alba:

Can we regard just SU(2) Yang-Mills theory?

- Bern:

Yes.

- Alba:

What will we obtain in the case of quiver gauge theory?

- Bern:

As long as this amplitude is generated by a non-abelian vertex, whatever this color factor is completely irrelevant; it just has to satisfy Jacobi identity. If you start modifying the states of the theory, like when you start adding matter and complicated global symmetries, than I'm not sure what exactly it corresponds to on the gravity side. But the way it works is that in the gravity theory the states you get are the tensor product of the Yang-Mills ones stripped of color.

- Alba:

If you write Lagrangian with six-point correction the number of items will be around one-hundred. Does it mean that we have no chance to write exact Lagrangian with any *n*-point correction?

- Bern:

That's a good question. I'd say you don't have any chance unless you are clever. Just because something is a mess doesn't mean that you can't reorganize it in a beautiful way. A perfect example is the Einstein action. If I gave you the Einstein action order by order in perturbation theory it would be a very unhappy mess, and you'd say, "what a mess, there is no logic to it," but in fact it is a very beautiful object if you look at it the right way. To me it seems that the situation with the Yang-Mills Lagrangian satisfying the duality with color is exactly the same, that there is a beautiful structure, but we just don't see it yet. I believe that, if you keep on adding terms and then at some point you realize that what you've added can be re-expressed in terms of something beautiful, the answer is out there, and if you want to look for it, that would be fantastic.

- Borsten:

The duality of color and kinematic factors suggests that the kinematic factors enjoy some symmetry, which one might expect to be related to the gauge symmetry. Does this symmetry carry over to gravity via the doubled copy-property and if so, how?

- Bern:

There is definitely a gauge symmetry, that I haven't talked about – a generalized gauge symmetry. Here's an amusing comment about the question of symmetry. You see this Lagrangian that we're adding? I would say, "Oh, there must be some symmetry to organize this thing in some reasonable way." But I'm adding zero, so I have to be careful about what exactly I mean by symmetry. It has to be something a little bit unusual, where you are imposing a symmetry, but then you're not allowed to use Jacobi identities that rearrange anything, because otherwise it's completely lost.

In the amplitude, however, there is a generalized gauge symmetry in Yang-Mills theory, which is induced in gravity. The invariance is the following: imagine all possible transformations of the gauge theory diagram numerators such that the amplitude is unchanged. Then if you come over to gravity, that means that one factor of the numerator here can be transformed in exactly the same way. This set of transformations is much larger than ordinary gauge symmetry. You can throw in arbitrary functions here, as long as you throw it in one diagram and cancel it out from another diagram, to leave the amplitude invariant. And you're allowed to use Jacobi identities to really make a mess of things. That invariance of the amplitude gets induced to the gravity amplitude. That's an example of a symmetry or invariance, that you could see in *this* way, by knowing about this double copy, that you will really not be able to see easily by thinking directly about the Einstein action or Feynman rules.

- *Arnowitt*:

Are the results automatically finite or do you have to regularize the amplitude first?

- *Bern:*

When making a claim at a certain loop order, that it is finite is automatic in the best representations we now have. The finiteness is apparent before we do any integral. Up to four loops we now have representations which are finite just by power counting alone in all terms, so there is no need for an ultraviolet regularization. There are infrared singularities if you're in four dimensions, but that's another story. When we investigate divergences in higher dimensions then we do introduce a form of dimensional regularization.

Another comment is that the double copy does not automatically imply that if N=4 sYM is finite, then N=8 supergravity (also) is finite. The reason why it doesn't do that is because, generically, you're squaring numerators, for example when there is a loop momentum in the numerator, and if you square it, it will have worse behavior. So then things can get more complicated to understand exactly what the right behavior of gravity is as the loop order increases.

- *Giddings:*

I noticed that in your "vague formula" for the metric in terms of a product of two gauge fields, actually that formula was non-local. Does this indicate any non-locality in the gravitational theory?

- *Bern:*

The reason why it became non-local is basically because the vertices are a product in momentum space, so then when you translate to the Lagrangian in momentum space, you get a product, so obviously if you have product in momentum space, you wind up with a convolution to coordinate space. But as to deep physical significance, for example as in your talk where you were discussing non-locality, first we should write down the precise formula. Right now I'd say it's a little bit too vague to conclude much.

- *Azmi:*

How can the theory of quantum mechanics be merged with the theory of general relativity/gravitational force and remain correct at macroscopic length scales?

- *Bern:*

The quantum corrections are so microscopically tiny. Let's assume that we are not going to worry about any infinities, or normally the way that we can think about it is the effective field theory. Those loop corrections are so tiny, for example, if you look at a tree scattering with two powers of gravitational coupling, then a loop diagram will have many more powers suppressing it. So basically it is irrelevant to the normal world that we're used to, because the quantum corrections are extremely tiny.

Scientific Secretaries: J. Jejelava, M. Kurkov

DISCUSSION III

- *Cerchiai:*

Is there some symmetry behind the on-shell recursion relations that you explained?

- *Bern:*

There is a paper from, Nima-Arkani-Hamed and Jarad Kaplan where they discuss an enhanced Lorentz symmetry that may be responsible for the good large z behavior needed to prove the recursion relation for gravity. For gauge theory I'm not aware of any symmetry that has been proposed, but it's certainly something people are looking at and there may be more to say soon. But it is fair to say that it is not as yet fully understood.

- *Kusmarski:*

Is this good behavior with z: does it work so for some sort of effective theories because they usually don't behave well in the high energy limit. Right?

- *Bern:*

Well, gravity is the really perfect example where things work where you would think it wouldn't because of bad high energy behavior. Naively you would think this is an effective field theory because it has bad high energy behavior. In fact, it behaves very well as far as the large z behavior is concerned. In other physical effective theories I would not be surprised if things behaved better than you might naively think, especially if you consider general complex shifts of multiple external legs.

- *Kusmarski:*

But for example four-Fermi theories.

- *Bern:*

Yes, four-Fermi theories are definitely unhappy theories with bad high energy behavior. But it turns out you can do something sneaky. Let's say you have some amplitude which behaves poorly under some z shift. There could, however, be more complicated shifts or shifts of multiple legs that give you good behavior. In many cases, if you had some kind of theory that has bad behavior, many times you can fix things just by shifting larger numbers of legs. The choice of shifts you made will drop out of the final answer, as long as there is good large z behavior, because we are just using the Cauchy's Theorem. So the ideas presented here are very powerful because they are very general.

- *Burda:*

Just wondering whether this technique could tell us something, maybe after simplifications, about non-planar limit of the Yang-Mills Theories or maybe Super Yang-Mills. So could it be possible to solve YM completely by using this technique?

- *Bern:*

In the current lecture, it was only about tree level, so it is very far away from planar and non-planar loop amplitudes. But the issue of non-planar is connected to my previous lectures. In those I showed you the duality relations between color and kinematics, and those relations precisely take you from planar to non-planar. But if you solve N=4 SYM in the planar limit I am not going to say "that means you solved the non-planar case as well". However, using the duality there is definitely information we can carry from planar to non-planar. In fact, today the way we carry out calculations in a non-planar $N=4$ SYM at loop level is first evaluate the planar contributions and then use that to determine non-planar integrands using the duality. This doesn't yet allow you to solve the theory because one would still need to do infinitely many nontrivial integrals.

- Boiskie:

So the way I understand it is that, in order to stay *on shell*, all you need is to use this complex shift. Is that right?

- Bern:

Yes, the complex shift was essential. It is how the recursion relation is constructed. If we did not use complex momenta it would mean you have a vanishing three-vertex. With real momenta the kinematic constraints force an on-shell gauge theory three vertex to vanish. If you were to start doing recursion starting from such a three vertex, you would get all amplitudes to vanish which would not be a happy situation. So yes, complex momenta and shifts are essential.

- Boiskie:

So is it appropriate to think of going in to the complex space, as being equivalent to going off-shell in the conventional formalism?

- Bern:

No, because a key feature about going off-shell is that the results are not gauge invariant in general. When you stay on shell with complex momenta the results are gauge invariant. So it's not the same thing. You can see this at the level of three vertices. If we go off shell, I have to tell you the gauge, but on shell there is a unique object that does not depend on the gauge.

- Boiskie:

Presumably there must be, even if it is terribly complicated, some way to find this recursion relations from the conventional off shell techniques.

- Bern:

Yes, there is some understanding of this in terms of a gauge called space-cone gauge. In this gauge it is possible to rearrange standard field theory into a form which can be interpreted in terms of the on-shell recursion relations. There are also off-shell recursion relations which were found a long time ago by Berend and Giele which give the recursion relations which are your off-shell recursion relations and those were used originally to prove the Parke-Taylor formula. However, the solution of those recursion relations are much more complicated than in the on shell case.

- 't Hooft:

I basically have two questions. The first is: do you see any special role played by conformal formal symmetry? As I explain in my lectures there is a very delicate feature that conformal symmetry places in gravity and actually as you know of course the theory has at least tree level conformal invariance.

- Bern:

There is conformal symmetry if you look at $N=4$ SYM theory. In $N=4$ SYM theory if you just look to pure gluonic sector at tree level it is equivalent to QCD, so even there the conformal symmetry is apparent. But if you ask me about gravity, it's not so obvious. The one path I can see for studying it is through the double copy. Certainly I don't see it.

- 't Hooft:

I explain in my lectures that if you first integrate over the conformal factor the results that remain are completely conformal invariant and so in indirect way conformal symmetry survive in a back-door.

- Bern:

Well, it's hard for us to see this since we are in flat space perturbation theory with on shell state so I do not think that we have an access to this at present. It's something we should try to think about in the future.

- 't Hooft:

Well, I have an example. Today, when we compute one loop counter terms, the expressions we should realize are completely conformal invariant objects, so the conformal invariance is going through diagrams in a very special way, which you may be interested in.

- Bern:

Certainly I am always interested in symmetries that can help explain structures. But this is something that we still have to think about.

- 't Hooft:

The other question is of course you have done spin 1 and spin 2, so why not include also the spin ½ and even spin 0? And a good question is to consider general Lagrangian containing everything: spin 1, spin 1/2, spin 0, spin 2 and even spin 3/2. Do you see any prospect of generalizing these theories ?

- Bern:

Well, when we do our computations in $N=4$ SYM or $N=8$ supergravity we use an on-shell super-space formalism so all the states are present there, from spin 0 to spin 2. Now you can impose projections, to keep some states or not keep states, so the methods should apply for any states. The on-shell recursion relations themselves work directly for any spin, as has been studied in many papers.

-'t Hooft:

I think that with the standard model you explained you can do pure QCD without any supersymmetry. But what if you have a whole bunch of the particle species as in the Standard Model: no supersymmetry and lots of particles. I am thinking about the question of generalizing the whole idea to do any calculation in the Standard Model.

- Bern:

We have devoted a lot of time to exactly this question, This tree level techniques have been applied very generally to many species and to amplitudes with massive particles as you would have in QCD. The technique is quite general because it relies on basic factorization properties in field theory as well as on Cauchy's Theorem. A number of people have studied massive amplitudes, including the massive fermions and vector bosons. At loop level when we do our calculations of $W+4$ jets obviously we must consider quarks, W-bosons and so on. I think the importance of these methods is precisely what you are pointing to: a general technique must be widely applicable to quarks, vector bosons and to massive particle. If it was a special technique that you can only use in pure gluon and pure gravitons then it's very nice but obviously of limited importance. I believe the methods I'm talking about belong to textbooks on quantum field theory precisely because of their generality.

- Schmidt-Sommerfield:

This is again physically a completely irrelevant question, but I'll anyway ask. The conformal field theories should not be defined on Minkowski space but its conformal compactification, right? So I am just wondering whether there are infrared singularities and could you define the S-matrix in principle?

- Bern:

First let's look at the tree level which was the subject of my previous lecture. At tree level there is essentially no difference between $N=4$ SYM theories and QCD if you ignore the masses. So obviously there is no issue because there is no issue in QCD. The problems you are referring to are at loop level. But even here I think this is a relatively minor technicality. When we perform our calculations of N=4 SYM amplitude and we do it to high loop orders and even to all loop orders, it exactly matches the string answer in Euclidean space. So, I think the question of whether this is this truly the "S-matrix" and how one might define it is an irrelevant question. It is well defined enough in dimensional regularization. This is the quantity we are talking about that's very closely related the same quantity for studying scattering in QCD, so I think that it's just a technicality. Maybe we don't understand everything about it. I don't think it is a crucial point.

- Korthals-Altes

From your lectures I came away with the impression that in order to have this double copy relation between gravity and gauge theory, you need these parallel relations between color factors and numerator momentum factors.

- Bern:

That is correct. Yes.

- Korthals-Altes

But it did not seem to play a role at all and that...

- Bern:

You mean in the homework example?

- Korthals-Altes

Exactly!

- Bern:

Yes. This is actually quite an interesting point. It turns out that there is something very special about four points. For four-point amplitudes it turns out that the relations automatically hold. That is, the relation between color and kinematic numerators and the double copy automatically hold in any gauge or representation in terms of diagrams. I don't know why this happens, but it does. So, the duality plays a role but you do not have to work for it. It is completely automatic. The place where this becomes quite nontrivial is when you go to five or more points. In these cases it plays a direct and essential role. You have to enforce this relationship of course maintaining you get the correct amplitude. It is only when the relation between color and kinematics is enforced that the double copy works properly to give gravity.

- Ferrara:

You say that in Einstein gravity there is some sort of unexpected better behavior?

- Bern:

Yes, there are unexpected ultraviolet cancellations.

- Ferrara:

Yes, but are there any counterterms in standard Einstein gravity which would be there and which are not there because of this.

- Bern:

When I say "it behaves better" I don't mean that pure Einstein gravity behaves well enough to be finite. It is still diverges, as we know from the explicit computations of Goroff and Sagnotti. What I'm expecting is that if we were able to compute the degree of divergence, it would not behave as badly as we would naively expect, even if you encounter divergences. To make gravity finite we need extra symmetries such as supersymmetry.

- Ferrara:

That's right.

- Pauk:

I'm not a specialist but how can you be sure that analytically continuation to complex z-plane is allowed? So, you made some assumption about analyticity or something like that?

- Bern:

Yes. When we make some claim that something is allowed we have a great advantage in that we have a very large number of results in gauge theories worked out previously that we can compare to. In addition we also have Feynman diagrams to help us resolve any potential subtleties along the lines you are worried about. An example of the type of confusion you can encounter where you do some analytic continuation is how to deal with the Feynman $i\varepsilon$ proscription. How do you know that we have dealt with all such subtleties properly? The bottom line is we have Feynman diagrams and people have performed an enormous number of checks. It really works! You may ask about loops because issues in loop calculations are more severe. It's something that people who do this pay very careful attention to. Basically you have to be paranoid precisely about analytic continuation issues, but it has been checked extensively at loop level as well.

- 't Hooft:

Are there any more questions? Then I thank Professor Bern very much for his lectures and discussions.

R. BOUSSO

"The Measure Problem in Cosmology"

This Lecture has already been published on the
Gen. Rel. Grav. 40 (2008) 607-637:
"TASI Lectures on the Cosmological Constant"

DISCUSSION I

- *Pagnutti:*

How special is this time in which the density of the vacuum energy is of the same order as the density of matter?

- *Bousso:*

You can think of this time in such a way. The density of matter behaves as one over expansion factor cubed. So when the universe is doubled in its size the density of matter is decreased by a factor of eight. It is already decreased by a factor of 3 to the density of the vacuum energy. So it would be comparable no longer.

- *Arnowitt:*

In your talk you were somewhat negative on quintessence models. Do you think that the planned measurements of the ratio of pressure to density will give -1?

- *Bousso:*

I should say that I do not think that there is a reason for not doing an experiment. I mean I am a stronger believer in doing incredibly precise experiments. What worries me is that we have designed a quality based on the completely suspicious parameters like w and w', which from any fundamental theoretical point of view correspond to insanely fine-tuned things. So what worries me is to decide which experiment we are going to do, this is based on how well we have constrained this parameter space, while these parameters have no theoretical foundation. May be, by some luck we pick the right experiment, but it is not the way you usually operate. What I would like to see is perhaps speaking about objects which certainly make sense, which tell us something about geometry. We would like to measure an expansion, to be even more precisely. We want to measure the Hubble constant, a bunch of red shifts and a clustering more precisely. It is very important. But due to lack of this information, planning the experiment based on w, w' is very dangerous.

- *Alberte:*

What are the reasons why the string landscape can be considered a better solution to the cosmological constant problem than anyone of modified gravity?

- *Bousso:*

There are two parts in this question. Why string landscape is a plausible solution to the cosmological constant problem? This will be highlighted in the future lectures. And the second part is why theories like modified gravity just add problems. The reason is that such kinds of theories are basically similar to the quintessence, in the sense that some miracle makes the Lambda-constant equal to zero. And there is something else suspicious that pushes the Lambda-constant away from zero, and this does not have something in common with what makes Lambda-constant equal to zero. It is true that any modified gravity theory still predicts enormous vacuum energy. This is one problem we definitely have, and these approaches do not even attempt to solve it.

- *Juansher:*

How would the limits on the Lambda-parameter change if we change the current density of the Universe?

- *Bousso:*

Actually, the density and the Lambda-parameter change rather independently with time. So, for example, their ratio is known to be changed in the Universe during its evolution. So there is no so obvious dependence between these quantities.

- *Inguglia:*

What is your own point of view on the possibility of interpretation of gravity not as a fundamental force, but as entropic force?

- *Bousso:*

I don't think that anyone seriously believes that gravity in a form of General Relativity is a last word. But there is one thing that I really want to emphasize; whatever your underlying theory is, it has to reproduce General Relativity in a limit where we tested it and a cosmological constant will arise. Hence, the assumption that everything in General Relativity is true in the limit where we tested it, is a basic reason why it is unclear how quantum gravity modifications of General Relativity could have anything to say about the cosmological constant problem, at least at any simple level.

- *Mironov:*

How could some recent research like screening of the Lambda-term or self-screening change our knowledge about cosmological constant problem?

- *Bousso:*

I am not going to go into technical details and discuss the very mechanism of screening in the de Sitter space. Even if such thing like screening is true, it is not going to work with the cosmological constant problem in our Universe. It might work for other Universe, which is empty. But it is not suitable in our case because of early stage of Universe evolution with huge amount of matter in addition to pure gravity. And in peculiar the solution of gravity equation of motion is not approximately de Sitter space. So you can see that the problem of dynamical reduction of the cosmological constant cannot work for our Universe.

- *Mironov:*

But in our Universe we can solve this problem even if we know there is matter. There will be processes of screening of the cosmological constant, in the early Universe it would not be a process of self screening, but screening by the particles which are emitted by matter.

- *Bousso:*

You see, this screening has to be a dynamical process. But you have to know what you are trying to cancel. Nobody, no observer, no matter system, nothing can measure the value of cosmological constant in early Universe with sufficient accuracy. Basically, with cosmological constant Lambda time scale and distance scale, proportional to the inverse square root of Lambda, are associated. And you have to do the experiment, which probes that distance scale, which lasts for at least that time scale, is the reason why it would be impossible for us to detect the cosmological constant if we were doing it 10 billions years ago.

CHAIRMAN: R. BOUSSO

Scientific Secretaries: M. Mulhearn, C. Pagnutti

DISCUSSION II

- Inguglia:

Today you were talking about the Big Bang as a decay of a parent vacuum. I'm not sure I have understood what you mean.

- Bousso:

When people talk about Big Bang cosmology, what they really mean is that when you let the clock run backwards, you extrapolate back in time from the expanding universe that we see today, you take into account what we know about the matter content that it has, and what we know about the standard model, and you run it backwards to an earlier era then it becomes more and more dense and more and more hot. If you just run it all the way back without thinking much about it you end up with a space time singularity that we call the Big Bang.

Now I don't think that very many people actually believe that it is really fair to extrapolate the evolution backward all the way to that singularity. One reason why that seems implausible is that it would be very difficult to understand the flatness and homogeneity of our universe if that was truly valid. So for that reason people have long considered the possibility that there was a period even earlier when there was a scalar field that rolled down some potential and insulated the universe and made it flat--this is a different kind of inflation from the eternal inflation that I was talking about today, it's not a meta-stable vacuum but just a slowly rolling scalar field.

But that still leaves open the question of how inflation began, or how this slow roll inflation began. You might be motivated to consider all sorts of different models for how the universe began and eventually ended up with this slow roll inflation. Here we started from the other way around, we started with string theory, we looked at the picture that it gave us, and we were forced to conclude that in this picture what happens is that this slow roll inflation is further preceded by a phase transition where a bubble formed inside a region with a much larger cosmological constant.

Of course that does not tell us how the universe started any more than before, we have simply post-poned the question a little bit further back. You can still ask how this eternally inflating universe got started. There are ways of thinking about eternal inflation which make the predictions that you are going to get rather insensitive to how you start the processes. So in that sense the picture that I was presenting perhaps reduces the urgency of having to find a theory of the initial conditions, but I think in principle it is still a very important problem.

- Ambrosetti:

I actually have two questions. One was in the context of eternal inflation, so you have the formation of this bubble, a mini-verse. How would we see the formation of a bubble in our own universe? And the other question which is also related is if we really had many many vacua even more than 10^{500}. Wouldn't it be quite likely that baby universes would form in our own?

- Bousso:

Let's start with your first question. If a bubble nucleated somewhere in our own visible universe then what would happen is pretty much what I described. For a very large set of relatively natural parameters this bubble would appear at a very small size. Remember that the size is basically controlled by how large I have to make it so that the energy is conserved when the bubble appears, and that tends to be a micro physical scale. Now the bubble expands approaching the speed of light very quickly, and the characteristic time-scale of its acceleration is also controlled by this micro-physical size divided by the speed of light. Because of the symmetry of the problem there is only one scale in this bubble. And for this reason, you can imagine that if a light ray gets emitted by something and approaches you, you get zero warning, you get hit by this light when you have no chance of anticipating that by causality. This thing is almost like a light ray, and the amount of warning that you can get at best is equal to the initial size of the bubble divided by the speed of light. That's the warning that you are going to get, so in practice none at all.

- Ambrosetti:
When it comes at you then...

- Bousso:
It will be a painless death.

- Ambrosetti:
So if something like this happens we could know about it...

- Bousso:
You won't be around to know about it I am afraid. In principle there is no fundamental physics obstruction to surviving a collision with such a bubble. It is just in practice extremely implausible, since not only will you be disintegrated but your actual electrons and quarks will be disintegrated into something new. So it is of course in practice not a realistic thing to hope to be able to see.

The second part of the question, is that if there are so many vacua with lower cosmological constant, why don't we decay into one of those? First of all, it is actually not relevant how many vacua there are with cosmological constant lower than ours but above zero. We can perfectly happily decay into the ones that have cosmological constant negative. So there is roughly half of the landscape that we can decay into while decreasing the vacuum energy. So that's quite a large number, about 10^{500} vacua, but the suppression rate for these decays can easily vastly exceed 10^{-500}. So even though there is a multiplicity that enhances the overall chance of vacuum decay, it is actually a relatively small number compared to the kind of suppressions that you can easily get from the potential. In particular, what you have to keep in mind, is that of the 10^{500} vacua only maybe 500 or so are going to be nearby in the landscape, and even those can already be very suppressed, and then the far away ones are going to be ridiculously suppressed, so you can basically forget about them.

- Alba:
You told that in principle it is possible to go from one vacuum to another vacuum, [that some kind of tunneling is possible]. And as far as I can tell, a different vacuum means a different universe.

- Bousso:
This is somewhat misleading language. Different vacuum really doesn't mean anything dramatically more different than for example the state of our vacuum before and after electroweak symmetry breaking. What you mean by that is that the perturbative physics around the minimum of the potential is different, and that means you have different particles, different forces and so on, different energy of empty space. But fundamentally, there is nothing dramatic happening here, it's the same space-time, it's the same fundamental theory, it's just that you've relocated to a different place in the landscape. When people say multi-verse or different universes it often makes it sound more mysterious than it really is, it's not some hypothetical different world that is sort of unconnected.

- Alba:

When I listened to lectures by Andre Linde, he also talked about multi-universes. He told us that there is a barrier between the different universes, and that it is impossible to get from one to another.

- Bousso:

Well, I can't speak for Andre of course. The only thing that I can imagine that he might have been referring to is in the situation where you have some other bubble that is behind your horizon and of course it is causally impossible to get to that place for the same reason that it is always causally impossible to reach a signal from a place that is beyond your event horizon. So of course that is a very conservative statement when you really think about it, and nothing terribly new. But what's always at least in principle possible is that if you have high enough energies available in your accelerators you should be able to go into these six extra dimensions and take some of these p-branes and put them somewhere else or tear them up and throw them away. This is just matter. It's just matter configurations which you in principle can reconfigure. Of course in practice it's not going to happen. But I'm just trying to say that there is nothing magical going on that goes beyond just moving around matter.

- Alba:

What do you mean by tunneling between different vacua? The tunelling of what?

- Bousso:

There are three ways that you can see a different vacuum. One is you can hope that a bubble forms in our vacuum that spontaneously by tunneling, overruns us, and mysteriously we survive this, and now we see a new vacuum. And of course, [this is] not likely to happen. The other way you could try to do it is by hoping that slow roll inflation didn't last so long as to wipe out all traces of the previous vacuum inside which we nucleated. For that you would have to get lucky, but I don't think it is an extremely long shot, and it is certainly worth working out what the signatures would be that we would see in the sky, and there are quite a lot of people working on that. And the third way, which, again, is totally impractical, is to make a little bubble that is going to re-collapse on some other vacuum in the laboratory. As a matter of principle that is another way you could possibly do it. The one thing you can't do is go somewhere beyond the event horizon and see what is going on there.

- Alba:

In your figure showing the different bubbles, why do none of the bubbles intersect?

- Bousso:

Oh, I'm sorry. They do! I just didn't draw any of these intersections. They are going to intersect. Let's go to the simplest picture that we can imagine where we just have one meta-stable vacuum which is de Sitter and then we can decay and produce other vacua which I'm just going to say have lambda less than zero so that inside nothing interesting anymore happens.

Here's your de Sitter space, you form these bubbles of vacua which have lambda less than zero inside so this is the new vacuum, so I have a potential that looks like that. This is lambda greater than zero, lambda less than zero. And the transition is this one, so in space time it looks like this. This is something that was shown by Alan Guth and Eric Weinberg back in 1982 or something. This is eternal in the sense that the old vacuum is never completely eaten up by these new vacua that form, but nevertheless it is true that every bubble of the new vacuum collides with infinitely many other such bubbles that form in its neighborhood, close enough that they can collide with it. So what ends up happening is that you form these sort of clusters. But what doesn't happen is that somehow the entire future infinity gets eaten up by the new vacuum. That does not happen as long as the rate, the number of vacua per unit volume and unit time at which these things form, is less than the Hubble rate. As long as fewer than one bubble form per Hubble time and Hubble volume in the old vacuum this will not happen. But you will form these clusters, and I just suppressed that in my drawing because I didn't want to make it look too complicated.

I just want to add one more thing. That's another thing that people are looking for as a possible smoking-gun sign of an apparent vacuum, as that would be the beginning of a multiverse. If inflation didn't last too long, and if our parent vacuum was sufficiently unstable that at lot of these nucleations happen, then there's a chance that we might see a collision in the sky. And you can calculate what signature that would have.

- Alberte:

You said that string theory is not really necessary for an explanation of the multiverse, that you could just do it from quantum field theory. I thought that maybe, somehow, this could be connected with the many-worlds interpretation of quantum mechanics.

- Bousso:

This is something I'm actually thinking about.

- Alberte:

It's actually quite similar. Their justification for that is that it can explain the paradox of the destruction of the wavefunction. Could the multiverse also handle these questions?

- Bousso:

Well, it's very tempting to think that the multiverse could shed some light on the many worlds interpretation of quantum mechanics. But I don't think we're at a point where we could make that speculation very precise yet. The basic idea behind this became a little more interesting in the context of the struggle to understand the measure problem. So whenever you have eternal inflation you have these infinities that I discussed today. You have to somehow cut them off. And there were two approaches that looked like they were completely different. One approach was to say, let's just stop at a particular time and count everything that happens prior to that time, t, and compute ratios, N1/N2, that give you the probabilities. By counting how many 1s and 2s happen below that time, and then later maybe take the limit as t goes to infinity and hope that this ratio converges to something nice. In fact, there are very many different measures that you can define in this way, even if there's many different ways of defining the time foliation. That's been considered one of the big problems in this subject. But, for each such way of defining this measure, there is actually a dual way of defining it, which looks completely different a priori, in which you say I'm going to follow a geodesic. So instead of keeping something below a certain time, we're going to keep everything that's in the neighbourhood of that geodesic, defined in a suitable way. How you have to define this neighbourhood depends on how you define this time variable. But there's a sort of duality which has been made the most precise in the pair of measures that's also been the most successful phenomenologically. So it's really quite an impressive result I think, which is telling us that there are two different ways that we can think about the multiverse.

Once you have that, it's very tempting to say the many-worlds branches of all the things that can happen in this region - if I measure an electron and it turns out up or down - somehow corresponds to actual regions in this global picture, so that the many different things that can happen in this local picture where you restrict to one local region and you go over all time... In the many-worlds interpretation you really have to let this local region branch into many worlds depending on what is happening. In the global picture you could hope to put this back together into definite things that are simply happening in different regions of the multiverse. This is something I probably shouldn't have said because it's wildly speculative and quite possibly wrong, but I find this question very hinting.

- Schmidt-Sommerfeld:

I'm not sure my question makes sense, so if not just tell me. What is the cosmological constant problem in the context of the holographic approach to gravity?

- Bousso:

Well, the dumb answer would be "the same" in the sense that it's not immediately clear what the holographic approach to quantum gravity would have to say about it. I mentioned in my first lecture that there's no obvious and simple way in which appealing to quantum gravity can help you get rid of this problem, because quantum gravity, whatever the theory is, has to reproduce classical gravity in the regime where we've tested it, and we've also tested quantum field theory quite well. On the other hand, it is true that holography tells us that there are far fewer degrees of freedom in a region of space than you might have naively thought, that they scale like the area and not like the volume. And when I compute the vacuum energy density at every point in space, I'm appealing to local point-like degrees of freedom of quantum field theory, which really aren't there. So there's a reason to be suspicious of the calculation as it is done. However, I haven't been able to find, or have seen, a good argument for why holography is a strong enough restriction to really get rid of the problem. In particular, it's difficult to see what would set the scale of the cosmological constant. Think of it this way. The universe we see is certainly compatible with the holographic principle, but if the cosmological constant was larger or smaller, we would again get a perfectly good solution of general relativity, which is again compatible with the holographic principle. So no matter what the value of the cosmological constant is, there's no conflict with the holographic principle. This is why it's been difficult to make progress by just appealing to the holographic principle.

- Azmi:

I have two questions. One is related to the present universe. The second one is related to dark energy. I will ask the first one. Why is the present universe so homogeneous?

- Bousso:

You mean why is the universe on large-scale so homogeneous?

- Azmi:

Yes.

- Bousso:

That's a loaded question. Most people that I know think that that's because of slow-roll inflation. It's because there was a period of exponential expansion in the early universe that essentially stretched out inhomogeneities and flattened out the universe. I honestly believe that that's by far the most plausible theory that's been put forward to explain this. Though, periodically, people try to find alternatives.

- Azmi:

The second question is, we know that dark energy and matter at present evolve differently over time. Why is the energy density of the dark energy component and matter, at present, are of the same order?

- Bousso:

That's the famous coincidence problem, or the "why now?" question. Why do we live at the time when the two are at the same order? And I hope I'll be able to convince you tomorrow that this can be explained in the context of modern measures combined with the multiverse of string theory. But I have not told you yet, how. So I want to wait until tomorrow before answering that question.

- 't Hooft:

Since we're running late, we can have one or two very short questions.

- Pagnutti:

On your last slide you had a Penrose diagram with these triangular sections along the top. I was wondering what those triangular sections represent.

- Bousso:

You mean these kind of triangles?

- Pagnutti:

No. They were going up.

- Bousso:

I just don't recall what my last slide was. Do you have an idea what it was about.

- Pagnutti:

It was similar to your first one.

- Bousso:

I'm afraid we might have to defer that to a private discussion.

- Inguglia:

Do you think that the asymmetry between matter and antimatter can be due to the fact that you're talking about the decay of the parent vacuum?

- Bousso:

I don't think there's any direct relation to that. But I think there are some intriguing proposals of how we might address this problem in the context of the landscape in novel ways.

CHAIRMAN: R. BOUSSO

Scientific Secretaries: B.L. Cerchiai, A. Marrani

DISCUSSION III

- Dennen:

What does the age of the universe (13.7 billion years) correspond to in the bubble picture? Could the multiverse go infinitely back in time?

- Bousso:

The age of the Universe corresponds to the nucleation time of our own bubble. There are no reliable theorems telling us whether eternal inflation had to have a beginning. There are some interesting arguments by Guth and Vilenkin, Borden and others, but we don't know how the Universe started. However, the picture is that regardless of the choice of initial conditions, the behavior at late times is actually universal. This is the main difference between the two methods of regulating the measure, i.e. the one with a local cut-off following a light-cone and the global one of taking slices at later and later times. The attractor behavior holds only in the global picture. For the choice of the causal patch you have to choose the longest lived metastable de Sitter vacuum. In practice for the computations, including the one of the cosmological constant, it would not make a lot of difference if I chose another random vacuum. In the global framework this is not so important, because the exponentially divergent numbers at late times are dominating when you have the universal attractor regime.

- Mulhearn:

Will observers in rare exotic universes reject the correct measure? Unlike experiments that may be repeated, they will use the incorrect measure for ever.

- Bousso:

This is a very important question. Your comment concerns a useful fiction but still a fiction nevertheless. Indeed, laboratory experiments cannot be repeated an infinite number of times in a finite universe. Let's consider CMB, which allegedly is perfectly consistent with slow roll inflation. Based on it we have successfully ruled out theories of structure formation, such as cosmic strings, cosmological defects etc. But you can imagine that the actual theory of structure formation is due to textures or cosmic strings. In fact, within such a framework the probability of observing what we see is not exactly zero, but rather very small. However, if we take this point of view, we could not use any cosmological data, or even any laboratory data, to construct our theories. This is ultimately due to the fact that our universe is a finite domain with a finite number of accessible quantum states.

- Krislock:

Is eternal inflation a perpetual motion machine?

- Bousso:

Yes, eternal inflation can seem like that if you look at it from the global perspective, because you are creating more and more from a finite starting point. This is one of the reasons why I am so skeptical about this viewpoint and why I think we should restrict ourselves to a causally connected patch. Of course, this doesn't mean that we are throwing out some of the information of the multiverse picture, but rather that we are not creating matter, entropy nor information out of nothing. These are questions that were already present in cosmology even before the multiverse, because even in our current cosmology it looks as if the universe is expanding forever, with continuously increasing entropy.

- Krislock:

Does the measure which predicts Λ to be 10^{-121} give a reason why Λ is so small compared to the value predicted by the standard model?

- Bousso:

The reasoning why the standard model seems to predict a large value for the cosmological constant is that there are several contributions from different large numbers, and therefore it is most likely that their sum is a large number. However, the landscape should be big enough that there should be universes where these contributions can cancel out and give a small number. The problem is that if we look at other quantities, the cosmological constant is the only one for which we have a precise confidence on the distribution in the string landscape. There are a few other quantities, which we could compute under certain circumstances, such as the distribution of dark matter in the presence of axions. But most of them are hard to determine, as e.g. the distribution of primordial density perturbations.

- Schmidt-Sommerfeld:

In the absence of a fundamental principle that selects a measure, in order to preserve some predictability you have to make sure that the number of possible measures is much smaller than the number of parameters. Is this true? What do we know about the number of possible measures?

- Bousso:

To write down a well-defined cosmological measure is an issue surprisingly hard. Once you have done that, you generally find a regime which gives rise to catastrophic predictions. In the end, I was able to write down only one consistent measure. But there is no theorem telling you that there aren't other measures which could work, there's no systematic way to construct them.

- Burda:

If the two concepts of black holes complementarity and holography give you the same answer for the cosmological constant problem, aren't they related perhaps?

- Bousso:

Yes, they both are aspects of holography, namely of the fact the number of degrees of freedom in the universe scales as the area.

Black holes and qubits

M. J. Duff[1]

[1]*The Blackett Laboratory, Imperial College London, Prince Consort Road, London SW7 2BZ, U.K.*

Quantum entanglement lies at the heart of quantum information theory, with applications to quantum computing, teleportation, cryptography and communication. In the apparently separate world of quantum gravity, the Hawking effect of radiating black holes has also occupied centre stage. Despite their apparent differences, it turns out that there is a correspondence between the two.

1. INTRODUCTION

Whenever two very different areas of theoretical physics are found to share the same mathematics, it frequently leads to new insights on both sides. Here we describe how knowledge of string theory and M-theory leads to new discoveries about Quantum Information Theory (QIT) and vice-versa [1–3].

2. BEKENSTEIN-HAWKING ENTROPY

Every object, such as a star, has a critical size determined by its mass, which is called the Schwarzschild radius. A black hole is any object smaller than this. Once something falls inside the Schwarzschild radius, it can never escape. This boundary in spacetime is called the event horizon. So the classical picture of a black hole is that of a compact object whose gravitational field is so strong that nothing, not even light, can escape.

Yet in 1974 Stephen Hawking showed that quantum black holes are not entirely black but may radiate energy, due to quantum mechanical effects in curved spacetime. In that case, they must possess the thermodynamic quantity called entropy. Entropy is a measure of how organized or disorganized a system is, and, according to the second law of thermodynamics, it can never decrease. Noting that the area of a black hole event horizon can never decrease, Jacob Bekenstein had earlier suggested such a thermodynamic interpretation implying that black holes must have entropy. This Bekenstein-Hawking black hole entropy is in fact given by one quarter the area of the event horizon. This is a remarkable fact relating a thermodynamic quantity, entropy, with a quantum mechanical origin, to a purely geometrical quantity, area, that is calculated in Einstein's classical theory of gravity.

Entropy also has a statistical interpretation as a measure of the number of quantum states available. However, it was not until 20 years later that string theory, as a theory of quantum gravity, was able to provide a microscopic explanation of this kind.

3. BITS AND PIECES

A classical bit is the basic unit of computer information and takes the value 0 or 1. A light switch provides a good analogy; it can either be off, denoted 0, or on, denoted 1. A quantum bit or "qubit" can also have two states but whereas a classical bit is either 0 or 1, a qubit can be both 0 and 1 until we make a measurement. In quantum mechanics, this is called a superposition of states. When we actually perform a measurement, we will find either 0 or 1 but we cannot predict with certainty what the outcome will be; the best we can do is to assign a probability.

There are many different ways to realize a qubit physically. Elementary particles can carry an intrinsic spin. So one example of a qubit would be a superposition of an electron with spin up, denoted 0, and an electron with spin down, denoted 1. Another example of a qubit would be the superposition of the left and right polarizations of a photon. So a single qubit state, usually called Alice, is a superposition of Alice-spin-up 0 and Alice-spin-down 1, represented by the line in figure 1. The most general two-qubit state, Alice and Bob, is a superposition of Alice-spin-up-Bob-spin-up 00, Alice-spin-up-Bob-spin-down 01, Alice-spin-down-Bob-spin-up 10 and Alice-spin-down-Bob-spin-down 11, represented by the square in figure 1.

Consider a special two-qubit state which is just 00 + 01. Alice can only measure spin up but Bob can measure either spin up or spin down. This is called a separable state; Bob's measurement is uncorrelated with Alice's. By contrast consider 00 + 11. If Alice measures spin up, so must Bob and if she measures spin down so must he. This is called an entangled state; Bob cannot help making the same measurement. Mathematically, the square in figure 1 forms a 2×2 matrix and a state is entangled if the matrix has a nonzero determinant.

This is the origin of the famous Einstein-Podolsky-Rosen (EPR) paradox put forward in 1935. Even if Alice is in the Vatican and Bob is millions of miles away in Alpha Centauri, Bob's measurement will still be determined by Alice's. No wonder Albert Einstein called it "spooky action at a distance". EPR concluded rightly that if quantum mechanics is correct then nature is nonlocal and if we insist on local "realism" then quantum mechanics must be incomplete. Einstein himself favoured the latter hypothesis. However, it was not until 1964 that CERN theorist John Bell proposed an experiment that could decide which

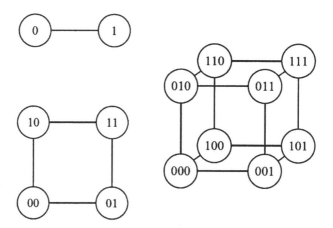

FIG. 1: A single quit is represented by a line, two qubits by a square and three qubits by a cube.

version was correct, and it was not until 1982 that Alain Aspect actually performed the experiment. Quantum mechanics was right, Einstein was wrong and local realism went out the window.

As QIT developed, the impact of entanglement went far beyond the testing of the conceptual foundations of quantum mechanics. Entanglement is now essential to numerous quantum information tasks such as quantum cryptography, teleportation and quantum computation.

4. CAYLEY'S HYPERDETERMINANT

As a high-energy theorist involved in research on quantum gravity, string theory and M-theory, I paid little attention to all this, even though as a member of staff at CERN in the 1980s my office was just down the hall from Bell's.

My interest was not aroused until 2006, when I attended a lecture by Hungarian physicist Peter Levay at a conference in Tasmania. He was talking about three qubits Alice, Bob and Charlie where we have eight possibilities 000, 001, 010, 011, 100, 101, 110, 111, represented by the cube in figure 1. Wolfgang Dur and colleagues at the University of Innsbruck have shown [4] that that three qubits can be entangled in several physically distinct ways: tripartite GHZ (Greenberger-Horne-Zeilinger), tripartite W, biseparable A-BC, separable A-B-C and null, as shown in the left hand diagram of figure 2.

The GHZ state is distinguished by a nonzero quantity known as the 3-tangle, which measures genuine tripartite entanglement. Mathematically, the cube in figure 1 forms what in 1845 the mathematician Arthur Cayley called a $2 \times 2 \times 2$ hypermatrix and the 3-tangle is given by the generalization of a determinant called Cayley's hyperdeterminant.

The reason this sparked my interest was that Levays equations reminded me of some work I had been doing on a completely different topic in the mid 1990s with my collaborators Joachim Rahmfeld and Jim Liu [5]. We found a particular black hole solution that carries eight charges (four electric and four magnetic) and involves three fields called S, T and U. When I got back to London from Tasmania I checked my old notes and asked what would happen if I identified S, T and U with Alice, Bob and Charlie so that the eight black-hole charges were identified with the eight numbers that fix the three-qubit state. I was pleasantly surprised to find that the Bekenstein-Hawking entropy of the black holes was given by the 3-tangle: both were described by Cayley's hyperdeterminant. This turned out to be the tip of an iceberg and there is now a growing dictionary between phenomena in the theory of black holes and phenomena in QIT.

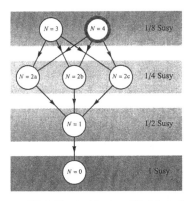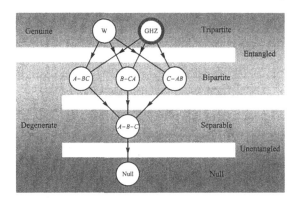

FIG. 2: The classification of black holes from N wrapped branes (left) exactly matches the classification of three-qubit entanglement (right). Only the GHZ state has non-zero 3-tangle and only the $N = 4$ black hole has non-zero entropy.

5. OCTONIONS

According to supersymmetry, for each known boson (integer spin 0, 1, 2 and so on), there is a fermion (half-integer spin 1/2, 3 /2, 5/2 and so on), and vice versa. CERN's Large Hadron Collider will be looking for these superparticles. The number of supersymmetries is denoted by $\mathcal{N}$ and ranges from 1 to 8 in four spacetime dimensions.

CERN's Sergio Ferrara and I have extended the STU model example, which has $\mathcal{N} = 2$, to the most general case of black holes in $\mathcal{N} = 8$ supergravity. We have shown that the corresponding system in quantum information theory is that of seven qubits (Alice, Bob, Charlie, Daisy, Emma, Fred and George), undergoing at most a tripartite entanglement of a very specific kind as depicted by the Fano plane of figure 3. The Fano plane has a strange mathematical property: it describes the multiplication table of a particular kind of number: the octonion. Mathematicians classify numbers into four types: real numbers, complex numbers (with one imaginary part A), quaternions (with three imaginary parts A, B, D) and octonions (with seven imaginary parts A, B, C, D, E, F, G). Quaternions are non-commutative because AB does not equal BA. Octonions are not only noncommutative but also non-associative since $(AB)C$ does not equal $A(BC)$.

Real, complex and quaternion numbers show up in many physical contexts. Quantum mechanics, for example, is based on complex numbers and Pauli's electron spin operators are quaternionic. Octonions have fascinated mathematicians and physicists for decades but have yet to find any physical application. In recent books both Roger Penrose and Ray Streater have characterized octonions as one of the great lost causes in physics. So we hope that the tripartite entanglement of seven qubits (which is just at the limit of what can be reached experimentally) will prove them wrong and provide a way of seeing the effects of octonions in the laboratory [6–8].

6. IMPLICATIONS FOR M-THEORY

We have also learned things about M-theory from QIT. The Fano plane suggests a whole new way of studying its symmetries based on the 7 imaginary octonions (completely different from the Jordan algebra approach that uses all 8 split octonions). Such expectations have recently been strengthened by the discovery of four supergravities with $16 + 16, 32 + 32, 64 + 64, 128 + 128$ degrees of freedom displaying some curious properties [9]. In particular they reduce to $\mathcal{N} = 1; 2; 4; 8$ theories all with maximum rank 7 in $D = 4$ which correspond to 0, 1, 3, 7 lines of the Fano plane and hence admit a division algebra ($\mathbb{R}; \mathbb{C}; \mathbb{H}; \mathbb{O}$) interpretation consistent with the black-hole/qubit correspondence. They exhibit unusual properties. For example they are all self-mirror with vanishing trace anomaly [10].

7. SUPERQUBITS

In another development, QIT has been extended to super-QIT with the introduction of the superqubit which can take on three values: 0 or 1 or $. Here 0 and 1 are bosonic and $ is fermionic [11]. Such values can be realised in condensed matter physics, such as the excitations of the t-J model of strongly correlated electrons, known as spinons and holons. The superqubits promise

60

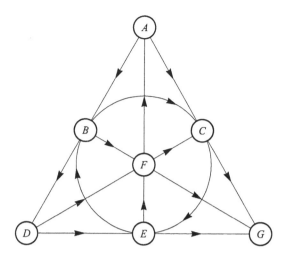

FIG. 3: The Fano plane. The vertices A, B, C, D, E, F, G represent the seven qubits and the seven lines ABD, BCE, CDF, DEG, EFA, FGB, GAC, represent the tripartite entanglement.

totally new effects, for example, could they be even more non-local than ordinary bits? Super quantum computing is also being investigated [12].

8. WRAPPED BRANES AS QUBITS

If current ideas are correct, a unified theory of all physical phenomena will require some radical ingredients in addition to supersymmetry. For example, there should be extra dimensions: supersymmetry places an upper limit of 11 on the dimension of spacetime. The kind of real, four-dimensional world that supergravity ultimately predicts depends on how the extra seven dimensions are rolled up, in a way suggested by Oskar Kaluza and Theodor Klein in the 1920s. In 1984, however, 11-dimensional supergravity was knocked off its pedestal by superstring theory in 10 dimensions. There were five competing theories: the E8 x E8 heterotic, the SO(32) heterotic, the SO(32) Type I, and the Type IIA and Type IIB strings. The E8 x E8 seemed, at least in principle, capable of explaining the elementary particles and forces, including their handedness. Moreover, strings seemed to provide a theory of gravity consistent with quantum effects.

However, the spacetime of 11 dimensions allows for a membrane, which may take the form of a bubble or a two-dimensional sheet. In 1987 Howe, Inami, Stelle and I were able to show [13] that if one of the 11 dimensions were a circle, we could wrap the sheet around it once, pasting the edges together to form a tube. If the radius becomes sufficiently small, the rolled-up membrane ends up looking like a string in 10 dimensions; it yields precisely the Type IIA superstring. In a landmark talk at the University of Southern California in 1995, Edward Witten [14] drew together all of this work on strings, branes and 11 dimensions under the umbrella of M-theory in 11 dimensions. Branes now occupy centre stage as the microscopic constituents of M-theory, as the higher-dimensional progenitors of black holes and as entire universes in their own right.

Such breakthroughs have led to a new interpretation of black holes as intersecting black-branes wrapped around the seven curled dimensions of M-theory or six of string theory. Moreover, the microscopic origin of the Bekenstein-Hawking entropy is now demystified. Using Polchinski's D-branes, Andrew Strominger and Cumrun Vafa were able to count the number of quantum states of these wrapped branes [15]. A p-dimensional D-brane (or Dp-brane) wrapped around some number p of the compact directions $(x_4, x_5, x_6, x_7, x_8, x_9)$ looks like a black hole (or D0-brane) from the four-dimensional (x_0, x_1, x_2, x_3) perspective. Strominger and Vafa found an entropy that agrees with Hawking's prediction, placing another feather in the cap of M-theory. Yet despite all these successes, physicists are glimpsing only small corners of M-theory; the big picture is still lacking. Over the next few years we hope to discover what M-theory really is. Understanding black holes will be an essential pre-requisite.

In string literature one may find D-brane intersection rules that tell us how N branes can intersect over one another and the fraction of supersymmetry that they preserve. In our black hole/qubit correspondence, my students Leron Borsten, Duminda

Dahanayake, Hajar Ebrahim, William Rubens and I showed that the microscopic description of the GHZ state, $000 + 011 + 101 + 110$ is that of the $\mathcal{N} = 4$, fraction 1/8, case of D3-branes of Type IIB string theory [16]. We denoted the wrapped circles by crosses and the unwrapped circles by noughts; 0 corresponds to xo and 1 to ox, as in table 1. So the number of qubits here is three because the number of extra dimensions is six. This also explains where the two-valuedness enters on the black-hole side. To wrap or not to wrap; that is the qubit.

4	5	6	7	8	9	macro charges	micro charges	$\lvert ABC \rangle$
x	o	x	o	x	o	p^0	0	$\lvert 000 \rangle$
o	x	o	x	x	o	q_1	0	$\lvert 110 \rangle$
o	x	x	o	o	x	q_2	$-N_3 \sin\theta \cos\theta$	$\lvert 101 \rangle$
x	o	o	x	o	x	q_3	$N_3 \sin\theta \cos\theta$	$\lvert 011 \rangle$
o	x	o	x	o	x	q_0	$N_0 + N_3 \sin^2\theta$	$\lvert 111 \rangle$
x	o	x	o	o	x	$-p^1$	$-N_3 \cos^2\theta$	$\lvert 001 \rangle$
x	o	o	x	x	o	$-p^2$	$-N_2$	$\lvert 010 \rangle$
o	x	x	o	x	o	$-p^3$	$-N_1$	$\lvert 100 \rangle$

TABLE I: Three qubit interpretation of the 8-charge $D = 4$ black hole from four D3-branes wrapping around the lower four cycles of T^6 with wrapping numbers N_0, N_1, N_2, N_3 and then allowing N_3 to intersect at an angle θ.

9. REPURPOSING STRING THEORY

In the forty years since its inception, string theory has undergone many changes of direction, in the light of new evidence and discovery:

1970s: Strong nuclear interactions
1980s: Quantum gravity; "theory of everything"
1990s: AdS/CFT: QCD (revival of 1970s); quark-gluon plasmas
2000s: AdS/CFT: superconductors
2000s: Cosmic strings
2010s: Fluid mechanics
2010s: Black hole/qubit correspondence: entanglement in Quantum Information Theory

For example, by stacking a large number of branes on top of one another, Juan Maldacena [17] showed that a (D+1)-dimensional spacetime with all its gravitational interactions, may be dual to a non-gravitational theory that resides on its D-dimensional boundary. If this so-called holographic picture is correct, our universe maybe like Plato' s cave and we are the shadows projected on its walls. Its technical name is the ADS/CFT correspondence. Maldacena's 1998 ADS/CFT paper has garnered an incredible 7000+ citations. Interestingly enough, this is partly because it has found applications outside the traditional "theory of everything" milieu that one normally associates with string and M-theory. These, frequently serendipitous, applications include quark-gluon plasmas, high temperature superconductors and fluid mechanics. ADS/CFT is not the only branch of string/M-theory that has found applications in different areas of physics. After all, as shown in the table, string theory was originally invented in the 1970s to explain the behaviour of protons, neutrons and pions under the influence of the strong nuclear force.

The partial nature of our understanding of string/M-theory has so far prevented any kind of smoking-gun experimental test in the fields of particle physics and cosmology. This has led some critics of string theory to suggest that it is not true science. This is easily refuted by studying the history of scientific discovery; the 30-year time lag between the EPR idea and Bells falsifiable prediction provides a nice example. Nevertheless it cannot be denied that a prediction in string theory would be very welcome. Here we describe a prediction, not in the fields of particle physics and cosmology, but in quantum information theory.

10. FOUR QUBIT ENTANGLEMENT: A FALSIFIABLE PREDICTION

More recently Borsten, Dahanayake, Rubens and I at Imperial College teamed up with Alessio Marrani at CERN. We invoked this black hole-qubit/correspondence to predict a new result in quantum information theory. Noting that the classification of stringy black holes puts them in 31 different families, we predicted that four qubits can be entangled in 31 different ways [18]. By the way, this particular aspect of the correspondence is not a guess or a conjecture but a consequence of the Kostant-Sekiguchi theorem:

Extremal black holes classification of STU model

$\updownarrow$

31 real nilpotent orbits of $SO(4,4)$ acting on the $\mathbf{28}$

$\updownarrow$

Kostant-Sekiguchi Correspondence

$\updownarrow$

31 complex nilpotent orbits of $SL(2)^4$ acting on the $(\mathbf{2},\mathbf{2},\mathbf{2},\mathbf{2})$

$\updownarrow$

4 qubits entanglement classification

This can, in principle, be tested in the laboratory and we are urging our experimental colleagues to find ways of doing just that. So the esoteric mathematics of string and M-theory might yet find practical applications.

Acknowledgements

I am grateful to my collaborators Leron Borsten, Duminda Dahanayake, Sergio Ferrara, Hajar Ibrahim, Alessio Marrani and William Rubens for their part in this research and especially to Leron Borsten for help with the manuscript.

[1] M. J. Duff, String triality, black hole entropy and Cayley's hyperdeterminant, Phys. Rev. D**76** (2007) 025017.

[2] R. Kallosh and A. Linde, Strings, black holes, and quantum information, Phys. Rev. D**73** (2006) 104033.

[3] P. Lévay, Stringy black holes and the geometry of entanglement, Phys. Rev. D**74** (2006) 024030.

[4] W. Dür, G. Vidal, J. I. Cirac, three-qubits can be entangled in two inequivalent ways, Phys. Rev. A**62** (2000) 062314.

[5] M. J. Duff, J. T. Liu, J. Rahmfeld, Four dimensional string-string-string triality, Nuclear Physics B**459** (1996) 125-159.

[6] M. J. Duff, S. Ferrara, E_7 and the tripartite entanglement of seven qubits, Phys. Rev. D**76** (2007) 025018.

[7] P. Lévay, Strings, black holes, and the tripartite entanglement of seven qubits and the Fano plane, Phys. Rev. D**75** (2007) 024024.

[8] L. Borsten, D. Dahanayake, M. J. Duff, H. Ebrahim, W. Rubens, Black holes, qubits and octonions, Physics Reports **471** (2009) 113-219.

[9] M. J. Duff, S. Ferrara, Four curious supergravities, Phys. Rev. D**83** (2011) 046007

[10] M. J. Duff, S. Ferrara, Generalized mirror symmetry and trace anomalies, Class.Quant.Grav. **28** (2011) 065005.

[11] L. Borsten, D. Dahanayake, M. J. Duff, W. Rubens, Superqubits, Phys. Rev. D**81** (2010) 105023.

[12] Leonardo Castellani, Pietro Antonio Grassi, Luca Sommovigo, Quantum Computing with Superqubits, e-Print: arXiv:1001.3753 [hep-th]

[13] M. J. Duff, Paul S. Howe, T. Inami, K.S. Stelle, Superstrings in D=10 from Supermembranes in D=11, Phys.Lett. B**191** (1987) 70.

[14] Edward Witten, String theory dynamics in various dimensions, Nucl. Phys. B**443** (1995) 85-126.

[15] A. Strominger and C. Vafa, Microscopic origin of the Bekenstein-Hawking entropy, 1996 Phys. Lett. B**379** 99.

[16] L. Borsten, D. Dahanayake, M. J. Duff, H. Ebrahim, and W. Rubens, Black Wrapped Branes as Qubits, Phys. Rev. Lett. **100** (2008) 251602.

[17] J. M. Maldacena, The Large N limit of superconformal field theories and supergravity, Adv. Theor. Math. Phys. **2** (1998) 231-252.

[18] L. Borsten, D. Dahanayake, M.J. Duff, A. Marrani, W. Rubens, Four-qubit entanglement from string theory, Phys. Rev. Lett. **105** (2010) 100507.

Scientific Secretaries: L. Borsten, F. Zaidi

DISCUSSION I

- Bousso:

You identified the state vector coefficients with the black hole charges. However, I would assume that quantum mechanically the coefficients of the state vectors |ABC> are small, order 1 or less, since the state is normalised, and certainly not integer as the black hole charges must be. Is there a contradiction in setting them to be equal?

- Duff:

That is a good question. Since the entanglement classification is concerned with only the SL(2) transformations it is not meaningful to normalise the state in this context as the norm is not an SL(2) invariant. So, on the qubit side we use un-normalised states, which does not effect the entanglement classification. As you point out for quantum string theory the black holes charges are integer, which is not what you typically find for qubits, and this is a difference.

- Busso:

Because you are not restricted to normalize you could then approximate a given qubit state with a black hole possessing very large charges?

- Duff:

On the black hole side we use the classical approximation valid for the large charge limit, in which case the charges are continuous.

- Burda:

On slide 28 why are there three minus signs on the coefficients a_{001}, a_{010}, a_{100} in the correspondence between the 8 black hole charges and the 8 state vector coefficients. Does the correspondence between the entropy and the hyperdeterminant only work with these signs?

- Duff:

This choice of signs is not unique, but it is required for the entropy to be given by the hyperdeterminant.

- Burda:

So, for four qubits will you have to have similar signs?

- Duff:

The four-qubit story will proceed in a different manner, but yes you need to be careful with the signs.

- Alba:

Does the relationship between black holes and qubits give any insight into the black hole information loss paradox?

- Duff:

Unfortunately, our work on the black hole qubit correspondence does not address the black hole information paradox.

Scientific Secretaries: P. Dunin-Barkowski, A. Popolitov

DISCUSSION II

- Alba:
How should we understand formula (1) on page 48, what is this bracket with three numbers there?

- Duff:
This is the funny thing about the Freudenthal triple system. It is represented by a ``2 by 2 matrix'', but there is a lack of democracy among the entries. Here we have 1 number in each diagonal entry and 3 numbers in each off-diagonal entry, so this is just a way of arranging 8 numbers in a convenient way. In the E_7 case, we have 1 number in each diagonal entry and 27 numbers in each off-diagonal entry, making up the 56. So it's not really a 2 by 2 matrix but it's convenient to call it a 2 by 2 matrix.

- Alba:
What corresponds to superqubits on the black hole side?

- Duff:
Good question. We don't have any black hole interpretation for superqubits, that's a separate game. Part of the reason is: on the supergravity side we are working with what is called the U-duality group, which is $SL(2)^3$ in the case of the 4-dimensional N=2 STU model and the exceptional group E_7 in the N=8 case. But whatever theory we're looking at, the U-duality group is always bosonic. When we come to superqubits, however, we are looking at supergroups. Strangely enough, Alessio Marrani was telling me today about some new work on new superalgebras involving exceptional groups.

- Marrani:
Yes, but unfortunately not involving derived exceptional groups.

- Duff:
So, for the time being anyway, superqubits and the black-holes/qubit correspondence are separate endeavours.

- Alba:
What corresponds to black hole charge and black hole angular momentum?

- Duff:
Angular momentum is a little mysterious. I don't think we fully understand the role of angular momentum. The closest we've come is something called the 4d-5d lift, that relates black holes in 5 dimensions with angular momentum to black holes in 4d without angular momentum. The black holes in 5d have angular momentum built from the 4d charges. So in that case the interpretation is that of a 4d black hole with no angular momentum that is lifted up to 5 dimensions. It's not an entirely satisfactory understanding, I think, and that's the only place where angular momentum entered our discussions. If you just work with Kerr black holes the 4d rotation parameter, as far as I can tell, would have no quantum information interpretation. But I may be wrong.

- Dennen:

Your student mentioned that scientists working in this field are focused on increasing the number of qubits that they entangle, but from your presentation it seems like there is an enormous increase of complexity in the analysis when going from 3 qubits to 4 qubits. First of all, do you have any hope to go beyond 4 qubits and, supposing you have the limit of maximum number of qubits going to infinity, are there any simplifications that might happen and is there any connection to black holes in that limit?

- Duff:

Two excellent questions and I have to tell I don't have the answer to either of them. 5 qubits is going to be several orders of magnitude more complicated than 4, so I wouldn't like to say we can't do it, but it is going to be very difficult. Generalizing to n qubits is a longstanding unsolved problem in quantum information theory. I might just add that Cayley's hyperdeteminant, which is the 2 by 2 by 2 entanglement measure for 3-qubits, generalizes to higher dimensions, but mathematicians rapidly run out of understanding it; it gets very very complicated. That's just one little indication of how difficult things can get. So on the theoretical side understanding the number of additional ways in which many qubits can be entangled seems like an enormous statistical problem, but maybe some clever student will figure out some neat way of generalizing it that we haven't thought of. Now the n going to infinity limit is another interesting question and you might hope for some simplifications, but I don't know of any.

- Dennen:

Does anybody work on any sort of approximation scheme?

- Duff:

Well, there are many attempts in the literature that focus on n-qubit generalizations of specific states, such as the GHZ states, rather than generalizing to more qubits the classification that we did, with 31 families of 4 qubits. They can say things about particular states without having the big picture.

Perturbative and Non-Perturbative Aspects
of $\mathcal{N} = 8$ Supergravity

Sergio Ferrara[1,2] **and Alessio Marrani**[1]

1 *Physics Department, Theory Unit, CERN,*
CH 1211, Geneva 23, Switzerland
sergio.ferrara@cern.ch
Alessio.Marrani@cern.ch

2 *INFN - Laboratori Nazionali di Frascati,*
Via Enrico Fermi 40, 00044 Frascati, Italy

Abstract

Some aspects of quantum properties of $\mathcal{N} = 8$ supergravity in four dimensions are discussed for non-practitioners.

At perturbative level, they include the Weyl trace anomaly as well as composite duality anomalies, the latter being relevant for perturbative finiteness. At non-perturbative level, we briefly review some facts about extremal black holes, their Bekenstein-Hawking entropy and attractor flows for single- and two- centered solutions.

1 Lecture I
On "Quantum" $\mathcal{N} = 8$, $d = 4$ Supergravity

$\mathcal{N} = 8$, $d = 4$ "quantum" supergravity may be defined by starting with the Einstein-Hilbert action, and setting "perturbative" Feynman rules as a *bona fide* gauge theory of gauge particles of spin 2, the gravitons. In supersymmetric gravity theories with $\mathcal{N}$-extended supersymmetry in $d = 4$ space-time dimensions, the massless particle content is given by

$$\binom{\mathcal{N}}{k} \equiv \frac{\mathcal{N}!}{k! \, (\mathcal{N} - k)!} \text{ particles of } \textit{helicity } \lambda = 2 - \frac{k}{2}, \tag{1.1}$$

where $k_{\max} = \mathcal{N}$, and $\mathcal{N} \leqslant 8$ if $|\lambda| \leqslant 2$ is requested (namely, no higher spin fields in the massless spectrum).

One possible approach to "quantum" supergravity is to consider it as it comes from M-theory restricted to the massless sector. The problem is that this theory, even if preserving maximal $\mathcal{N} = 8$ supersymmetry in $d = 4$ space time dimensions (corresponding to $32 = 8 \times 4$ supersymmetries), is *not* uniquely defined, because of the multiple choice of internal compactification manifolds and corresponding duality relations:

$$\text{I. } M_{11} \longrightarrow M_4 \times T_7 \qquad (GL^+(7, \mathbb{R}) \text{ and } SO\,(7) \text{ manifest});$$

$$\text{II. } M_{11} \longrightarrow AdS_4 \times S^7 \qquad (SO\,(8) \text{ manifest, } \textit{gauged}); \tag{1.2}$$

$$\text{III. } M_{11} \longrightarrow M_4 \times T_{7,\mathcal{R}} \qquad (SL(8, \mathbb{R}) \text{ and } SO\,(8) \text{ manifest}),$$

where T_7 is the 7-torus and S^7 is the 7-sphere. $T_{7,\mathcal{R}}$ denotes the case in which, according to Cremmer and Julia [1], the dualization of 21 vectors and 7 two-forms makes $SL(8, \mathbb{R})$ (in which $GL^+(7, \mathbb{R})$ is maximally embedded) manifest as maximal non-compact symmetry of the Lagrangian. Note that in case **III** one can further make $E_{7(7)}$ (and its maximal compact subgroup $SU\,(8)$) manifest *on-shell*, by exploiting a Cayley transformation supplemented by a rotation through $SO\,(8)$ gamma matrices on the vector 2-form field strengths [1, 2]. As we discuss further below, $E_{7(7)}$ can be promoted to a Lagrangian symmetry if one gives up manifest diffeomorphism invariance, as given by treatment in [3], then used in the anomaly study of [4].

It is worth remarking that $\mathcal{N} = 8$, $d = 4$ *gauged* supergravity with gauge group $SO(8)$ cannot be used for electroweak and strong interactions model building, because

$$SO\,(8) \not\supseteq SU\,(3) \times SU\,(2) \times U\,(1). \tag{1.3}$$

Furthermore, also the cosmological term problem arises out: the vacuum energy in anti De Sitter space AdS_4 is much higher than the vacuum energy in Standard Model of non-gravitational interactions (see *e.g.* the discussion in [5]). However, by exploiting the AdS_4/CFT_3 correspondence, theory **II** of (1.2) recently found application in $d = 3$ condensed matter physics (see *e.g.* [6] for a review and list of Refs.). Furthermore, the recently established fluid-gravity correspondence was object of many studies (see *e.g.* [7] for recent reviews and lists of Refs.).

The fundamental massless fields (and the related number $\sharp$ of degrees of freedom) of M-theory in $d = 11$ flat space-time dimensions are

$$g_{\mu\nu} \ (graviton): \qquad \sharp = \frac{(d-1)(d-2)}{2} - 1, \qquad in \ d = 11 : \sharp = 44;$$

$$\Psi_{\mu\alpha} \ (gravitino): \qquad \sharp = (d-3)2^{(d-3)/2}, \qquad in \ d = 11 : \sharp = 128; \qquad (1.4)$$

$$A_{\mu\nu\rho} \ (three\text{-}form): \qquad \sharp = \frac{(d-2)(d-3)(d-4)}{3!}, \qquad in \ d = 11 : \sharp = 84.$$

Because a $(p + 1)$-form ("Maxwell-like" gauge field) A_{p+1} couples to p-dimensional extended objects, and its "magnetic" dual B_{d-p-3} couples to $(d - p - 4)$-dimensional extended objects, it follows that the fundamental (massive) objects acting as sources of the theory are $M2$- and $M5$-branes.

In general, a compactification on an n-torus T_n has maximal manifest non-compact symmetry $GL^+(n, \mathbb{R}) \sim \mathbb{R}^+ \times SL(n, \mathbb{R})$. The metric g_{IJ} of T_n parametrizes the $n(n+1)/2$-dimensional coset $\mathbb{R}^+ \times \frac{SL(n,\mathbb{R})}{SO(n)}$, whereas the Kaluza-Klein vectors $g_{\hat{\mu}}^I$ are in the $\mathbf{n'}$ irrep. of $GL^+(n, \mathbb{R})$ itself. By reducing M-theory on T_7 a $d = 4$ theory with maximal ($\mathcal{N} = 8$) local supersymmetry arises. By splitting the $d = 11$ space-time index $\mu = 0, 1, ..., 10$ as $\mu = (\hat{\mu}, I)$, where $\hat{\mu} = 0, 1, ...3$ is the $d = 4$ space-time index, and $I = 1, ..., 7$ is the internal manifold index, the bosonic degrees of freedom of M-theory split as follows (below (1.6), for simplicity's sake we will then refrain from hatting the $d = 4$ curved indices):

$$g_{\mu\nu} \longrightarrow \begin{cases} g_{\hat{\mu}\hat{\nu}} \ (d = 4 \ graviton), & 1 + 1; \\\\ g_{\hat{\mu}}^I \ (vectors), & \mathbf{7'}; \\\\ g_{IJ} \ (scalars), & \mathbf{28}; \end{cases} \qquad (1.5)$$

$$A_{\mu\nu\rho} \longrightarrow \begin{cases} A_{\hat{\mu}\hat{\nu}\hat{\rho}} \ (d = 4 \ domain \ wall), & \sharp = 0; \\\\ A_{\hat{\mu}\hat{\nu}I} \ (antisymmetric \ tensors : strings), & \mathbf{7}; \\\\ A_{\hat{\mu}IJ} \ (vectors), & \mathbf{21}; \\\\ A_{IJK} \ (scalars), & \mathbf{35}, \end{cases} \qquad (1.6)$$

where the indicated irreps. pertain to the maximal manifest non-compact symmetry $GL^+(7, \mathbb{R})$, whose maximal compact subgroup is $SO(7)$. The 28 scalars g_{IJ} (metric of T_7) parametrize the coset $\mathbb{R}^+ \times \frac{SL(7,\mathbb{R})}{SO(7)}$.

By switching to formulation **III** of (1.2) [1], the 7 antisymmetric rank-2 tensors $A_{\hat{\mu}\hat{\nu}I}$ (sitting in the $\mathbf{7}$ of $GL^+(7, \mathbb{R})$) can be dualized to scalars ϕ^I (in the $\mathbf{7'}$ of $GL^+(7, \mathbb{R})$), and therefore one obtains $35 + 28 + 7 = 70$ scalar fields. It is worth remarking that in Cremmer and Julia's [1] theory the gravitinos ψ_I and the gauginos χ_{IJK} respectively have the following group theoretical assignment[1] (I in $\mathbf{8}$ of $SU(8)$):

$$\text{theory III [1]}: \begin{cases} \psi_I: \underset{\mathbf{8}}{SO(7)} \subset \underset{\mathbf{8}_s}{SO(8)} \subset \underset{\mathbf{8}}{SU(8)}; \\[2ex] \chi_{IJK}: \underset{\mathbf{8+48}}{SO(7)} \subset \underset{\mathbf{56}_s}{SO(8)} \subset \underset{\mathbf{56}}{SU(8)}. \end{cases} \tag{1.7}$$

Thus, in this theory the 70 scalars arrange as

$$\text{theory III [1]}: \underset{(\sharp=70)}{s=0} \; dofs: \underset{\mathbf{1+7+21+35}}{SO(7)} \subset \underset{\mathbf{35}_v+\mathbf{35}_c}{SO(8)} \subset \underset{\mathbf{70}}{SU(8)}, \tag{1.8}$$

where $\mathbf{70}$ is the rank-4 completely antisymmetric irrep. of $SU(8)$, the maximal compact subgroup of the U-duality group $E_{7(7)}$ (also called $\mathcal{R}$-symmetry).

On the other hand, also the vector fields $A_{\hat{\mu}IJ}$ (sitting in the $\mathbf{21}$ of $GL^+(7,\mathbb{R})$) can be dualized to $A_{\hat{\mu}}^{IJ}$ (sitting in the $\mathbf{21}'$ of $GL^+(7,\mathbb{R})$). Together with $g_{\hat{\mu}}^I$, the "electric" and "magnetic" vector degrees of freedom can thus be arranged as follows:

$$\underset{(\sharp=56)}{s=1} \; dofs: \begin{cases} \underset{\mathbf{7'+21'+7+21}}{GL^+(7,\mathbb{R})} \subset \underset{\mathbf{28'+28}}{SL(8,\mathbb{R})} \subset \underset{\mathbf{56}}{E_{7(7)}}; \\[2ex] \underset{\mathbf{7+21+7+21}}{SO(7)} \subset \underset{\mathbf{28+28}}{SO(8)} \subset \underset{\mathbf{28+\overline{28}}}{SU(8)}. \end{cases} \tag{1.9}$$

The counting of degrees of freedom is completely different in the *gauged* maximal supergravity theory II of (1.2), based on the $AdS_4 \times S^7$ solution of $d = 11$, $\mathcal{N} = 1$ M-theory field equations; in this framework, rather than using torus indices as in theories I and III of (1.2), Killing vector/spinor techniques are used (for a discussion, see *e.g.* [8], and the lectures [9], and Refs. therein). However, the 70 scalars still decompose[2] as $\mathbf{35}_v + \mathbf{35}_c$ of $SO(8)$ but, with respect to the chain of branchings (1.8), they lack of any $SO(7)$ interpretation. It is worth recalling that a formulation of this theory directly in $d = 4$ yields to the de Wit and Nicolai's $\mathcal{N} = 8$, $d = 4$ *gauged* supergravity [10].

Since the 70 scalar fields fit into an unique irrep. of $SU(8)$, it follows that they parameterize a non-compact coset manifold $\frac{G}{SU(8)}$. Indeed, the $SU(8)$ under which both the scalar fields and the fermion fields transform is the *"local"* $SU(8)$, namely the stabilizer of the scalar manifold. On the other hand, the $SU(8)$ appearing in the second line of (1.9), *i.e.* the one under which the vector 2-form self-dual/anti-self-dual field strengths transform, is the *"global"* $SU(8)$ ($\mathcal{R}$-symmetry group). Roughly speaking, the physically relevant group $SU(8)$ is the diagonal one in the product $SU_{\text{local}}(8) \times SU_{\text{global}}(8)$ (see also discussion below).

---

[1]As evident from (1.7), we use a different convention with respect to [5] (see *e.g.* Table 36 therein). Indeed, we denote as $\mathbf{8}_v$ of $SO(8)$ the irrep. which decomposes into $\mathbf{7+1}$ of $SO(7)$, whereas the two spinorial irreps. $\mathbf{8}_s$ and $\mathbf{8}_c$ both decompose into $\mathbf{8}$ of $SO(7)$. The same change of notation holds for $\mathbf{35}$ and $\mathbf{56}$ irreps..

[2]There are three distinct 35-dimensional $SO(8)$ irreps., usually denoted as $\mathbf{35}_v$, $\mathbf{35}_s$ and $\mathbf{35}_c$, obeying the relations:

$$(ab) \leftrightarrow [ABCD]_+ \; ; \; [abcd]_+ \leftrightarrow [ABCD]_- \; ; \; [abcd]_- \leftrightarrow (AB),$$

where $a, b = 1, ..., 8$ are in $\mathbf{8}_v$, $A, B, C, D = 1, ..., 8$ are in $\mathbf{8}_s$ (or in $\mathbf{8}_c$), and "$\pm$" denotes self-dual/anti-self-dual irreps.. For a discussion, see *e.g.* [1] and [9].

Remarkably, there exists an *unique* simple, non-compact Lie group with real dimension $70 + 63 = 133$ and which embeds $SU(8)$ as its maximal compact subgroup: this is the real, non-compact split form $E_{7(7)}$ of the exceptional Lie group E_7, thus giving rise to the symmetric, rank-7 coset space

$$\frac{E_{7(7)}}{SU(8)/\mathbb{Z}_2}, \tag{1.10}$$

which is the scalar manifold of $\mathcal{N} = 8$, $d = 4$ supergravity ($\mathbb{Z}_2$ is the kernel of the $SU(8)$-representations of even rank; in general, spinors transform according to the double cover of the stabilizer of the scalar manifold; see *e.g.* [11, 12]).

$E_{7(7)}$ acts as electric-magnetic duality symmetry group [13], and its maximal compact subgroup $SU(8)$ has a chiral action on fermionic as well as on (the vector part of the) bosonic fields. While the chiral action of $SU(8)$ on fermions directly follows from the chirality (complex nature) of the relevant irreps. of $SU(8)$ (as given by Eq. (1.7)), the chiral action on vectors is a crucial consequence of the electric-magnetic duality in $d = 4$ space-time dimensions. Indeed, this latter allows for "self-dual/anti-self-dual" complex combinations of the field strengths, which can then fit into complex irreps. of the stabilizer H of the coset scalar manifold G/H itself. For the case of maximal $\mathcal{N} = 8$ supergravity, the relevant chiral complex irrep. of $H = SU(8)$ is the rank-2 antisymmetric **28**, as given by Eq. (1.9).

Note that if one restricts to the $SL(8,\mathbb{R})$-covariant sector, the chirality of the action of electric-magnetic duality is spoiled, because the maximal compact subgroup of $SL(8,\mathbb{R})$, namely $SO(8)$, has not chiral irreps.

Composite (sigma model G/H) anomalies can arise in theories in which G has a maximal compact subgroup with a *chiral* action on bosons and/or fermions (see *e.g.* [14, 15, 4]). Surprising cancellations among the various contributions to the composite anomaly can occur as well. An example is provided by $\mathcal{N} = 8$, $d = 4$ supergravity itself, in which standard anomaly formulæ yield the remarkable result [15, 4]

$$3Tr_8 X^3 - 2Tr_{28} X^3 + Tr_{56} X^3 = (3 - 8 + 5) Tr_8 X^3 = 0, \tag{1.11}$$

where X is any generator of the Lie algebra $\mathfrak{su}(8)$ of the *rigid (i.e. global)* $SU(8)$ group ($\mathcal{R}$-symmetry). In light of the previous considerations, the first and third contributions to (1.11) are due to fermions: the 8 gravitinos ψ_A and the 56 spin-$\frac{1}{2}$ fermions χ_{ABC}, respectively, whereas the second contribution is due to the 28 chiral vectors. Note that, for the very same reason, the *local* $SU(8)$ (stabilizer of the non linear sigma-model of scalar fields), under which only fermions do transform[3], would be *anomalous* [14]. In an analogous way, in [15] it was discovered that $\mathcal{N} = 6$ and $\mathcal{N} = 5$ "pure" supergravities are *composite anomaly-free*, whereas $\mathcal{N} \leqslant 4$ theories are not.

A crucial equivalence holds at the homotopical level:

$$E_{7(7)} \cong (SU(8)/\mathbb{Z}_2) \times \mathbb{R}^{70}, \tag{1.12}$$

implying that the two group manifolds have the same De Rham cohomology. This is a key result, recently used in [4] to show that the aforementioned absence of $SU(8)$ current

---

[3] Also scalar fields transform under *local* $SU(8)$, but they do not contribute to the composite anomaly, because they sit in the *self-real* (and thus *non-chiral*) rank-4 antisymmetric irrep. **70** of $SU(8)$.

anomalies yield to the absence of anomalies for the non-linearly realized $E_{7(7)}$ symmetry, thus implying that the $E_{7(7)}$ continuous symmetry of classical $\mathcal{N} = 8$, $d = 4$ supergravity is preserved at all orders in perturbation theory (see e.g. [16, 17, 18, 19, 20, 21, 22, 23]). This implies the perturbative finiteness of supergravity at least up to seven loops; Bern, Dixon et al. explicitly checked the finiteness up to four loops included [16] (computations at five loops, which might be conclusive, are currently in progress; for a recent review, see e.g. [24]).

In order to achieve the aforementioned result on the anomalies of $E_{7(7)}$, in [4] the manifestly $E_{7(7)}$-covariant Lagrangian formulation of $\mathcal{N} = 8$, $d = 4$ supergravity [3] was exploited, by using the ADM decomposition of the $d = 4$ metric, namely[4]:

$$g_{\mu\nu}dx^\mu dx^\nu = -N^2 dt^2 + h_{ij}\left(dx^i + N^i dt\right)\left(dx^j + N^j dt\right),\qquad(1.13)$$

with lapse N and shift N^i (h_{ij} is the metric on the spatial slice). Within this approach [25, 3], the diffeomorphism symmetry is not realized in the standard way on the vector fields: the 28 vector fields A_μ^Λ of the original formulation [1, 10] are replaced by 56 vector fields $A_i^{\mathbb{B}}$ ($\mathbb{B} = 1, ..., 56$) with only spatial components, which recover the number of physical degrees of freedom by switching to an Hamiltonian formulation. Besides the 56×56 symplectic metric Ω:

$$\Omega^T = -\Omega, \ \Omega^2 = -\mathbb{I},\qquad(1.14)$$

a crucial quantity is the scalar field-dependent 56×56 symmetric matrix $\mathcal{M}$ (see Eq. (2.20) below), which is symplectic (see e.g. [26]):

$$\mathcal{M}\Omega\mathcal{M} = \Omega,\qquad(1.15)$$

and negative definite due to the positivity of the vector kinetic terms (see also discussion below). $\mathcal{M}$ allows for the introduction of a symplectic, scalar field-dependent complex structure:

$$\mathcal{J} \equiv \mathcal{M}\Omega \Rightarrow \mathcal{J}^2 = \mathcal{M}\Omega\mathcal{M}\Omega = -\mathbb{I}.\qquad(1.16)$$

Thus, the equations of motion of the 56 vector fields $A_i^{\mathbb{B}}$ can be expressed as a twisted self-duality condition[5] [1] for their super-covariant fields strengths $\hat{F}_{\mu\nu}^{\mathbb{A}}$, namely (see [3, 4] for further elucidation)

$$\hat{F}_{\mu\nu}^{\mathbb{A}} = -\frac{1}{2\sqrt{|g|}}\epsilon_{\mu\nu}{}^{\rho\sigma}\mathcal{J}^{\mathbb{A}}{}_{\mathbb{B}}\hat{F}_{\mu\nu}^{\mathbb{B}}.\qquad(1.17)$$

Although the time components $A_0^{\mathbb{B}}$ do not enter the Lagrangian, they appear when solving the equations of motion for the spatial components $A_i^{\mathbb{B}}$, and diffeomorphism covariance is recovered on the solutions of the equations of motion [3, 4].

From power counting arguments in quantum gravity, an n-loop counterterm contains $2n + 2$ derivatives, arranged such that it does not vanish on-shell. In $\mathcal{N} = 8$ supergravity the first (non-BPS) full superspace integral which is $E_{7(7)}$-invariant is the super-Vielbein superdeterminant, which may contain as last component a term $\sim \partial^8 R^4$ (see e.g. [28], and also [29]), then possibly contributing to a divergence in the four-graviton amplitude.

---

[4]We use units in which the Newton gravitational constant G and the speed of light in vacuum c are all put equal to 1.

[5]For interesting recent developments on twisted self-duality, see [27].

However, in [22] R. Kallosh argued that, by exploiting the light-cone formulation, candidate counterterms can be written in chiral, but *not* in real, light-cone superspace. This would then imply the ultraviolet finiteness of $\mathcal{N} = 8$, $d = 4$ supergravity, *if* supersymmetry and $E_{7(7)}$ symmetry are non-anomalous. Recently, in [30] the latter symmetry was advocated by the same author to imply ultraviolet finiteness of the theory to all orders in perturbation theory.

A puzzling aspect of these arguments is that string theory certainly violates continuous $E_{7(7)}$ symmetry at the perturbative level, as it can be easily realized by considering the dilaton dependence of loop amplitudes (see *e.g.* [23]). However, this is not the case for $\mathcal{N} = 8$ supergravity. From this perspective, two (perturbatively finite) theories of quantum gravity would exist, with 32 local supersymmetries; expectedly, they would differ at least in their non-perturbative sectors, probed *e.g.* by black hole solutions. String theorists [31, 32, 33] claim that $\mathcal{N} = 8$, $d = 4$ supergravity theory is probably not consistent at the non-perturbative level. From a purely $d = 4$ point of view, their arguments could be overcome by excluding from the spectrum, as suggested in [18], black hole states which turn out to be singular or ill defined if interpreted as purely four-dimensional gravitational objects. Inclusion of such singular states (such as $\frac{1}{4}$-BPS and $\frac{1}{2}$-BPS black holes) would then open up extra dimensions, with the meaning that a non-perturbative completion of $\mathcal{N} = 8$ supergravity would lead to string theory [31]. Extremal black holes with a consistent $d = 4$ interpretation may be defined as having a Bertotti-Robinson [34] $AdS_2 \times S^2$ near-horizon geometry, with a non-vanishing area of the event horizon. In $\mathcal{N} = 8$ supergravity, these black holes are[6] $\frac{1}{8}$-BPS or non-BPS (for a recent review and a list of Refs., see *e.g.* [35]). The existence of such states would in any case break the $E_{7(7)}$ $(\mathbb{R})$ continuous symmetry, because of Dirac-Schwinger-Zwanziger dyonic charge quantization conditions. The breaking of $E_{7(7)}$ $(\mathbb{R})$ into an arithmetic subgroup $E_{7(7)}$ $(\mathbb{Z})$ would then manifest only in exponentially suppressed contributions to perturbative amplitudes (see *e.g.* the discussion in [4], and Refs. therein), in a similar way to instanton effects in non-Abelian gauge theories.

The composite anomaly concerns the gauge-scalar sector of the supergravity theories. Another anomaly, originated in the gravitational part of the action is the so-called *gravitational anomaly*, which only counts the basic degrees of freedom associated to the field content of the theory itself [38, 39] (see also [40] for a review):

$$g_{\mu\nu} \langle T^{\mu\nu} \rangle_{1-loop} = \frac{\mathcal{A}}{32\pi^2} \int d^4x \sqrt{|g|} \left(R^2_{\mu\nu\lambda\rho} - 4R^2_{\mu\nu} + R^2 \right), \qquad (1.18)$$

where $\langle T^{\mu\nu} \rangle_{1-loop}$ is the 1-loop *vev* of the gravitational stress-energy tensor. In general, this trace anomaly is a total derivative and therefore it can be non-vanishing only on topologically non-trivial $d = 4$ backgrounds. Furthermore, as found long time ago by Faddeev and Popov [41], $(p + 1)$-form gauge fields have a complicated quantization procedure, due to the presence of ghosts; thus, their contribution to the parameter $\mathcal{A}$ appearing in the formula (1.18) vary greatly depending on the field under consideration. This is because at the quantum level different field representations are generally inequivalent [38]. Consequently, one may expect that different formulations of $\mathcal{N} = 8$, $d = 4$ supergravity (1.2), give rise to different gravitational anomalies. This is actually what happens:

---

[6]We also remark that these are the only black holes for which the *Freudenthal duality* [36, 37] is well defined.

- in the formulation **III** of (1.2) [1], with maximal manifest compact symmetry $SO\,(8)$, the antisymmetric tensors $A_{\mu\nu I}$ are dualized to scalars, and $\mathcal{A} \neq 0$.

- in the formulation **I** of (1.2) [1], with maximal manifest compact symmetry $SO\,(7)$, obtained by compactifying $d = 11$ M-theory on T_7, the antisymmetric tensors $A_{\mu\nu I}$ are not dualized, and, as found some time ago in [38], the gravitational anomaly vanishes: $\mathcal{A} = 0$. Recently, a wide class of models has been shown to have $\mathcal{A} = 0$, by exploiting *generalized mirror symmetry* for seven-manifolds [42].

- in the formulation **II** of (1.2) [10, 39] (see also [8, 9] and the discussion above), with maximal manifest compact *gauged* symmetry $SO\,(8)$, the gravitational anomaly is the sum of two contributions: one given by (1.18), and another one related to the non-vanishing cosmological constant Λ, given by

$$\frac{\mathcal{B}}{12\pi^2} \int d^4x \sqrt{|g|} \Lambda^2, \tag{1.19}$$

where $\mathcal{B}$, through the relation $\Lambda \sim -e^2$ [39], vanishes whenever the charge e normalization beta function[7] [43]

$$\beta_e\,(s) = \frac{\hbar}{96\pi^2} e^3 C_s \left(1 - 12s^2\right) (-1)^{2s} \tag{1.20}$$

vanishes, namely in $\mathcal{N} > 4$ supergravities (compare *e.g.* Table II of [39] with Table 1 of [43]). The contribution to the coefficients $\mathcal{A}$ and $\mathcal{B}$ of (1.18) and (1.19) depends on the spin s of the *massless* particle, but also, as mentioned above, on the its field representation ([39]; see also *e.g.* Table 1 of [42]):

$s:$	0 (ϕ)	0 $(A_{\mu\nu\rho})$	$\frac{1}{2}$	1 (A_μ)	1 $(A_{\mu\nu})$	$\frac{3}{2}$	2
$360\mathcal{A}:$	4	-720	7	-52	364	-233	848
$60\mathcal{B}:$	-1	0	-3	-12	0	137	$-522.$

$$\tag{1.21}$$

---

[7]C_s is the appropriate (positive) *quadratic invariant* for the gauge group representation in which the particle of spin s sits (see *e.g.* Table 1 of [43], and Refs. therein).

2 Lecture II
(Multi-Center) Black Holes and Attractors

If $E_{7(7)}$ is a continuous non-anomalous symmetry of $\mathcal{N} = 8$ supergravity, then it is likely that non-perturbative effects are exponentially suppressed in perturbative amplitudes.

Black holes (BHs) are examples of non-perturbative states which, in presence of Dirac-Schwinger-Zwanziger dyonic charge quantization, would break $E_{7(7)}(\mathbb{R})$ to a suitable (not unique) arithmetic subgroup of $E_{7(7)}(\mathbb{Z})$ (see *e.g.* [44, 36, 19, 35], and Refs. therein).

Here we confine ourselves to recall some very basic facts on extremal BHs (for further detail, see *e.g.* [45], and Refs. therein), and then we will mention some recent developments on multi-center solutions.

For simplicity's sake, we consider the particular class of *extremal* BH solutions constituted by static, asymptotically flat, spherically symmetric solitonic objects with dyonic charge vector $\mathcal{Q}$ and scalars ϕ describing trajectories (in the radial evolution parameter r) with[8] *fixed points* determined by the *Attractor Mechanism* [46]:

$$\begin{cases} \lim_{r \to r_H^+} \phi(r) = \phi_H(\mathcal{Q}); \\ \\ \lim_{r \to r_H^+} \frac{d\phi(r)}{dr} = 0. \end{cases} \qquad (2.1)$$

At the horizon, the scalars lose memory of the initial conditions (*i.e.* of the asymptotic values $\phi_\infty \equiv \lim_{r \to \infty} \phi(r)$), and the fixed (attractor) point $\phi_H^a(\mathcal{Q})$ only depends on the BH charges $\mathcal{Q}$. In the supergravity limit, for $\mathcal{N} > 2$ superymmetry, the attractor behavior of such BHs is now completely classified (see *e.g.* [47, 48] for a review and list of Refs.).

The classical BH entropy is given by the Bekenstein-Hawking entropy-area formula [49]

$$S(\mathcal{Q}) = \frac{A_H(\mathcal{Q})}{4} = \pi V_{BH}(\phi_H(\mathcal{Q}), \mathcal{Q}) = \pi \sqrt{|\mathcal{I}_4(\mathcal{Q})|}. \qquad (2.2)$$

where V_{BH} is the effective BH potential [50] (see Eq. (2.22) below).

The last step of (2.2) holds[9] for those theories admitting a quartic polynomial invariant $\mathcal{I}_4$ in the (symplectic) representation of the electric-magnetic duality group in which $\mathcal{Q}$ sits. This is the case *at least* for the *"groups of type E_7"* [51], which are the electric-magnetic duality groups of supergravity theories in $d = 4$ with *symmetric* scalar manifolds (see *e.g.* [37] for recent developments, and a list of Refs.). These include all $\mathcal{N} \geqslant 3$ supergravities as well as a broad class of $\mathcal{N} = 2$ theories in which the vector multiplets' scalar manifold is a *special Kähler* symmetric space (see *e.g.* [52, 53, 54, 68], and Refs. therein). In the D-brane picture of type IIA supergravity compactified on Calabi-Yau threefolds CY_3, charges can be denoted by q_0 ($D0$), q_a ($D2$), p^a ($D4$) and p^0 ($D6$), and

---

[8]The subscript "H" denotes the evaluation at the BH event horizon, whose radial coordinate is r_H (see treatment below).

[9]Incidentally, the last step of (2.2) also holds for arbitrary cubic scalar geometry if particular charge configurations are chosen.

the quartic invariant polynomial $\mathcal{I}_4$ is given by [55]

$$\mathcal{I}_4 = -\left(p^0 q_0 + p^a q_a\right)^2 + 4\left(-p^0 \mathcal{I}_3\left(q\right) + q_0 \mathcal{I}_3\left(p\right) + \frac{\partial \mathcal{I}_3\left(p\right)}{\partial p^a}\frac{\partial \mathcal{I}_3\left(q\right)}{\partial q_a}\right); \quad (2.3)$$

$$\mathcal{I}_3\left(p\right) \equiv \frac{1}{3!}d_{abc}p^a p^b p^c; \quad \mathcal{I}_3\left(q\right) \equiv \frac{1}{3!}d^{abc}q_a q_b q_c, \quad (2.4)$$

where d_{abc} and d^{abc} are completely symmetric rank-3 invariant tensors of the relevant electric and magnetic charge irreps. of the U-duality group in $d = 5$. A typical (single-center) BPS configuration is (q_0, p^a), with all charges positive (implying $\mathcal{I}_4 > 0$), while a typical non-BPS configuration is (p^0, q_0) (implying $\mathcal{I}_4 < 0$), see $e.g.$ the discussion in [56] (other charge configurations can be chosen as well). In the dressed charge basis, manifestly covariant with respect to the $\mathcal{R}$-symmetry group, the charges arrange into a complex skew-symmetric central charge matrix Z_{AB}. This latter can be skew-diagonalized to the form [57]

$$Z_{AB} = \text{diag}\left(z_1, z_2, z_3, z_4\right) \otimes \begin{pmatrix} 0 & -1 \\ 1 & 0 \end{pmatrix}, \quad (2.5)$$

and the quartic invariant can be recast in the following form [58]:

$$\mathcal{I}_4 = \sum_{i=1}^{4}|z_i|^4 - 2\sum_{i<j=1}^{4}|z_i|^2|z_j|^2 + 4\prod_{i=1}^{4}z_i + 4\prod_{i=1}^{4}\overline{z}_i. \quad (2.6)$$

In such a basis, a typical BPS configuration is the one pertaining to the Reissner-Nördstrom BH, with charges $z_1 = (q + ip)$ and $z_2 = z_3 = z_4 = 0$ (implying $\mathcal{I}_4 = (q^2 + p^2)^2 > 0$), whereas a typical non-BPS configuration has (at the event horizon) $z_i = \rho e^{i\pi/4} \,\forall i = 1, ..., 4$ (implying $\mathcal{I}_4 = -16\rho^4 < 0$); see $e.g.$ the discussion in [59, 60, 61].

The simplest example of BH metric is the Schwarzschild BH:

$$ds^2 = -\left(1 - \frac{2M}{r}\right)dt^2 + \left(1 - \frac{2M}{r}\right)^{-1}dr^2 + r^2 d\Omega^2, \quad (2.7)$$

where M is the ADM mass [62], and $d\Omega^2 = d\theta^2 + sin^2\theta d\psi^2$. This BH has no $naked$ $singularity$, $i.e.$ the singularity at $r = 0$ is $covered$ by the event horizon at $r_H = 2M$.

The metric (2.7) can be seen as the neutral $q, p \rightarrow 0$ limit of the Reissner-Nordström (RN) BH:

$$ds_{RN}^2 = -\left(1 - \frac{2M}{r} + \frac{q^2 + p^2}{r^2}\right)dt^2 + \left(1 - \frac{2M}{r} + \frac{q^2 + p^2}{r^2}\right)^{-1}dr^2 + r^2 d\Omega^2. \quad (2.8)$$

Such a metric exhibits two horizons, with radii

$$r_\pm = M \pm \sqrt{M^2 - q^2 - p^2}. \quad (2.9)$$

In the $extremal$ case $r_+ = r_-$, and it holds that

$$M^2 = q^2 + p^2, \quad (2.10)$$

thus a unique event horizon exists at $r_H = M$. Notice that for RN BHs the extremality condition coincides with the saturation of the *BPS bound* [63]

$$M^2 \geqslant q^2 + p^2. \tag{2.11}$$

By defining $\rho \equiv r - M = r - r_H$, the extremal RN metric acquires the general static Papapetrou-Majumdar [64] form

$$ds^2_{RN,extr} = -\left(1 + \frac{M}{\rho}\right)^{-2} dt^2 + \left(1 + \frac{M}{\rho}\right)^2 \left(d\rho^2 + \rho^2 d\Omega^2\right) = -e^{2U} dt^2 + e^{-2U} d\vec{x}^2, \tag{2.12}$$

where $U = U(\vec{x})$ is an harmonic function satisfying the $d = 3$ Laplace equation

$$\Delta e^{-U(\vec{x})} = 0. \tag{2.13}$$

In order to determine the near-horizon geometry of an extremal RN BH, let us define a new radial coordinate as $\tau = -\frac{1}{\rho} = \frac{1}{r_H - r}$. Thus, after a further rescaling $\tau \to \frac{\tau}{M^2}$, the near-horizon limit $\rho \to 0^+$ of extremal metric (2.12) reads

$$\lim_{\rho \to 0^+} ds^2_{RN,extr} = \frac{M^2}{\tau^2} \left(-dt^2 + d\tau^2 + \tau^2 d\Omega^2\right), \tag{2.14}$$

which is nothing but the $AdS_2 \times S^2$ Bertotti-Robinson metric [34], both *flat* and *conformally flat*.

In presence of scalar fields coupled to the BH background, the BPS bound gets modified, and in general extremality does not coincide with the saturation of BPS bound (and thus with supersymmetry preservation) any more. Roughly speaking, the charges $\mathcal{Q}$ gets "dressed" with scalar fields ϕ into the central extension of the local $\mathcal{N}$-extended supersymmetry algebra, which is an antisymmetric complex matrix $Z_{AB}(\phi, \mathcal{Q})$, named *central charge matrix* $(A, B = 1, ..., \mathcal{N})$:

$$\begin{cases} \left\{\mathbf{Q}_{\alpha A}, \overline{\mathbf{Q}}_{\dot\alpha}^B\right\} = \delta_A^B \sigma_{\alpha\dot\alpha}^\mu P_\mu; \\[2mm] \left\{\mathbf{Q}_{\alpha A}, \mathbf{Q}_{\beta B}\right\} = \epsilon_{\alpha\beta} Z_{AB}(\phi, \mathcal{Q}). \end{cases} \tag{2.15}$$

In general

$$Z_{AB}(\phi, \mathcal{Q}) = L_{AB}^{\mathbf{\Lambda}}(\phi) \mathcal{Q}_{\mathbf{\Lambda}}, \tag{2.16}$$

where $L_{AB}^{\mathbf{\Lambda}}(\phi)$ are the scalar field-dependent symplectic sections of the corresponding *(generalized) special geometry* (see *e.g.* [26, 59, 37], and Refs. therein).

In the BH background under consideration, the general *Ansätze* for the vector 2-form field strengths $F_{\mu\nu}^\Lambda$ of the n_V vector fields $(\Lambda = 1, \ldots, n_V)$ and their duals $G_{\Lambda\mu\nu} = \frac{\delta \mathcal{L}}{\delta F_{\mu\nu}^\Lambda}$ are given by [50]

$$F = e^{2U} \mathbb{C} \mathcal{M}(\phi) \mathcal{Q} dt \wedge d\tau + \mathcal{Q} \sin\theta d\theta \wedge d\psi; \tag{2.17}$$

$$F = \begin{pmatrix} F_{\mu\nu}^\Lambda \\ G_{\Lambda\mu\nu} \end{pmatrix} \frac{dx^\mu dx^\nu}{2}, \tag{2.18}$$

and electric and magnetic charges $\mathcal{Q} \equiv \left(p^{\Lambda}, q_{\Lambda}\right)^{T}$ are defined by

$$q_{\Lambda} \equiv \frac{1}{4\pi} \int_{S_{\infty}^{2}} G_{\Lambda}, \qquad p^{\Lambda} \equiv \frac{1}{4\pi} \int_{S_{\infty}^{2}} F^{\Lambda}, \qquad (2.19)$$

where S_{∞}^{2} is the 2-sphere at infinity. $\mathcal{M}(\phi)$, already discussed in Sec. 1, is a $2n_{V} \times 2n_{V}$ real symmetric $Sp(2n_{V}, \mathbb{R})$ matrix (see Eq. (1.15)) whose explicit form reads [26]

$$\mathcal{M}(\phi) = \begin{pmatrix} I + RI^{-1}R & -RI^{-1} \\ -I^{-1}R & I^{-1} \end{pmatrix}, \qquad (2.20)$$

with $I \equiv \mathrm{Im}\,\mathcal{N}_{\Lambda\Sigma}$ and $R \equiv \mathrm{Re}\,\mathcal{N}_{\Lambda\Sigma}$, where $\mathcal{N}_{\Lambda\Sigma}$ is the (scalar field dependent) kinetic vector matrix entering the $d = 4$ Lagrangian density

$$\mathcal{L} = -\frac{R}{2} + \frac{1}{2}g_{ij}(\phi)\partial_{\mu}\phi^{i}\partial^{\mu}\phi^{j} + I_{\Lambda\Sigma}F^{\Lambda} \wedge^{*} F^{\Sigma} + R_{\Lambda\Sigma}F^{\Lambda} \wedge F^{\Sigma}. \qquad (2.21)$$

The black hole effective potential [46] is given by

$$V_{BH}\left(\phi, \mathcal{Q}\right) = -\frac{1}{2}\mathcal{Q}^{T}\mathcal{M}\left(\phi\right)\mathcal{Q}, \qquad (2.22)$$

This is the effective potential which arises upon reducing the general $d \geqslant 4$ Lagrangian on the BH background to the $d = 1$ almost geodesic action describing the radial evolution of the $n_{V} + 1$ scalar fields $(U(\tau), \phi^{i}(\tau))$ [65]:

$$S = \int \mathcal{L}d\tau = \int (\dot{U} + g_{ij}\dot{\phi}^{i}\dot{\phi}^{j} + e^{2U}V_{BH}(\phi(\tau), p, q)d\tau. \qquad (2.23)$$

In order to have the same equations of motion of the original theory, the action must be complemented with the Hamiltonian constraint, which in the extremal case reads [50]

$$\dot{U}^{2} + g_{ij}\dot{\phi}^{i}\dot{\phi}^{j} - e^{2U}V_{BH}(\phi(\tau), p, q) = 0. \qquad (2.24)$$

The black hole effective potential V_{BH} can generally be written in terms of the super-potential $W(\phi)$ as

$$V_{BH} = W^{2} + 2g^{ij}\partial_{i}W\partial_{j}W. \qquad (2.25)$$

This formula can be viewed as a differential equation defining W for a given V_{BH}, and it can lead to multiple choices, one corresponding to BPS solutions, and the others associated to non-BPS ones. W allows to rewrite the ordinary second order supergravity equations of motion

$$\ddot{U} = e^{2U}V_{BH}; \qquad (2.26)$$

$$\ddot{\phi}^{i} = g^{ij}\frac{\partial V_{BH}}{\partial\phi_{j}}e^{2U}, \qquad (2.27)$$

as first order flow equations, defining the radial evolution of the scalar fields ϕ^{i} and the warp factor U from asymptotic (radial) infinity towards the black hole horizon [66] :

$$\dot{U} = -e^{U}W, \qquad \dot{\phi}^{i} = -2e^{U}g^{ij}\partial_{j}W. \qquad (2.28)$$

At the prize of finding a suitable "fake" first order superpotential W, one only has to deal with these first order flow equations even for non-supersymmetric solutions, where one does not have Killing spinor equations [66, 67].

For $\frac{1}{N}$-BPS supersymmetric BHs in $\mathcal{N} \geqslant 2$ supergravity theories (with central charge matrix Z_{AB}), W is given by the square root[10] $\sqrt{\lambda_{h}}$ of the largest of the eigenvalues of

---

[10]The subscript "h" stands for "the highest".

$Z_{AB} Z^{\dagger BC}$ [66, 67]. Furthermore, $\mathcal{W}$ has a known analytical expression for all $\mathcal{N} \geqslant 2$ charge configurations with $\mathcal{I}_4 > 0$ (for $\mathcal{N} = 2$, this applies to special Kähler geometry based on symmetric spaces, see *e.g.* [68]) [67]. For $\mathcal{I}_4 < 0$, $\mathcal{W}^2$ has an analytical expression for rank-1 and rank-2 cosets [69, 70, 71], while it is known to exist in general as a solution of a sixth order algebraic equation [70, 71, 72].

The Bekenstein-Hawking BH entropy [49] (2.2) can be written in terms of W as follows:

$$S(\mathcal{Q}) = \pi \left. W^2 \right|_{\partial W = 0}, \tag{2.29}$$

where the critical points of the suitable W reproduce a class of critical points of V itself. It is worth remarking that the value of the superpotential W at radial infinity also encodes other basic properties of the extremal black hole, namely its ADM mass [62], given by ($\phi_\infty^i \equiv \lim_{r \to \infty} \phi^i(r)$)

$$M_{ADM}(\phi_\infty, \mathcal{Q}) = \dot{U}(\tau = 0) = W(\phi_\infty, \mathcal{Q}), \tag{2.30}$$

and the scalar charges

$$\Sigma^i(\phi_\infty, \mathcal{Q}) = 2g^{ij}(\phi_\infty) \frac{\partial W}{\partial \phi^i}(\phi_\infty, \mathcal{Q}). \tag{2.31}$$

Multi-center BHs are a natural extension of single-center BHs, and they play an important role in the dynamics of quantum theories of gravity, such as superstrings and M-theory.

In fact, interesting multi-center solutions have been found for BPS BHs in $d = 4$ theories with $\mathcal{N} = 2$ supersymmetry, in which the *Attractor Mechanism* [46, 50] is generalized by the so-called *split attractor flow* [73]. This name comes from the existence, for 2-center solutions, of a co-dimension one region (named *marginal stability (MS) wall*) in the scalar manifold, where in fact a stable 2-center BH configuration may decay into two single-center constituents, whose scalar flows then separately evolve according to the corresponding attractor dynamics.

The study of these phenomena has recently progressed in many directions. By combining properties of $\mathcal{N} = 2$ supergravity and superstring theory, a number of interesting phenomena, such as split flow tree, entropy enigma, bound state recombination walls, and microstate counting have been investigated (see *e.g.* [74, 75, 76, 77]).

The MS wall is defined by the condition of stability for a marginal decay of a 2-center BH compound solution with charge $\mathcal{Q} = \mathcal{Q}_1 + \mathcal{Q}_2$ into two single-center BHs (respectively with charges $\mathcal{Q}_1$ and $\mathcal{Q}_2$):

$$M(\phi_\infty, \mathcal{Q}_1 + \mathcal{Q}_2) = M(\phi_\infty, \mathcal{Q}_1) + M(\phi_\infty, \mathcal{Q}_2). \tag{2.32}$$

As mentioned, after crossing the MS wall each flow evolves towards its corresponding attractor point, and the classical entropy of each BH constituent follows the Bekenstein-Hawking formula (2.2). It should be noted that the entropy of the original compound (conceived as a *single-center* BH with total charge $\mathcal{Q} = \mathcal{Q}_1 + \mathcal{Q}_2$) can be smaller, equal, or larger than the sum of the entropies of its constituents:

$$S(\mathcal{Q}_1 + \mathcal{Q}_2) \gtreqless S(\mathcal{Q}_1) + S(\mathcal{Q}_2). \tag{2.33}$$

For $\mathcal{N} = 2$ BPS compound and constituents in $\mathcal{N} = 2$, $d = 4$ supergravity (in which $Z_{AB} = \epsilon_{AB}Z$), (2.32) can be recast as a condition on the central charge ($Z_i \equiv M\left(\phi_\infty, \mathcal{Q}_i\right)$, $i = 1, 2$, and $Z_{1+2} \equiv Z\left(\phi_\infty, \mathcal{Q}_1 + \mathcal{Q}_2\right) = Z_1 + Z_2$):

$$|Z_1 + Z_2| = |Z_1| + |Z_2|. \tag{2.34}$$

Furthermore, before crossing the MS wall, the relative distance $|\overrightarrow{x_1} - \overrightarrow{x_2}|$ of the two BH constituents with *mutually non-local* charges $\langle \mathcal{Q}_1, \mathcal{Q}_2 \rangle \neq 0$ is given by [74]

$$|\overrightarrow{x_1} - \overrightarrow{x_2}| = \frac{1}{2} \frac{\langle \mathcal{Q}_1, \mathcal{Q}_2 \rangle |Z_1 + Z_2|}{\mathrm{Im}\left(Z_1 \overline{Z_2}\right)}, \tag{2.35}$$

where

$$2\left|\mathrm{Im}\left(Z_1 \overline{Z_2}\right)\right| = \sqrt{4|Z_1|^2 |Z_2|^2 - \left(|Z_1 + Z_2|^2 - |Z_1|^2 - |Z_2|^2\right)^2}. \tag{2.36}$$

Correspondingly, the 2-center BH has an intrinsic (orbital) angular momentum, given by [74]

$$\overrightarrow{J} = \frac{1}{2} \langle \mathcal{Q}_1, \mathcal{Q}_2 \rangle \frac{\overrightarrow{x_1} - \overrightarrow{x_2}}{|\overrightarrow{x_1} - \overrightarrow{x_2}|}. \tag{2.37}$$

Note that when the charge vectors $\mathcal{Q}_1$ and $\mathcal{Q}_2$ are *mutually local* (*i.e.* $\langle \mathcal{Q}_1, \mathcal{Q}_2 \rangle = 0$), $|\overrightarrow{x_1} - \overrightarrow{x_2}|$ is not constrained at all, and $J = 0$. Actually, this is always the case for the scalarless case of extremal Reissner-Nördstrom double-center BH solutions in $\mathcal{N} = 2$ *pure* supergravity. Indeed, in this case the central charge simply reads (see also discussion above)

$$Z_{RN}\left(p, q\right) = q + ip, \tag{2.38}$$

and it is immediate to check that the marginal stability condition (2.34) implies $\langle \mathcal{Q}_1, \mathcal{Q}_2 \rangle = q_1 p_2 - p_1 q_2 = 0$.

It is here worth observing that $\mathrm{Im}\left(Z_1 \overline{Z_2}\right) = 0$ both describes marginal and *anti-marginal* stability [76]. *Marginal stability* further requires

$$\mathrm{Re}\left(Z_1 \overline{Z_2}\right) > 0 \Leftrightarrow |Z_1 + Z_2|^2 > |Z_1|^2 + |Z_2|^2. \tag{2.39}$$

The other (unphysical) branch, namely

$$\mathrm{Re}\left(Z_1 \overline{Z_2}\right) < 0 \Leftrightarrow |Z_1 + Z_2|^2 < |Z_1|^2 + |Z_2|^2, \tag{2.40}$$

pertains to *anti-marginal stability*, reached for $|Z_1 + Z_2| = ||Z_1| - |Z_2||$.

Eq. (2.35) implies the stability region for the 2-center BH solution to occur for

$$\langle \mathcal{Q}_1, \mathcal{Q}_2 \rangle \, \mathrm{Im}\left(Z_1 \overline{Z_2}\right) > 0, \tag{2.41}$$

while it is forbidden for $\langle \mathcal{Q}_1, \mathcal{Q}_2 \rangle \mathrm{Im}\left(Z_1 \overline{Z_2}\right) < 0$. The scalar flow is directed from the stability region towards the instability region, crossing the MS wall at $\langle \mathcal{Q}_1, \mathcal{Q}_2 \rangle \mathrm{Im}\left(Z_1 \overline{Z_2}\right) = 0$. This implies that the stability region is placed *beyond* the MS wall, and *on the opposite side* of the split attractor flows.

By using the fundamental identities of $\mathcal{N} = 2$ special Kähler geometry in presence of two (mutually non-local) symplectic charge vectors $\mathcal{Q}_1$ and $\mathcal{Q}_2$ (see *e.g.* [73, 78, 59]), one can compute that at BPS attractor points of the centers 1 *or* 2:

$$\langle \mathcal{Q}_1, \mathcal{Q}_2 \rangle = -2\mathrm{Im}\left(Z_1 \overline{Z_2}\right) \Rightarrow 2 \langle \mathcal{Q}_1, \mathcal{Q}_2 \rangle \, \mathrm{Im}\left(Z_1 \overline{Z_2}\right) = -\langle \mathcal{Q}_1, \mathcal{Q}_2 \rangle^2 < 0. \tag{2.42}$$

By using (2.35) and (2.42), one obtains $|\vec{x_1} - \vec{x_2}| < 0$: this means that, as expected, the BPS attractor points of the centers 1 *or* 2 do not belong to the stability region of the 2-center BH solution. Furthermore, the result (2.42) also consistently implies:

stability region :
$$\langle \mathcal{Q}_1, \mathcal{Q}_2 \rangle \operatorname{Im} \left(Z_1 \overline{Z_2} \right) = |\langle \mathcal{Q}_1, \mathcal{Q}_2 \rangle| \sqrt{4 |Z_1|^2 |Z_2|^2 - \left(|Z_1 + Z_2|^2 - |Z_1|^2 - |Z_2|^2 \right)^2} > 0; \tag{2.43}$$

instability region :
$$\langle \mathcal{Q}_1, \mathcal{Q}_2 \rangle \operatorname{Im} \left(Z_1 \overline{Z_2} \right) = -|\langle \mathcal{Q}_1, \mathcal{Q}_2 \rangle| \sqrt{4 |Z_1|^2 |Z_2|^2 - \left(|Z_1 + Z_2|^2 - |Z_1|^2 - |Z_2|^2 \right)^2} < 0, \tag{2.44}$$

where a particular case of (2.44), holding at the attractor points, is given by (2.42).

As shown in [77], by exploiting the theory of *matrix norms*, all above results can be extended *at least* to $\mathcal{N} = 2$ non-BPS states with $\mathcal{I}_4 > 0$, as well as to BPS states in $\mathcal{N} > 2$ supergravity.

For two-center BHs, by replacing $|Z|$ with $\sqrt{\lambda_h}$, the generalization of (2.35) *e.g.* to $\mathcal{N} = 8$ maximal supergravity reads

$$|\vec{x_1} - \vec{x_2}| = \frac{|\langle \mathcal{Q}_1, \mathcal{Q}_2 \rangle| \sqrt{\lambda_{1+2,h}}}{\sqrt{4\lambda_{1,h}\lambda_{2,h} - (\lambda_{1+2,h} - \lambda_{1,h} - \lambda_{2,h})^2}}, \tag{2.45}$$

where $\lambda_{1+2,h} \equiv \lambda_h (\phi_\infty, \mathcal{Q}_1 + \mathcal{Q}_2)$ and $\lambda_{i,h} \equiv \lambda_h (\phi_\infty, \mathcal{Q}_i)$.

Analogously, also result (2.42) can be generalized *e.g.* to suitable states in $\mathcal{N} = 8$ supergravity. Indeed, by exploiting the $\mathcal{N} = 8$ generalized special geometry identities [59] ($\mathbf{Z}_i \equiv Z_{AB} (\phi_\infty, \mathcal{Q}_i)$)

$$\langle \mathcal{Q}_1, \mathcal{Q}_2 \rangle = -\operatorname{Im} \left(Tr \left(\mathbf{Z}_1 \mathbf{Z}_2^\dagger \right) \right), \tag{2.46}$$

one can compute that at the $\frac{1}{8}$-BPS attractor points of the centers 1 *or* 2 it holds

$$|\langle \mathcal{Q}_1, \mathcal{Q}_2 \rangle| = \sqrt{4\lambda_{h,1}\lambda_{h,2} - (\lambda_{1,h} + \lambda_{2,h} - \lambda_{1+2,h})^2}. \tag{2.47}$$

Analogously to the $\mathcal{N} = 2$ case treated above, note that $\frac{1}{8}$-BPS attractor points of the centers 1 *or* 2 do not belong to the stability region of the two-center BH solution, but instead they are placed, with respect to the stability region, on the opposite side of the MS wall.

Acknowledgments

The work of S. F. is supported by the ERC Advanced Grant no. 226455, *"Supersymmetry, Quantum Gravity and Gauge Fields" (SUPERFIELDS)*.

82

References

[1] E. Cremmer and B. Julia, *The SO(8) Supergravity*, Nucl. Phys. **B159**, 141 (1979).

[2] C. M. Hull and N. P. Warner, *The Structure of the Gauged $\mathcal{N}= 8$ Supergravity Theories*, Nucl. Phys. **B253**, 650 (1985). C. M. Hull and N. P. Warner, *The Potentials of the Gauged $\mathcal{N}= 8$ Supergravity Theories*, Nucl. Phys. **B253**, 675 (1985).

[3] C. Hillmann, *$E_{7(7)}$ Invariant Lagrangian of $d = 4$ $\mathcal{N}= 8$ Supergravity*, JHEP **1004**, 010 (2010), arXiv:0911.5225 [hep-th].

[4] G. Bossard, C. Hillmann and H. Nicolai, *$E_{7(7)}$ Symmetry in Perturbatively Quantised $\mathcal{N}= 8$ Supergravity*, JHEP **1012**, 052 (2010), arXiv:1007.5472 [hep-th].

[5] R. Slansky, *Group Theory for Unified Model Building*, Phys. Rep. **79**, 1 (1981).

[6] S. A. Hartnoll, *Lectures on Holographic Methods for Condensed Matter Physics*, Class. Quant. Grav. **26**, 224002 (2009), arXiv:0903.3246 [hep-th].

[7] M. Rangamani, *Gravity & Hydrodynamics : Lectures on Fluid-Gravity Correspondence*, Class. Quant. Grav. **26**, 224003 (2009), arXiv:0905.4352 [hep-th].

[8] B. Biran, F. Englert, B. de Wit and H. Nicolai, *Gauged $\mathcal{N}= 8$ Supergravity and its Breaking from Spontaneous Compactification*, Phys. Lett. **B124**, 45 (1983); erratum ibidem, **B128**, 461 (1983).

[9] M. J. Duff and C. N. Pope, *Kaluza-Klein Supergravity and the Seven Sphere*, Lectures given at September School on Supergravity and Supersymmetry, ICTP Trieste (Italy), 1982.

[10] B. de Wit and H. Nicolai, *$\mathcal{N}= 8$ Supergravity*, Nucl. Phys. **B208**, 323 (1982).

[11] I. Yokota, *Subgroup $SU(8)/\mathbb{Z}_2$ of compact simple Lie group E_7 and non-compact simple Lie group $E_{7(7)}$ of type E_7*, Math. J. Okoyama Univ. **24**, 53 (1982).

[12] P. Aschieri, S. Ferrara and B. Zumino, *Duality Rotations in Nonlinear Electrodynamics and in Extended Supergravity*, Riv. Nuovo Cim. **31**, 625 (2008), arXiv:0807.4039 [hep-th].

[13] M. K. Gaillard and B. Zumino, *Duality Rotations for Interacting Fields*, Nucl. Phys. **B193**, 221 (1981).

[14] P. Di Vecchia, S. Ferrara and L. Girardello, *Anomalies of Hidden Local Chiral Symmetries in Sigma Models and Extended Supergravities*, Phys. Lett. **B151**, 199 (1985).

[15] N. Marcus, *Composite Anomalies in Supergravity*, Phys. Lett. **B157**, 383 (1985).

[16] Z. Bern, J. J. Carrasco, L. J. Dixon, H. Johansson and R. Roiban, *The Ultraviolet Behavior of $\mathcal{N}= 8$ Supergravity at Four Loops*, Phys. Rev. Lett. **103**, 081301 (2009), arXiv:0905.2326 [hep-th].

[17] R. Kallosh, *On UV Finiteness of the Four Loop $\mathcal{N}= 8$ Supergravity*, JHEP **0909**, 116 (2009), arXiv:0906.3495 [hep-th].

[18] M. Bianchi, S. Ferrara and R. Kallosh, *Perturbative and Non-Perturbative $\mathcal{N}= 8$ Supergravity*, Phys. Lett. **B690**, 328 (2010), arXiv:0910.3674 [hep-th].

[19] M. Bianchi, S. Ferrara and R. Kallosh, *Observations on Arithmetic Invariants and U-Duality Orbits in $\mathcal{N}= 8$ Supergravity*, JHEP **1003**, 081 (2010), ArXiv:0912.0057 [hep-th].

[20] P. Vanhove, *The Critical Ultraviolet Behaviour of $\mathcal{N}= 8$ Supergravity Amplitudes*, arXiv:1004.1392 [hep-th].

[21] L. J. Dixon, *Ultraviolet Behavior of $\mathcal{N}= 8$ Supergravity*, Lectures presented at ISSP 2009, Aug 29 - Sep 7 '09, Erice (Italy), arXiv:1005.2703 [hep-th].

[22] R. Kallosh, *The Ultraviolet Finiteness of $\mathcal{N}= 8$ Supergravity*, JHEP **1012**, 009 (2010), arXiv:1009.1135 [hep-th].

[23] N. Beisert, H. Elvang, D. Z. Freedman, M. Kiermaier, A. Morales and S. Stieberger, *$E_{7(7)}$ Constraints on Counterterms in $\mathcal{N}= 8$ Supergravity*, Phys. Lett. **B694**, 265 (2010), arXiv:1009.1643 [hep-th].

[24] Z. Bern, J. J. Carrasco, L. Dixon, H. Johansson and R. Roiban, *Amplitudes and Ultraviolet Behavior of $\mathcal{N}= 8$ Supergravity*, arXiv:1103.1848 [hep-th].

[25] M. Henneaux and C. Teitelboim, *Dynamics of chiral (selfdual) p-forms*, Phys. Lett. **B206**, 650 (1988).

[26] A. Ceresole, R. D'Auria and S. Ferrara, *The Symplectic Structure of $\mathcal{N}= 2$ Supergravity and its Central Extension*, Nucl. Phys. Proc. Suppl. **46**, 67 (1996), hep-th/9509160.

[27] C. Brunster and M. Henneaux, *The Action of Twisted Self-Duality*, arXiv:1103.3621 [hep-th].

[28] P. S. Howe and U. Lindström, *Higher Order Invariants in Supergravity*, Nucl. Phys. **B181**, 487 (1981).

[29] J. Björnsson and M. B. Green, *5 Loops in 24/5 Dimensions*, JHEP **1008**, 132 (2010), arXiv:1004.2692 [hep-th].

[30] R. Kallosh, *$E_{7(7)}$ Symmetry and Finiteness of $\mathcal{N}= 8$ Supergravity*, arXiv:1103.4115 [hep-th].

[31] M. Green, H. Ooguri and J. H. Schwarz, *Nondecoupling of Maximal Supergravity from the Superstring*, Phys. Rev. Lett. **99**, 041601 (2007), arXiv:0704.0777 [hep-th].

[32] N. Arkani-Hamed, F. Cachazo and J. Kaplan, *What is the Simplest Quantum Field Theory?*, JHEP **09**, 016 (2010), arXiv:0808.1446 [hep-th].

[33] T. Banks, *Why I don't Believe $\mathcal{N}=8$ SUGRA is Finite*, talk at the Workshop *"Supergravity versus Superstring Theory in the Ultraviolet"*, PennState Univ, PA USA, August 27-30 2009.

[34] B. Bertotti, *Uniform Electromagnetic Field in the Theory of General Relativity*, Phys. Rev. **116**, 1331 (1959). I. Robinson, *A solution of the Maxwell-Einstein equations*, Bull. Acad. Pol. Sci. Ser. Sci. Math. Astron. Phys. **7**, 351 (1959).

[35] L. Borsten, D. Dahanayake, M. J. Duff, S. Ferrara, A. Marrani and W. Rubens, *Observations on Integral and Continuous U-Duality Orbits in $\mathcal{N}=8$ Supergravity*, Class. Quant. Grav. **27**, 185003 (2010), arXiv:1002.4223 [hep-th].

[36] L. Borsten, D. Dahanayake, M. J. Duff and W. Rubens, *Black Holes admitting a Freudenthal Dual*, Phys. Rev. **D80**, 026003 (2009), arXiv:0903.5517 [hep-th].

[37] S. Ferrara, A. Marrani and A. Yeranyan, *Freudenthal Duality and Generalized Special Geometry*, arXiv:1102.4857 [hep-th].

[38] M. J. Duff and P. van Nieuwenhuizen, *Quantum Inequivalence of Different Field Representations*, Phys. Lett. **B94**, 179 (1980).

[39] S. M. Christensen, M. J. Duff, G. W. Gibbons and M. Rocek, *Vanishing One-Loop Beta Function in Gauged $\mathcal{N}>4$ Supergravity*, Phys. Rev. Lett. **45**, 161 (1980).

[40] M. J. Duff, *Twenty Years of the Weyl Anomaly*, Class. Quant. Grav. **11**, 1387 (1994), hep-th/9308075.

[41] L. D. Faddeev and V. N. Popov, *Feynman Diagrams for the Yang-Mills Field*, Phys. Lett. **B25**, 29 (1967).

[42] M. J. Duff and S. Ferrara, *Generalized Mirror Symmetries and Trace Anomalies*, Class. Quant. Grav. **28**, 065005 (2011), arXiv:1009.4439 [hep-th].

[43] T. L. Curtright, *Charge Renormalization and High Spin Fields*, Phys. Lett. **B102**, 17 (1981).

[44] A. Sen, *Arithmetic of $\mathcal{N}=8$ Black Holes*, JHEP **1002**, 090 (2010), arXiv:0908.0039 [hep-th].

[45] S. Ferrara, K. Hayakawa and A. Marrani, *Lectures on Attractors and Black Holes*, Fortsch. Phys. **56**, 993 (2008), arXiv:0805.2498 [hep-th].

[46] S. Ferrara, R. Kallosh and A. Strominger, *$\mathcal{N}=2$ extremal black holes*, Phys. Rev. **D52**, 5412 (1995), hep-th/9508072. A. Strominger, *Macroscopic entropy of $\mathcal{N}=2$ extremal black holes*, Phys. Lett. **B383**, 39 (1996), hep-th/9602111. S. Ferrara and R. Kallosh, *Supersymmetry and attractors*, Phys. Rev. **D54**, 1514 (1996), hep-th/9602136. S. Ferrara and R. Kallosh, *Universality of supersymmetric attractors*, Phys. Rev. **D54**, 1525 (1996), hep-th/9603090.

[47] L. Andrianopoli, R. D'Auria, S. Ferrara and M. Trigiante, *Extremal Black Holes in Supergravity*, Lect. Notes Phys. **737**, 661 (2007), hep-th/0611345.

[48] S. Bellucci, S. Ferrara, R. Kallosh and A. Marrani, *Extremal Black Holes and Flux Vacua Attractors*, Lect. Notes Phys. **755**, 115 (2008), arXiv:0711.4547 [hep-th].

[49] S. W. Hawking: *Gravitational Radiation from Colliding Black Holes*, Phys. Rev. Lett. **26**, 1344 (1971). J. D. Bekenstein: *Black Holes and Entropy*, Phys. Rev. **D7**, 2333 (1973).

[50] S. Ferrara, G. W. Gibbons and R. Kallosh, *Black Holes and Critical Points in Moduli Space*, Nucl. Phys. **B500**, 75 (1997), hep-th/9702103.

[51] R. B. Brown, *Groups of Type E_7*, J. Reine Angew. Math. **236**, 79 (1969).

[52] E. Cremmer, J. P. Derendinger, B. de Wit, S. Ferrara, L. Girardello, C. Kounnas and A. Van Proeyen, *Vector Multiplets Coupled to $\mathcal{N}= 2$ Supergravity : SuperHiggs Effect, Flat Potentials and Geometric Structure*, Nucl. Phys. **B250**, 385 (1985).

[53] E. Cremmer and A. Van Proeyen, *Classification of Kähler Manifolds in $\mathcal{N}= 2$ Vector Multiplet Supergravity Couplings*, Class. Quant. Grav. **2**, 445 (1985).

[54] M. Günaydin, G. Sierra and P. K. Townsend, *Exceptional Supergravity Theories and the Magic Square*, Phys. Lett. **B133**, 72 (1983). M. Günaydin, G. Sierra and P. K. Townsend, *The Geometry of $\mathcal{N}= 2$ Maxwell-Einstein Supergravity and Jordan Algebras*, Nucl. Phys. **B242**, 244 (1984).

[55] S. Ferrara and M. Günaydin, *Orbits of Exceptional Groups, Duality and BPS States in String Theory*, Int. J. Mod. Phys. **A13**, 2075 (1998), hep-th/9708025.

[56] A. Ceresole, S. Ferrara and A. Marrani, *4d/5d Correspondence for the Black Hole Potential and its Critical Points*, Class. Quant. Grav. **24**, 5651 (2007), arXiv:0707.0964 [hep-th].

[57] L. K. Hua, *On the Theory of Automorphic Functions of a Matrix Variable. I: Geometrical Basis*, Amer. J. Math. **66**, 470 (1944). C. Bloch and A. Messiah, *The Canonical Form of an Antisymmetric Tensor and its Application to the Theory of Superconductivity*, Nucl. Phys. **B39**, 95 (1962). B. Zumino, *Normal Forms of Complex Matrices*, J. Math. Phys. **3**, 1055 (1962).

[58] R. Kallosh and B. Kol, *E_7 Symmetric Area of the Black Hole Horizon*, Phys. Rev. **D53**, 5344 (1996), hep-th/9602014.

[59] S. Ferrara and R. Kallosh, *On $\mathcal{N}= 8$ Attractors*, Phys. Rev. **D73**, 125005 (2006), hep-th/0603247.

[60] A. Ceresole, S. Ferrara, A. Gnecchi and A. Marrani, *More on $\mathcal{N}= 8$ Attractors*, Phys. Rev. **D80**, 045020 (2009), arXiv:0904.4506 [hep-th].

[61] A. Ceresole, S. Ferrara and A. Gnecchi, *5d/4d U-Dualities and $\mathcal{N}= 8$ Black Holes*, Phys. Rev. **D80**, 125033 (2009), arXiv:0908.1069 [hep-th].

[62] R. Arnowitt, S. Deser and C. W. Misner: *Canonical Variables for General Relativity*, Phys. Rev. **117**, 1595 (1960).

[63] G.W.Gibbons and C.H.Hull: *A Bogomol'ny bound for general relativity and solitons in $\mathcal{N}= 2$ supergravity*, Phys. Lett. **B109**, 190 (1982).

[64] A. Papapetrou, Proc. R. Irish Acad. **A51**, 191 (1947). S. D. Majumdar, Phys. Rev. **72**, 930 (1947).

[65] P. Breitenlohner, D. Maison and G. W. Gibbons, *Four-Dimensional Black Holes from Kaluza-Klein Theories*, Commun. Math. Phys. **120**, 195 (1988).

[66] A. Ceresole and G. Dall'Agata, *Flow Equations for non-BPS Extremal Black Holes*, JHEP **0703**, 110 (2007), hep-th/0702088.

[67] L. Andrianopoli, R. D'Auria, E. Orazi and M. Trigiante, *First Order Description of Black Holes in Moduli Space*, JHEP **0711**, 032 (2007), arXiv:0706.0712 [hep-th].

[68] B. de Wit, F. Vanderseypen and A. Van Proeyen, *Symmetry Structure of Special Geometries*, Nucl. Phys. **B400**, 463 (1993), hep-th/9210068.

[69] A. Ceresole, G. Dall'Agata, S. Ferrara and A. Yeranyan, *First Order Flows for $\mathcal{N}= 2$ Extremal Black Holes and Duality Invariants*, Nucl. Phys. **B824**, 239 (2010), arXiv:0908.1110 [hep-th].

[70] G. Bossard, Y. Michel and B. Pioline, *Extremal Black Holes, Nilpotent Orbits and the True Fake Superpotential*, JHEP **1001**, 038 (2010), arXiv:0908.1742 [hep-th].

[71] A. Ceresole, G. Dall'Agata, S. Ferrara and A. Yeranyan, *Universality of the Superpotential for $d = 4$ Extremal Black Holes*, Nucl. Phys. **B832**, 358 (2010), arXiv:0910.2697 [hep-th].

[72] S. Ferrara, A. Marrani and E. Orazi, *Maurer-Cartan Equations and Black Hole Superpotentials in $\mathcal{N}= 8$ Supergravity*, Phys. Rev. **D81**, 085013 (2010), arXiv:0911.0135 [hep-th].

[73] F. Denef, *Supergravity flows and D-brane stability*, JHEP **0008**, 050 (2000), hep-th/0005049. F. Denef, B. R. Greene and M. Raugas, *Split attractor flows and the spectrum of BPS D-branes on the quintic*, JHEP **0105**, 012 (2001), hep-th/0101135.

[74] B. Bates and F. Denef, *Exact solutions for supersymmetric stationary black hole composites*, arXiv:hep-th/0304094.

[75] R. Kallosh, N. Sivanandam, and M. Soroush, *Exact Attractive Non-BPS STU Black Holes*, Phys. Rev. **D74**, 065008 (2006), hep-th/0606263. F. Denef and G. W. Moore, *Split States, Entropy Enigmas, Holes and Halos*, hep-th/0702146. F. Denef, D. Gaiotto, A. Strominger, D. Van den Bleeken and X. Yin, *Black Hole Deconstruction*, hep-th/0703252. F. Denef and G. W. Moore, *How many black holes fit on the head of a pin?*, Gen. Rel. Grav. **39**, 1539 (2007), arXiv:0705.2564 [hep-th]. D. Gaiotto, W. W. Li and M. Padi, *Non-Supersymmetric Attractor Flow in Symmetric Spaces*, JHEP **0712**, 093 (2007), arXiv:0710.1638 [hep-th]. M. C. N. Cheng and E. P. Verlinde, *Wall Crossing, Discrete Attractor Flow, and Borcherds Algebra*,

SIGMA **4**, 068 (2008), arXiv:0806.2337 [hep-th]. E. G. Gimon, F. Larsen and J. Simon, *Constituent Model of Extremal non-BPS Black Holes*, JHEP **0907**, 052 (2009), arXiv:0903.0719 [hep-th]. A. Castro and J. Simon, *Deconstructing the D0-D6 system*, JHEP **0905**, 078 (2009), arXiv:0903.5523 [hep-th]. J. R. David, *On walls of marginal stability in $\mathcal{N}=2$ string theories*, JHEP **0908**, 054 (2009), arXiv:0905.4115 [hep-th]. J. Manschot, *Stability and Duality in $\mathcal{N}=2$ Supergravity*, Commun. Math. Phys. **299**, 651 (2010), arXiv:0906.1767 [hep-th]. P. Galli and J. Perz, *Non-supersymmetric extremal multicenter black holes with superpotentials*, JHEP **1002**, 102 (2010), arXiv:0909.5185 [hep-th]. A. Sen, *Walls of Marginal Stability and Dyon Spectrum in $\mathcal{N}=4$ Supersymmetric String Theories*, JHEP **0705**, 039 (2007), hep-th/0702141. A. Sen, *Two Centered Black Holes and $\mathcal{N}=4$ Dyon Spectrum*, JHEP **0709**, 045 (2007), arXiv:0705.3874 [hep-th]. A. Sen, $\mathcal{N}=8$ *Dyon Partition Function and Walls of Marginal Stability*, JHEP **0807**, 118 (2008), arXiv:0803.1014 [hep-th]. A. Sen, *Wall Crossing Formula for $\mathcal{N}=4$ Dyons: A Macroscopic Derivation*, JHEP **0807**, 078 (2008), arXiv:0803.3857. S. Ferrara, A. Marrani and E. Orazi, *Split attractor Flow in $\mathcal{N}=2$ Minimally Coupled Supergravity*, Nucl. Phys. **B846**, 512 (2011), arXiv: 1010.2280 [hep-th]. J. Manschot, B. Pioline and A. Sen, *Wall Crossing from Boltzmann Black Holes*, arXiv:1011.1258 [hep-th]. S. Ferrara, A. Marrani, E. Orazi, R. Stora and A. Yeranyan, *Two-Center Black Holes Duality-Invariants for stu Model and its lower-rank Descendants*, arXiv: 1011.5864 [hep-th]. L. Andrianopoli, R. D'Auria, S. Ferrara, A. Marrani and M. Trigiante, *Two-Centered Magical Charge Orbits*, arXiv.1101.3496 [hep-th].

[76] E. Andriyash, F. Denef, D. L. Jafferis and G. W. Moore, *Wall-crossing from supersymmetric galaxies*, arXiv:1008.0030 [hep-th]. E. Andriyash, F. Denef, D. L. Jafferis and G. W. Moore, *Bound state transformation walls*, arXiv:1008.3555 [hep-th].

[77] S. Ferrara and A. Marrani, *Matrix Norms, BPS Bounds and Marginal Stability in $\mathcal{N}=8$ Supergravity*, JHEP **1012**, 038 (2010), arXiv:1009.3251 [hep-th].

[78] S. Bellucci, S. Ferrara and A. Marrani, *On some properties of the attractor equations*, Phys. Lett. **B635**, 172 (2006), hep-th/0602161.

Scientific Secretaries: N. Ambrosetti, A. Zahabi

DISCUSSION I

- *Schmidt-Sommerfeld:*
    In what sense is N=8 supergravity not unique? What are the free parameters?

- *Ferrara:*
    At the classical level there are two versions of N=8 supergravity: one with a cosmological constant and one without. Both can be interpreted as compactifications of 11-dimensional supergravity (M-theory): compactifying on a 7-torus gives N=8 supergravity in four dimensions without cosmological constant, whereas compactifying on $AdS_4 \times S^7$ yields a gauged supergravity in four dimensions with a negative cosmological constant, and there is a relation between the gauge coupling and the gravitational coupling constant fixed by supersymmetry.
    In N=8 supergravity without a cosmological constant one can dualise antisymmetric tensors into scalars. On the other hand, a calculation by Duff and van Nieuwanhuizen shows that the trace anomaly vanishes in the case of antisymmetric tensors but not when they are dualised into scalars. This means that the two theories are different at the quantum level.
    In general, local supersymmetry always allows a negative cosmological constant in four dimensions, while it is forbidden in eleven dimensions.

- *Dennen:*
    Are there any values of the cosmological constant that are forbidden by N=8 supergravity?

- *Ferrara:*
    No, there are no conditions imposed by supersymmetry. In string theory there might be some restrictions on the possible values, like in the string landscape, but not in supergravity.

- *Burda:*
    What can you say about the finiteness problem of N=8 supergravity? In particular what happens at five loops?

- *Ferrara:*
    This is a long story. It was known that there is the N=8 supersymmetrisation of the R^3 counterterm, but it turned out that for a miraculous reason it vanished. We now understand that this comes from the fact that it is not invariant under the $E_{7,7}$ global symmetry of N=8, which is not anomalous and must be preserved order by order in perturbation theory. Then the next counterterm that is allowed both by supersymmetry and by $E_{7,7}$ is a D^8R^4 term – corresponding to four graviton scattering at seven loops - which should also arise as the N=8 superspace measure. In fact the superspace measure should give a term with sixteen derivatives on dimensional grounds, but it's not clear that it would be exactly the D^8R^4 term. This issue is still open for debate.

- *Schmidt-Sommerfeld:*
    What does N=8 superspace say about which counterterms are allowed by N=8 supersymmetry?

- Ferrara:

We don't have an explicit formulation of N=8 superspace, but in principle it would let us construct explicitly counterterms invariant under N=8 supersymmetry and the $E_{7,7}$ global symmetry. In particular the superspace measure should be invariant under these symmetries and would contain sixteen derivatives.

CHAIRMAN: S. FERRARA

Scientific Secretaries: N .Ambrosetti, A. Zahabi

DISCUSSION II

- *Schmidt-Sommerfeld:*
    Does the split attractor flow you showed us occur in space-time or in field space?

- *Ferrara:*
    It occurs in field space.

- *Galakhov:*
    Is it possible to apply the results of your analysis of BPS states and marginal stability to other theories?

- *Ferrara:*
    Yes, I presented these results in a model independent way which can be applied to any theory with these symmetries.

- *Alba:*
    Why did you first consider the Reissner-Nordstrom black hole, which is classical, and then use it to discuss quantum black holes?

- *Ferrara:*
    I just considered the quantum black holes in a simple way as being quantum particles characterized by their mass and their charges. The entropy of a black hole in N=8 supergravity is given by the square root of the absolute value of the quartic Cartan invariant depending on the electric and magnetic charges. Thus, the entropy is invariant under the $E_{7,7}$ symmetry, acting on the 56 dimensional fundamental representation in which the electric and magnetic charges of the black hole sit.

- *Ambrosetti:*
    Which BPS black hole states can possibly be reached through decay?

- *Ferrara:*
    The possible black hole decays are constrained by the relations among masses and charges, and it is believed that by considering all possible decays one can construct all BPS states. Therefore, the BPS states that are related by decay should form a closed system.

- *Galakhov:*
    Is it possible to construct BPS states charged under a non-Abelian group?

- *Ferrara:*
    The black holes I considered were in asymptotically flat space. In N=8 supergravity in AdS there are non-Abelian symmetries and in principle one should be able to construct non-Abelian BPS states, but as far as I know this has been done only for an Abelian subgroup of the non-Abelian symmetries. I think Michael Duff wrote a paper on the subject.

-*Duff:*
    Yes, that's correct.

The gravitational S-matrix: Erice lectures

Steven B. Giddings[*]

Department of Physics, University of California, Santa Barbara, CA 93106

Abstract

These lectures discuss an S-matrix approach to quantum gravity, and its relation to more local spacetime approaches. Prominent among the problems of quantum gravity are those of unitarity and observables. In a unitary theory with solutions approximating Minkowski space, the S-matrix (or, in four dimensions, related inclusive probabilities) should be sharply formulated and physical. Features of its perturbative description are reviewed. A successful quantum gravity theory should in particular address the questions posed by the ultrahigh-energy regime. Some control can be gained in this regime by varying the impact parameter as well as the collision energy. However, with decreasing impact parameter gravity becomes strong, first eikonalizing, and then entering the regime where in the classical approximation black holes form. Here one confronts what may be the most profound problem of quantum gravity, that of providing unitary amplitudes, as seen through the information problem of black hole evaporation. Existing approaches to quantum gravity leave a number of unanswered questions in this regime; there are strong indications that new principles and mechanisms are needed, and in particular there is a good case that usual notions of locality are inaccurate. One approach to these questions is investigation of the approximate local dynamics of spacetime, its observables, and its limitations; another is to directly explore properties of the gravitational S-matrix, such as analyticity, crossing, and others implied by gravitational physics.

_____

[*]giddings@physics.ucsb.edu

1 The problem(s) of quantum gravity

Reconciliation of quantum mechanics with gravity may well be the most profound theoretical problem left unsolved by twentieth-century physics. Both general relativity and quantum-mechanics became well established in the first quarter of the last century, yet their clash continues to this day. One might simply declare the problem too difficult, and move on to more modest goals. But, there are glimmers of hope, both from new approaches to the problem, and perhaps more importantly, through a better understanding of the nature of the underlying difficulty.

There are a number of sub-problems to that of quantizing gravity. A list includes:

1. The problem of UV divergences and nonrenormalizability: what structure cures an apparent infinite proliferation of perturbative ultraviolet divergences, and needed counterterms?

2. The problem of singularities: how can a quantum theory resolve singularities of the classical theory, such as are found in black holes and in many cosmologies?

3. The problem of observables/time: how can one formulate gauge invariant observables in the quantum theory, given that diffeomorphisms are gauge transformations in the classical theory?

4. The problem of unitarity: what dynamics unitarizes amplitudes, in the high-energy regime?

5. The conundrums of cosmology: this includes the problem of defining infrared finite observables in inflationary cosmology – an extreme example is the measure problem; or, perhaps we need an alternative to inflation.

Historically, there has been a lot of focus on problems one and two, and they have strongly motivated approaches to quantum gravity such as string theory and loop quantization. However, there are increasing indications that problems three through five are both more profound, and are linked; they seemingly drive more deeply at the heart of the central problem of quantum gravity.

Indeed, there has been a growing appreciation of the fact that there are deep *long-distance* issues in quantum gravity, which do not appear easily resolved by short-distance modifications of the theory. This suggests that a successful theory must incorporate more radical departures from local quantum field theory (QFT), relevant over even very long-distance scales, in certain circumstances. These lectures will explore one way to probe these questions – through study of the gravitational S-matrix.

2 The gravitational S-matrix

2.1 Observables in quantum gravity?

Given our difficulties with gravity, it is important to understand what are sharply formulated questions. One might naïvely expect that a theory of quantum gravity should calculate correlation functions – as in QFT – such as

$$\langle \phi(x)\phi(y)\cdots\rangle \ , \ \langle g_{\mu\nu}(x)\rangle \ , \ \langle \phi(x)\phi(y)g_{\mu\nu}(z)\cdots\rangle \ , \quad \text{etc.} \ , \qquad (2.1)$$

where ϕ is some matter field, and $g_{\mu\nu}$ is the metric. However, such correlators cannot be gauge invariant, and hence physical. The reason is that the gauge symmetries of gravity, at least at the semiclassical level, are diffeomorphisms, whose infinitesimal forms act on local fields by

$$\delta\phi(x) = \xi^\mu \partial_\mu \phi(x) \ , \ \delta g_{\mu\nu}(x) = \nabla_{(\mu}\xi_{\nu)}(x) \ , \text{etc.} \qquad (2.2)$$

In field theory, local observables are given in terms of the fundamental fields; but in gravity, by this reasoning, no local observables can be gauge invariant.

Clearly our actual observations are governed by a theory incorporating both quantum mechanics and gravity, and we apparently make at least approximately local observations, certainly as compared to cosmological scales – so how is this problem resolved? A proposed answer goes back to Leibniz and Einstein, and began to be concretely developed by De-Witt: an important class of observables in gravity are *relational*, that is give location with relation to other features of the particular quantum state of the system. In particular, [1] explored formulation, within a semiclassical/perturbative approximation, of a class of *protolocal* observables, which are relational and *approximately* reduce to local observables in certain states. Very concrete examples of this construction can be given, in the context of two-dimensional gravity, using world-sheet methods of string theory [2]. Other approaches to implementing relational ideas have been explored in a number of places in the literature;* moreover, such relational ideas are concretely used to handle gauge invariance in inflationary cosmology, where, for example, fluctuations are computed on time slices defined by the inflaton field taking on a particular value, like that corresponding to "reheating."

While relational observables are important for quantum cosmology, there are many subtleties in their proper definition, which ultimately appears to require a deeper knowledge of the quantum theory, and also encounters various infrared issues. For that reason, in the present discussion I will primarily focus on another approach to formulation of sharp physical quantities: the *gravitational S-matrix.*[†]

2.2 The S-matrix and the ultraplanckian regime

A starting point for discussing the gravitational S-matrix is the observation that, to a very good approximation, Minkowski space looks like a good solution of the theory of quantum

---

*For one list of references, see [1].

[†]Important early references on gravitational scattering include [3–8].

gravity. Indeed, the curvature radius of the observed Universe is of order 10^{60} in Planck units – flat space is an excellent approximation for all but the longest-scale questions, and we therefore assume that it makes sense to consider the exactly flat limit.

Next, we also observe that there are excitations about this which we describe as "particle" states, such as electrons, photons, *etc.*, whatever their more fundamental description may be. Their asymptotic states are described by their momenta p_i, and other quantum numbers. Moreover, we can consider asymptotic multi-particle states, consisting of widely separated particles. And, in a quantum theory, we can ask for the amplitudes to transition from a given such "in" state, in the far past, to another "out" state, in the far future. This collection of amplitudes, which we might denote

$$S(p_{i'}, p_i) = {}_{out}\langle p_{i'} | p_i \rangle_{in} , \qquad (2.3)$$

is the S-matrix.

I should note there are two technicalities that we are glossing over. First, for gravity in $D = 4$ spacetime dimensions, these amplitudes aren't well-defined: instead we must perform a sum over soft gravitons, like with soft photons in QED. However, as we will see, this technicality can be avoided in higher dimensions. Secondly, a more precise definition of the S-matrix involves a map between normalizable Hilbert-space states; momentum-space states give a singlular limit of these, though one whose handling is well-understood in asymptotically flat spacetimes.

Such an S-matrix, which is tightly constrained by properties such as unitarity and analyticity, can be a very powerful way to summarize our ignorance of a theory. The S-matrix approach is so powerful that it lead to the *discovery* of string theory: Veneziano [9] guessed amplitudes satisfying additional "duality" properties, and they were then found to describe excitations of one-dimensional objects. We might anticipate that such study in the context of gravity, supplemented by additional physical input, could bear important fruit.

There are multiple proposed approaches to quantizing gravity, and the S-matrix also provides a very important test for these: any theory of quantum gravity should give us a means to calculate S, at least approximately, or must provide us with some alternative physical quantities.

This test becomes particularly incisive in the ultraplanckian regime, at center of mass energies $E \gg M_D$, where M_D denotes the D-dimensional Planck mass. That any theory of quantum gravity must describe this regime follows from very general principles, and avoiding this regime appears to require radical assumptions. The first principle is that of Lorentz invariance: the Minkowski-space solution is invariant under boosts, and we can perform an arbitrarily large boost on a single-particle excitation of it to get a state with an arbitrarily large energy.[‡] The second principle is a very weak notion of locality: one can

---

[‡]There are advocates of Lorentz-invariance violation. This won't be considered here because a) Lorentz invariance is hard to violate by a small amount; b) there are very stringent constraints on such violation; c) it is hard to reconcile such violation with basic properties of black holes, *e.g.* in the astrophysical context; and d) if black holes are still possible, they would form in other contexts, *e.g.* that of high-mass collapse, and we would still face similar puzzles to those I will describe.

consider independently boosted particles at very large separations, and so prepare a state with a large center of mass energy – and the then allow the particles to collide.

Plausibly during the big bang particles reached such energies, and it is interesting to ask whether such collisions are likely to take place anywhere in nature today. We observe that cosmic ray accelerators (argued to be active galactic nuclei) produce particles up to $\sim 10^{12}$ GeV, and now and then they could collide; that is only short by $\mathcal{O}(10^7)$. Even more interestingly, in scenarios with large and/or strongly warped extra dimensions, the fundamental Planck mass M_D could be as low as the TeV range, and in the most optimistic scenarios black holes could give important signatures at LHC [10, 11]! While I won't talk about this phenomenology, there are multiple reviews, including [12].

Whether or not this physics is experimentally accessible, the gedanken experiment and theoretical problem remain: we need a calculational framework that makes predictions in the ultraplanckian regime. As I will describe, here an apparently critical and fundamental problem is *unitarity*. However, as viewed through the lens of earlier attempts to quantize gravity, a more basic concern is that of *nonrenormalizability*. We will begin to understand this problem, and then explore the apparently more fundamental issue of unitarity, starting with the perturbative approach to gravity.

3 Review of perturbative gravity

3.1 Perturbative quantization and the Born amplitude

A starting point for perturbative quantization is the action,

$$S = \int d^D x \sqrt{-g} \left[\frac{1}{16\pi G_D} \mathcal{R} + \mathcal{L}_M \right] \tag{3.1}$$

where $\mathcal{R}$ is the Ricci scalar, $\mathcal{L}_M$ the matter lagrangian, and Newton's constant G_D is related to M_D by the *Particle Data Book* [13] convention:

$$M_D^{D-2} = (2\pi)^{D-4}/(8\pi G_D) . \tag{3.2}$$

For simplicity we will typically consider for matter a minimally-coupled scalar of mass m,

$$\mathcal{L}_m = -\frac{1}{2} (\nabla \phi)^2 - \frac{1}{2} m^2 \phi^2 . \tag{3.3}$$

We investigate perturbations about Minkowski space,

$$g_{\mu\nu} = \eta_{\mu\nu} + \sqrt{32\pi G_D}\, h_{\mu\nu} \tag{3.4}$$

and expand in powers of h. This gives an action of the form

$$S = S_{m,0} + \int d^D x \left\{ \frac{1}{2} h^{\mu\nu} (\hat{\mathcal{L}} h)_{\mu\nu} + \sqrt{8\pi G_D} [h^{\mu\nu} T_{\mu\nu}^{m,0} + \mathcal{L}_3(h)] + \cdots \right\} . \tag{3.5}$$

Here, quantities evaluated with $g = \eta$ are labeled with zeros, $\hat{\mathcal{L}}$ is a second-order differential operator (a tensor analog of the laplacian), indices are raised and lowered with η, T is the stress tensor, and $\mathcal{L}_3$ is the third-order term in h. The ellipses contain higher powers of ϕ and h. The propagators are given by inverting the kinetic operators for ϕ and h; the latter requires gauge fixing. There are two approaches to this, both using the quantity

$$\bar{h}_{\mu\nu} = h_{\mu\nu} - \frac{1}{2}\eta_{\mu\nu}h_\lambda^\lambda \ . \tag{3.6}$$

The first is to apply the gauge fixing condition

$$\partial^\mu \bar{h}_{\mu\nu} = 0 \ . \tag{3.7}$$

Alternately, one may add a gauge fixing term

$$\mathcal{L}_{gf} = \frac{1}{16\pi G_D \alpha} \left(\partial^\mu \bar{h}_{\mu\nu}\right)^2 \tag{3.8}$$

to the lagrangian; particularly nice simplifications occur for $\alpha = -1$. In either case, one straightforwardly finds the propagator, and using that can immediately calculate a first amplitude, that for tree-level scattering of two matter particles. This is the Born amplitude. It is most conveniently given in terms of the Mandelstam invariants; let p_1, p_2 denote the momenta of the incoming particles, and $-p_3$, $-p_4$ those for outgoing, and define

$$s = -(p_1 + p_2)^2 = E^2 \ , \ t = -(p_3 + p_1)^2 = -q^2 \ , \ u = -(p_4 + p_1)^2 \ . \tag{3.9}$$

Here E is the total center-of-mass (CM) energy, q is the "momentum transfer," and one finds $s + t + u = 4m^2$. If we then define the reduced transition matrix element by

$$S = 1 + iT(s,t)(2\pi)^D \delta^D\left(\sum p\right) \tag{3.10}$$

and take the limit of large $s = E^2$, we find the high-energy approximation of the Born amplitude T_0:

$$T_0(s,t) \simeq -8\pi G_D s^2/t \ . \tag{3.11}$$

Exercise: *Derive this formula from the Feynman rules of the action (3.5).*

This formula is easy to understand; see fig. 1. There is a coupling constant $\sim \sqrt{8\pi G_D}s$ at each vertex, and the graviton propagator gives a pole $\sim 1/q^2$. So, we're on our way to computing the S-matrix! The question is where T_0 gives a good approximation to the full S-matrix. As I will explain further, the answer is apparently for small enough t, or equivalently, large enough impact parameter b. To see this, we need to consider possible corrections to (3.11), from loops or radiation.

99

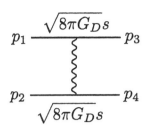

Figure 1: A Feynman diagram representing scattering in the Born approximation.

3.2 Radiative corrections, nonrenormalizability, and the problem of ultraplanckian energies

First, note that an accelerated particle will radiate gravitational bremsstrahlung, analogous to that of QED. To estimate how it corrects a given process, say with amplitude T', we attach a graviton of momentum k to one of the legs of T', as in fig. 2. Then there is an additional external propagator factor, and the amplitude becomes

$$T_{soft} \propto \frac{T'}{(p-k)^2 + m^2} = -\frac{T'}{2p \cdot k} \tag{3.12}$$

where we use the on-shell condition for the leg's momentum p, and for the graviton momentum. The sum over final graviton states thus gives

$$\int d^D k \, \delta(k^2) \, |T_{soft}|^2 \sim \int \frac{d^D k}{k^4} \ . \tag{3.13}$$

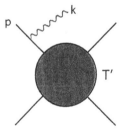

Figure 2: Addition of a soft graviton to a diagram.

Exercise: *Work out the details of this calculation, checking the form of the preceding expression.*

This is infrared divergent for $D \leq 4$, and is an indicator of the soft-graviton problem alluded to before. Like with QED, the S-matrix will not be defined, but the resolution appears straightforward [14]: include IR divergences from loop corrections, and perform an inclusive sum over scattering probabilities with gravitons below a momentum corresponding to, *e.g.*, experimental resolution, to get an IR finite result. For $D > 4$ there is no such IR divergence obstructing the definition of S, and rough estimates [15] indicate that the emission probabilities are small at small t/large b.

Next consider loops. Since the coupling constant G_D has negative mass dimension, there will be UV divergences requiring counterterms of the form

$$\Delta \mathcal{L} \propto G_D^a p^b \hbar^c M^d \, , \tag{3.14}$$

where M is a UV cutoff scale, and the exponents satisfy $a(D-2) + D = b + \frac{c}{2}(D-2) + d$. Thus, as one considers arbitrarily high-order loops (increasing a), there can be arbitrarily many different higher-dimension counterterms, unless there is some new magic to prevent them. The theory is *nonrenormalizable*, and really this means nonpredictive. Once scattering energies approach the Planck scale, amplitudes can depend on all of these counterterms, and thus there is no finite set of measurements which can determine the theory.

This of course does not mean that the action (3.1) is useless! Instead, we take the viewpoint of effective field theory: the Einstein action makes good (and in some cases, very well-tested) predictions in "low energy" regimes, for example in scattering at $E \ll M_D$. The loop corrections indeed renormalize G_D and the cosmological constant Λ, which must be adjusted to match their observed low-energy values. But, corrections corresponding to higher-dimension operators, such as $\mathcal{R}^2$, are *irrelevant*, that is, make negligible contribution at low energy. In particular, the Born amplitude remains a good approximation for scattering in this regime.

As E approaches M_D, we at first suspect that nonrenormalizability is severely problematical, and that we lose ability to say anything about scattering. However, a little reflection raises a question: the scattering of the Earth and the Moon (here ignoring the side issue that they form a bound state) is certainly a problem with large energy, $E >>> M_D$, and where the Einstein action makes excellent predictions – how can this be? A reason for this is the weakening of gravity with distance, which compensates for the large energy. Likewise, we expect that in the relativistic regime, we can learn more if we allow ourselves to specify the impact parameter b, which is a measure of the distance scale probed, along with E. A schematic plot of the different scattering regimes in the $E - b$ plane is shown in fig. 3. Let's discuss some of its features.

First, as we raise the energy, we can probe to shorter distances, staying above a line determined by the uncertainty principle. This is the story of high-energy accelerators probing smaller scales, and followed to its end reaches the uncontrolled, nonrenormalizable Planck regime, $E \sim M_D$, $b \sim 1/M_D$, and marked "NR". But, suppose we control the impact parameter – like one really does in accelerators – and try to exceed $E = M_D$. For large enough impact parameter, there is no obvious obstacle to still using the Born approximation.

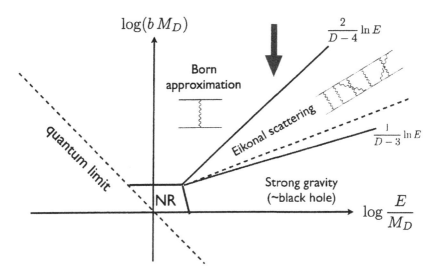

Figure 3: A proposed "phase diagram" of different regimes for gravitational scattering. In particular, we consider the effect of decreasing impact parameter, at fixed ultraplanckian energy, as indicated by the arrow. NR indicates the regime where higher-dimension operators are expected to be important.

Next, let us investigate the scattering behavior in such a "phase diagram" by fixing the energy $E \gg M_D$ and lowering the impact parameter. The Born approximation ultimately does fail, due to the contribution of higher-loop diagrams, but those of a very specific kind – the ladder diagrams, as pictured, which add up to give an eikonal approximation to the amplitude, closely related to the classical approximation. Earth-Moon scattering in fact lies in this eikonal regime, which we will study shortly.

But first, what happens for smaller impact parameters? While not all details are worked out, and ultimately a profound puzzle remains, an apparently consistent picture is the following. The scattered objects are deflected in each other's gravitational fields, and thus radiate. Moreover, if they are composite (e.g. hydrogen atoms, or strings), their internal degrees of freedom can become excited once tidal gravitational forces pass a threshhold. Thus the picture changes in a model-dependent way at some impact parameter depending on which "internal" dynamics becomes important first; this is indicated by the dotted line.

However, with present knowledge it appears that this kind of model-dependent behavior doesn't interfere with consideration of even smaller impact parameters, where something apparently very model-independent happens. Namely, once the impact parameter reaches the scale of the Schwarzschild radius corresponding to E, gravity becomes strong, and classically a black hole would form [16]. Our ultimate and central question is how to give a quantum description of this scattering regime.

Given the nonrenormalizability of gravity, mention of the word "loops" may immediately ring an alarm bell – how can we trust any of the story, once they become important? But, despite these concerns, our approach to the problem seems to make sense. First off, nonrenormalizability and the corresponding large contributions of higher-dimension operators appear to be *short distance* issues, and we are describing phenomena at large, even macroscopic, distances. Moreover, we can investigate this story rather explicitly, using various candidate regulators for the theory. One would be a naïve cutoff on loop momentum, but more sophisticated cutoff prescriptions are provided by supergravity (at least for sufficiently low loop level), or string theory! I will summarize some of the things we can say so far, and give a picture that while provisional, appears to hang together well.

3.3 Eikonal regime, and classical scattering

I have claimed that ladder diagrams (together with crossed ladder diagrams) are the first loop diagrams to become important as the impact parameter is lowered; let's investigate them more closely. One is shown in fig. 4. The nice thing is that they can be approximately computed, and summed up, for small enough momentum transfer $q = \sqrt{-t}$.

Consider for example the one-loop ladder. As shown in fig. 5, this can be thought of as two tree diagrams, sewn together. Specifically, if the momentum transfers k_i by the individual gravitons are small, we can replace the scalar propagators using

$$\frac{-i}{(p - k_i)^2 + m^2 - i\epsilon} \approx \frac{i}{2p \cdot k_i + i\epsilon} \, . \tag{3.15}$$

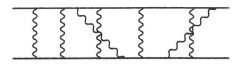

Figure 4: A ladder diagram, with multiple graviton exchange. (Note that the crossed lines do not connect.)

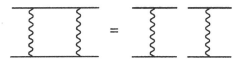

Figure 5: The one-loop ladder diagram is obtained by combining two tree diagrams.

Adding the crossed diagram, it is a straightforward exercise to compute the combined one-loop amplitude. This is most easily written in terms of the Fourier transform of the tree amplitude, with respect to the momentum transfer $q_\perp$ perpindicular to the CM momentum,

$$
\begin{aligned}
\chi(b, s) \quad &= \tfrac{1}{2s} \int \frac{d^{D-2}q_\perp}{(2\pi)^{D-2}} \, e^{i\mathbf{q}_\perp \cdot \mathbf{b}} T_{\text{tree}}(s, -q_\perp^2) \\
&= \frac{4\pi}{(D-4)\Omega_{D-3}} \frac{G_D s}{b^{D-4}} \, ,
\end{aligned}
\tag{3.16}
$$

where

$$
\Omega_n = \frac{2\pi^{(n+1)/2}}{\Gamma\left[(n+1)/2\right]}
\tag{3.17}
$$

is the volume of the unit n-sphere. We refer to χ as the *eikonal phase*. In the first line, we could have considered the tree-level amplitude for some other theory, for example string theory. The second line is the result using the Born amplitude for Einstein's action, (3.11), but for small momentum transfers string or other corrections to this would be very small anyways. The variable $\mathbf{b}$ is a vector-version of the impact parameter. The combined one-loop amplitude has a very simple form in terms of χ:

$$
T_1 \approx 2s \int d^{D-2}b \, e^{-i\mathbf{q}_\perp \cdot \mathbf{b}} \frac{[i\chi(b, s)]^2}{2!} \, .
\tag{3.18}
$$

Exercise: *By combining the one-loop ladder and crossed-ladder diagrams, and using the approximation (3.15), derive this equation.*

Once you've seen the derivation of this, and its simple form, generalization to the N-loop ladder diagram is fairly clear. The powers of $i\chi$ give an exponential sum, and the full *eikonal amplitude* is

$$iT_{\text{eik}}(s,t) = 2s \int d^{D-2}b\, e^{-i\mathbf{q}_\perp \cdot \mathbf{b}} [e^{i\chi(b,s)} - 1] \,. \tag{3.19}$$

And this now shows the origin of the top line on the right side of fig. 3. For $\chi \ll 1$, the sum is approximated by the linear term, which is just the Born amplitude. For $\chi \gtrsim 1$, one must sum higher terms in the series. Indeed, in the latter regime, the Born amplitude becomes large; the full sum then *unitarizes* the amplitude, as can be seen from the partial-wave expansion [15]. The crossover line given by $\chi \sim 1$ corresponds to

$$b \sim (G_D E^2)^{1/(D-4)} \,, \tag{3.20}$$

as follows from (3.16).

But now, we encounter the question of short distances. Expanding out (3.19) in powers of χ yields terms that are increasingly badly divergent as $\mathbf{b} \to 0$ – in fact, the approximation (3.15) has worsened the bad short-distance behavior responsible for nonrenormalizability. From this viewpoint, the eikonal amplitude (3.19) seems nonsensical. Fortunately, this is not a good viewpoint.

To see this, consider a simplified version of (3.19),

$$I = \int_{M^{-1}}^{1} db\, b^3 e^{ig/b^2} \,. \tag{3.21}$$

This looks a lot like (3.19) for $D = 6$, but the exponential of $q_\perp \cdot b$, which makes the integral finite at large b, has been replaced by an upper limit, and an explicit cutoff $1/M$ has been introduced to regulate the short-distance behavior. Expanding the exponential gives

$$I(M) = \frac{1 - M^{-4}}{4} + ig\frac{(1 - M^{-2})}{2} - \frac{g^2}{2}\log M + \frac{ig^3}{12}(1 - M^2) + \cdots \,, \tag{3.22}$$

which is terribly divergent as $M \to \infty$. *But*, plugging into Mathematica (or using a table of integrals, for the old-fashioned) gives

$$I(M) = \frac{g^2}{4}\left[Ei(ig) + \frac{1}{g}\left(i + \frac{1}{g}\right)e^{ig} - Ei(igM^2) - \frac{1}{gM^2}\left(i + \frac{1}{gM^2}\right)e^{igM^2}\right] \,, \tag{3.23}$$

with Ei the exponential integral, and this approaches a finite limit, with controllable small corrections, as $M \to \infty$!

So, expanding the integral (3.19) was not a good thing to do, and this carries an important lesson. Instead, we can see that (3.19) has a saddlepoint at

$$q_\perp \sim \partial\chi/\partial b \,, \tag{3.24}$$

and the saddlepoint approximation is a good way to approximate the integral. For large E and $q/E \ll 1$, this saddlepoint is at *long distances*

$$b \sim [G_D E(E/q)]^{1/(D-3)} \tag{3.25}$$

– so the short distance behavior doesn't play a role, and likewise short distance corrections to the integrand are not expected to be significant. Apparently, *short distance dynamics is essentially irrelevant.* Moreover, for a good approximation to the amplitude, we clearly need the full behavior of the summed exponential.

This sum introduces a connection with classical scattering.[§] Consider the classical metric of a particle of mass m that has been boosted to very high energy, $E/m = \gamma \gg 1$, but take m to zero such that E is fixed. The resulting metric is most easily explored in light-cone coordinates, $x^{\pm} = t \pm z$, $x_{\perp}$, where it takes the form found by Aichelburg and Sexl [17]:

$$ds^2 = -dx^+ dx^- + dx_{\perp}^2 + \Phi(x_{\perp})\delta(x^-)dx^{-2} , \qquad (3.26)$$

with

$$\Phi = -8G_D E \log(x_{\perp}) , \ D = 4 \ ; \quad \Phi = \frac{16\pi G_D E}{\Omega_{D-3}(D-4)x_{\perp}^{D-4}} , \ D > 4 . \qquad (3.27)$$

Exercise: *Derive this metric, by considering the stated limit of the Schwarzschild solution (3.34). Hint: work in isotropic coordinates.*

The departure from a flat metric is completely supported in a shock wave at $x^- = 0$; like with a boosted charge, the field lines pancake up into the transverse plane. Despite the delta function in the metric, one may solve the geodesic equation for a particle incident along $-z$ that scatters at impact parameter b, and derive a finite scattering angle. One finds that this scattering angle matches the momentum transfer given by the eikonal saddlepoint; note also $\chi \propto E\Phi(b)$.

Exercise: *Check these statements.*

Returning to the question of short vs. long distance physics, one can understand why even very ultraplanckian scattering is only probing long-distance physics, in terms of a phenomenon we will call "momentum fractionation [18]." For a given momentum transfer q, which can also be ultraplanckian, we expect the scattering to be dominated by impact parameters near that given by (3.25); if the corresponding eikonal phase χ is large, then from the exponential in (3.19) we see that the dominant loop order in the sum is $N \sim \chi$. Then, in the ladder diagram fig. 4, one estimates a typical momentum $k \sim q/N \sim (\partial\chi/\partial b)/\chi$ to flow through each rung. This gives $k \sim 1/b$, which is very small as long as $b \gg M_D$. Large momentum transfer fractionates into many small transfers, and one effectively only has a soft probe of the dynamics.

3.4 Match to supergravity amplitudes

If one still harbors suspicion about such cavalier treatment of divergent loop amplitudes, these can be examined in a case where they are manifestly convergent – for low enough loop level (or possibly to all order [19]), in supergravity. This was done in [18]. In particular, the explicit, finite one loop amplitudes given in [20] take the form (up to polarization dependence)

$$M_1(s,t) = -i(8\pi G_D)^2 s^4 \left[I^1(s,t) + I^1(t,u) + I^1(s,u) \right] , \qquad (3.28)$$

[§]For some early discussion of this connection, see [3,5].

given in terms of the integral

$$I^1(s,t) = \int \frac{d^D k}{(2\pi)^D} \frac{1}{k^2(p_1 - k)^2(p_2 + k)^2(p_1 + p_3 - k)^2} \; .$$

(3.29)

For non-supersymmetric gravity, the numerator in the one-loop integrals would have additional terms $\sim (k/E)^p$, giving divergent behavior – thus, we see that the effective cutoff on k provided by supergravity is at $k \sim \sqrt{s}$. Interestingly, the expression (3.28) captures *all* of the complicated one loop structure, in a small set of diagrams easily matched with those of the eikonal approximation! These are shown in fig. 6; the first two are clearly identifiable with the ladder and crossed ladder diagrams. The third is suppressed in an expansion in $q^2/E^2 \sim -t/s$, as it requires a large momentum exchange through a single line. Examining the explicit expressions, we see that

$$M_1(s,t) = T_{eik}^{1\,loop} + \mathcal{O}(t/s) + \text{cutoff dependent} \; .$$

(3.30)

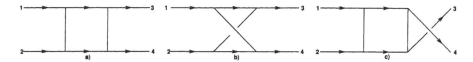

Figure 6: All one-loop, four-point supergravity diagrams can be written in terms of the three scalar box diagrams shown.

Going further, the two-loop supergravity diagrams are also finite. We again find [18], using the explicit SUGRA results of [20],

$$M_2^{SUGRA}(s,t) = T_{eik}^{2\,loop} + \mathcal{O}(t/s) + \text{cutoff dependent} \; .$$

(3.31)

The two-loop diagrams of [20] are shown in fig. 7; those in a), c), g), h), i), and j) correspond to the ladder and crossed ladder diagrams, and the rest are subleading in t/s.

This story illustrates another important phenomenon, that of *graviton dominance* [3, 18]. Specifically, all the states of the supergravity multiplet contribute to the amplitudes, yet we have just found that the gravitational exchange dominates at high-energies and long distances. This is generic, and can be explained [18] by noting that the coupling of an exchanged particle with a given spin grows like

$$E^{\text{spin}} \; ,$$

(3.32)

basically from couplings to $\partial_\mu^{\text{spin}}$.

Exercise: *Explain this more carefully, investigating explicit couplings.*

Thus, the high-energy, long-distance behavior is expected to be relatively generic to any theory of gravity. One caveat is that the emitted radiation will depend on the spectrum of light states – though this is expected to be a subleading effect [15].

107

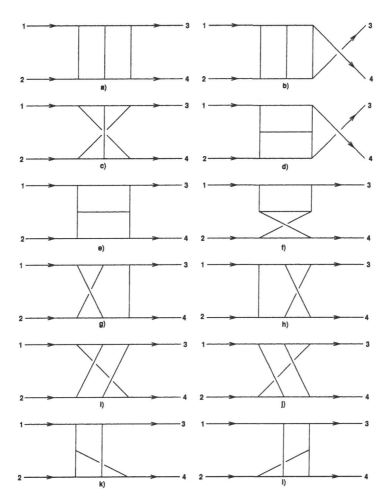

Figure 7: All two-loop, four-point supergravity diagrams can be written in terms of the scalar diagrams shown.

3.5 Momentum fractionation and strong gravity

Returning to momentum fractionation, we see it has some apparently important consequences [18]. First, it apparently presents a significant challenge to a meaningful formulation of the proposed phenomenon of *asymptotic safety* [21, 22] in terms of physical amplitudes [18, 23]. Specifically, if one wished to find a physical probe of this phenomenon, which is basically running of the gravitational coupling to a non-perturbative fixed-point at large energy, ultraplanckian scattering is the obvious place to look. But, the preceding arguments indicate that the gravitational coupling constant only enters the ultrahigh-energy scattering at low momentum transfer; for example, the argument involving eikonal diagrams shows that they depend on $G_D(q)$ at momentum transfer $q \sim k \sim 1/b$, and are not sensitive to Newton's constant at $q \gg M_D$. The same logic obstructs a meaningful probe of hypothesized Regge behavior in gravity.

This UV/IR behavior, telling us that in high-energy gravitational scattering we are really probing long distances, appears important and profound. In particular, we have encountered the problem of nonrenormalizability in other incomplete theories, like the four-Fermi theory of weak interactions. In these previous examples, one could also see the trouble with the theory through a breakdown of unitarity, and these problems are linked, as one finds by considering growth of amplitudes near the cutoff energy. But, in gravity we encounter a unitarity problem of a very different character, apparently arising purely in *long distance* physics, and delinked from the issues of short-distance behavior, nonrenormalizability, and related considerations.

To begin to understand the origin of this problem, let us imagine lowering the impact parameter further, specifically to the value

$$b \sim R(E) \sim (G_D E)^{1/(D-3)}. \tag{3.33}$$

This is the parametric dependence of the D-dimensional Schwarzschild radius given by the center-of-mass energy, and corresponds to the line demarcating the strong-gravity region in fig. 3. The former statement is seen explicitly from the form of the D-dimensional Schwarzschild solution,

$$ds^2 = -\left[1 - \frac{k_D M}{M_D^{D-2} r^{D-3}}\right] dt^2 + \frac{dr^2}{1 - \frac{k_D M}{M_D^{D-2} r^{D-3}}} + r^2 d\Omega_{D-2}^2 \ , \tag{3.34}$$

with k_D a numerical constant. At impact parameters below the value (3.33), one classically expects a black hole to form.

First, let us investigate this problem from the perturbative perspective. Specifically, there are subleading loop diagrams we have so far neglected, like that shown in fig. 8. These are subleading in an expansion in $[R(E)/b]^{2(D-3)} \sim -t/s$, as can be seen from the Feynman rules.

Exercise: *Using the Feynman rules, infer that if one of the rungs of an eikonal diagram with typical momentum $k \sim 1/b$ is replaced with the H diagram of fig. 8, the resulting expression is indeed suppressed by a factor of $[R(E)/b]^{2(D-3)}$ relative to the original amplitude.*

Figure 8: The "H-diagram" is one of a family of diagrams with graviton trees attached to the high-energy source particles; these produce subleading terms in an expansion in t/s.

Indeed, we saw the presence of such subleading behavior both in the corrections to the eikonal amplitudes, and in the supergravity amplitudes. These corrections are small as long as the scattering angle $\theta \sim \sqrt{-t/s} \sim [R(E)/b]^{(D-3)}$ is small. But, they suggest that the perturbation series is no longer an expansion in a small parameter once $b \lesssim R(E)$, and in fact diverges.

A check on this comes from a simpler problem; in [24], Duff showed that an analogous sum over graviton tree diagrams sourced by a point mass produce the Schwarzschild metric. The corresponding sum is indeed divergent at the Schwarzschild radius. We likewise expect that the sum of tree diagrams attached to the high-energy sources gives the classical collision geometry, if we trust the picture to this point; there have been a variety of tests of this picture, though more are still being performed. It was shown in [16] that this classical geometry contains a black hole. Specifically, for $b \lesssim R(E)$, a closed-trapped surface, or apparent horizon forms; singularity theorems plus cosmic censorship then impliy a corresponding black hole.

There are three further comments on this picture. First, there will also be some classical radiation. But, cosmic censorship gives a lower bound on the mass of the resulting black hole, since the horizon area can only grow from its initial value; such lower bounds are in the range of half the collision energy. Second, as with Duff's simpler calculation, one expects a divergent perturbation expansion associated with black hole formation here: a black hole is *not* a small perturbation of flat space. Moreover, there is no indication of how features of string theory or supergravity will save us from this breakdown. Now, we could take the resulting classical geometry as a new starting point, and try to quantize in a perturbation expansion about this. The third comment is that this has been done, in cases with more symmetry, beginning with Hawking's classic demonstration of black hole evaporation [25]. And, here we encounter the most serious problem: such a quantization led to the prediction of massive unitarity violation [26].

To summarize this section, at $E \gg M_D$ we can control the impact parameter by scattering wavepackets. We can also see a correspondence with the semiclassical picture. Classically, collisions at sufficiently small impact parameters produce black holes, plus some radiation. Quantum corrections to this picture lead to Hawking radiation. And, this leads to a unitarity crisis.

To understand the last statement, we turn to a deeper examination of black hole evaporation.

4 The gravitational unitarity crisis

The situation I have presented leads to an apparently critical challenge to the foundations of present-day physics, through what has been called the black hole information paradox. I will give a lightening review of this; more details appear in other reviews such as [27, 28].

As I have noted, Hawking found that perturbative quantization about a black hole geometry leads to its evaporation, with a final state that is mixed so that information is destroyed and unitary evolution fails [26]. The analysis can be carried out even more explicitly in two-dimensional models [29, 30]. This story represents, at the least, a breakdown of quantum mechanics, but we find that the situation is even worse.

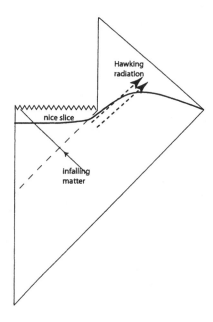

Figure 9: A Penrose diagram of an evaporating black hole, with one of a family of nice slices pictured.

Let's start by giving an updated version of Hawking's argument – the *nice slice* argument. First, a family of spatial slices can be drawn through the black hole, avoiding the strong curvature near the classical singularity, crossing the horizon, and asymptoting to a constant-time slice in the asymptotic flat geometry. One such slice is sketched in fig. 9. These "nice slices" were described in [31], with one explicit construction in [32].

Locality in field theory tells us that the state on such a slice can be represented as a sum of products of states from two Hilbert spaces, one corresponding to inside the black hole, and one outside. Schematically, this can be written (for more detail, see *e.g.* [30])

$$|\psi_{NS}\rangle \sim \sum_i p_i |i\rangle_{in} |i\rangle_{out} \ . \tag{4.1}$$

A description of the state outside the horizon is given by the density matrix formed by tracing over the inside Hilbert space:

$$\rho_{HR} \sim \mathrm{Tr}_{in} |\psi_{NS}\rangle \langle \psi_{NS}| \ . \tag{4.2}$$

This is manifestly a mixed state. In fact, one can trace over all degrees of freedom that have not left the vicinity of the black hole by a given time; *e.g.* one can describe the density matrix on future null infinity as a function of the corresponding retarded time x^-. This density matrix has an entropy $S(x^-) = -\mathrm{Tr}[\rho(x^-) \log \rho(x^-)]$, and one straightforwardly estimates that it grows with x^- to a value of order the Bekenstein-Hawking entropy, proportional to the original horizon area, $S(E) \propto G_D [R(E)]^{D-2}$, at a retarded time corresponding to the time the black hole evaporates to close to the Planck size. This represents a huge missing information, which is also largely independent of what was thrown into the black hole. This gives a modern update of Hawking's original argument [26] for the breakdown of unitarity.

The problem with the story is that quantum mechanics is remarkably robust. Hawking proposed a linear evolution law for density matrices

$$\rho \to \$\rho \ , \tag{4.3}$$

generalizing the S-matrix. But Banks, Peskin, and Susskind [33] argued that this causes severe problems with energy conservation. A basic argument is that information transfer and loss requires energy loss, and that once allowed, energy non-conservation will pollute all of physics through virtual effects. In particular, [33] leads to the conclusion that the world would appear to be thermal at a temperature $T \sim M_D$.

If locality tells us that there is no information escape during Hawking evaporation, and energy conservation or quantum mechanics tell us that information is conserved, the obvious alternative is that information does not escape the black hole until it has reached the Planck size, where the nice-slice argument manifestly fails. However, with an available decay energy $\sim M_D$ and an information $\Delta I \sim S(E)$ to transmit to restore purity, on very general grounds this must take a very long time, $\sim S^2$. This implies very long-lived, or perhaps stable, remnants, which come in $\mathcal{O}(\exp S(E))$ species (*i.e.* internal states) to encode the large information content.

Since one can in principle consider an arbitrarily large initial black hole, the number of species is unboundedly large. This leads to unboundedly large inclusive pair production in generic processes with total available energy $E \gg M_D$, as well as problems with inconsistent renormalization of the Planck mass [34], *etc.* (The former problem can be seen particularly clearly in the charged black hole sector [35], where one considers Schwinger production.)

So, information apparently cannot get out of a black hole, cannot be lost, and cannot be left behind, and this is the essence of the "paradox." It represents a deep conflict between basic principles of Lorentz/diffeomorphism invariance (on a macroscopic scale), quantum mechanics, and locality (also on a macroscopic scale). These are the foundation stones of quantum field theory, which therefore must apparently fall.

Both quantum mechanics and Lorentz invariance appear very robust; simple modifications of them lead to violent contradiction with experiment. On the other hand, locality is a concept that, as we have seen, is not even easily formulated in quantum gravity. So, it is natural to propose that locality is not sharply defined in this context, and that this ultimately underlies an explanation of how unitarity is restored. If this is true, and black hole formation and evaporation is a unitary process, Page [36] has argued on general information-theoretic grounds that information must begin to be emitted by the time scale

$$t_{Page} \sim R(E)S(E) , \qquad (4.4)$$

where the black hole has radiated an $\mathcal{O}(1)$ fraction of its mass. This indicates a needed breakdown of the nice-slice argument, and some departure from locality as described in the semiclassical picture of fig. 9, over distances comparable to the black hole size, $\sim R(E)$ – which can be a macroscopic scale.

While other surprises in the context of high-energy gravitational scattering are not unfathomable – we are still checking aspects of the picture outlined above – it seems clear that black holes also form in collapse of massive bodies, yielding a version of the preceding argument for a conflict between basic principles.

5 Nice slices and the local spacetime perspective

The essence of the nice-slice argument can be summarized by:

$$|\psi\rangle \quad \rightarrow \quad \rho = \text{Tr}|\psi\rangle\langle\psi| \quad \rightarrow \quad S = -\text{Tr}\rho ln\rho = \Delta I . \qquad (5.1)$$

Specifically, evolution on the nice slices leads to a quantum state $|\psi\rangle$ describing the black hole plus surroundings. Locality tells us this can be decomposed in terms of a product Hilbert space, representing distinct degrees of freedom inside and outside the black hole, and that moreover these degrees of freedom become entangled, and cannot become disentangled through escape of information from the black hole. Finally, the von Neumann entropy S gives a quantitative measure of the information that is "stuck" inside the black hole. Since this argument leads to an apparent paradox, and moreover it rests on the apparent "weak link" of locality, it is worth examining more closely, to see if it has a loophole [32, 37].

To begin, note that the way the state $|\psi\rangle$ has been specified is gauge-dependent; it depends on some rather arbitrary, and in the case of the nice slices, extreme and artificial, choice of time slicing. The question of a gauge-invariant specification encounters precisely the problem of gauge-invariant local observables outlined in section 2.1. Specifically, we need to talk about information in localized degrees of freedom, either inside or outside the black hole, and so we need to describe this in terms of local observables, which we don't have.

Of course we know that local QFT has a built-in notion of local observables, and since our fundamental theory of gravity has a QFT approximation it should have an approximate notion of local observables. The question of how these emerge from a gauge-invariant description is still being investigated, but we can outline a picture. This picture is given in terms of the (effective) field theory description of gravity, which we don't believe to be the fundamental description, so it is incomplete. But, nonetheless, it seems to give a suggestive guide, and moreover exploring its limitations plausibly hints at outlines of the more fundamental theory.

The basic idea has been outlined: approximate local observables emerge in *relation* to features of the state. This is because gauge transformations tell us that there is no intrinsic notion of location, but one can imagine locating something with respect to another feature of the state – such as, *e.g.*, planet Earth! In fact, this is also how current treatments of inflationary theory handle the problem of diffeomorphism invariance, which implies arbitrariness in the time slicing: one commonly computes the perturbations, which we ultimately observe, at the reheating time. This is defined in the same kind of relational approach, as the slice where the inflaton field takes a specific value where reheating takes place.

Ref. [1] investigated construction of such gauge-invariant relational observables, that approximately reproduce local QFT observables in certain states. An example of such a "proto-local" observable is an operator of the form

$$\mathcal{O} = \int d^D x \sqrt{-g}\, O(x) B(x) \ . \tag{5.2}$$

Here, $O(x)$ could be the local operator that we wish to discuss, and $B(x)$ is another operator acting on some "reference background" fields with respect to which we localize. Such an expression is diffeomorphism invariant. If we are in a state where the operator $B(x)$ is strongly peaked in some small region, then the integral (5.2) essentially gives the operator $O(x)$ localized to that region. In this sense, locality is emergent, and only in an approximation. This approximation can break down, when fluctuations of the reference background are important, and/or it has a large backreaction. An explicit toy model realizing such ideas can be given by working in 1+1 -dimensional gravity [2] (aka: "on the string world sheet"). One way to think of the reference background is in terms of inclusion of a measuring apparatus. The necessity of including such a background appears in harmony with a viewpoint enunciated by Bohr: "no phenomenon is a real phenomenon until it is an observed phenomenon."

If one needs a reference background in order to give a proper gauge-invariant description of the state, one can ask what kind of reference background is needed to resolve the features of the nice slice state. A simple estimate is the following. Hawking quanta are emitted with a typical energy $\sim 1/R(E)$, and roughly one such quantum is emitted every time $\sim R(E)$.

These have partner excitations, represented in (4.1), that fall into the black hole and should be registered in the internal part of the nice-slice state. Thus, a reference background needs a spatio-temporal resolution at least to scales $\sim R(E)$, over time scales (4.4). This means the reference background carries a minimum energy $S(E)/R(E) \sim E$, indicating that this would have a large backreaction on the black hole.

One might ask if the approximately local quantities like (5.2) characterizing the nice-slice state might be defined using the existing background of the black hole, and thus avoid the need to include other excitations. If so, one might ignore the preceding argument, and just try to give a QFT description of the nice slice state, working in a perturbative expansion in G_D about the semiclassical black hole geometry. The leading correction to this geometry, due to the expectation value of the stress tensor, just causes the black hole to shrink, as in [29], and one can then try to compute the quantum state on slices in this evaporating geometry. Here, though, one encounters an apparent issue as well [37]. If one tries to compute the state $|\psi\rangle$ perturbatively, fluctuations about the evaporating geometry ultimately appear to have a large backreaction on $|\psi\rangle$, leading to a breakdown of the perturbative expansion. This also occurs on the time scale (4.4).

Thus, a proposed resolution [32, 37] of the information paradox is the following. The arguments leading to the conclusion that information is lost are not sharp; the semiclassical/perturbative nice slice picture is not an accurate representation of the detailed quantum state of a black hole. In particular, it is not clear that the arguments can be made in a gauge-invariant fashion, and moreover a naïve attempt to quantize on nice slices encounters backreaction indicating a breakdown of such a perturbative treatment. This means that we have not given a controlled perturbative argument for information loss, and indicates that if one wants to examine the ultimate fate of the information, one needs to do so within a nonperturbative framework.

With this proposed resolution, the paradox becomes a problem, that of determining the correct *nonperturbative* mechanics governing the theory, and in particular unitarizing the description of an evaporating black hole. This may well not have the usual notions of locality when described with respect to the semiclassical evaporating geometry. Moreover, if locality breaks down, or is not a "sharp" notion, this statement needs to be relevant over distance scales that are large, of size $\sim R(E)$, which can be macroscopic, in order for information to begin to escape a big evaporating black hole by the time (4.4).

We would clearly like to understand these issues better. It is always useful to have a simplifying toy model, and fortuitously we have one with a number of parallel features: inflationary cosmology. This of course has its own intrinsic interest(!), but also is a simpler place to examine closely related questions regarding dynamical evolution of geometry, and the role of observables, since it has greater symmetry. Space does not permit covering developments in this area here, but some highlights can be mentioned. First of all, some investigation of approximately local observables, and limitations to recovering precisely local observables, are described in [38] (see also [1]). Moreover, ref. [37] explores the parallel between quantization on nice slices in a black hole background, and quantization on spatial slices in an inflationary background. There an apparently analagous problem with perturbation theory was noted, due to backreaction of fluctuations, appearing on the timescale RS, where now R and S are the Hubble radius and corresponding entropy.¶ More explicit calculations have been performed, and additional arguments provided for breakdown of a perturbative treatment at such times [41, 42] (see also [43]), associated with growth of the effect of fluctuations. It is clearly worthwhile to explore and sharpen this parallel further, particularly with an eye on the needed nonperturbative completion and its implications.

6 Lessons from loops or strings?

There are two particularly popular approaches to quantum gravity, string theory and loop quantum gravity: what do they say about these issues?

If we focus on the S-matrix, this is still largely problematic for loop quantum gravity, where people are working to recover a complete description of configurations closely-approximating flat space, and to give a description of scattering of perturbations about it. A modest first goal for this program would be, for example, to rederive the Born and eikonal amplitudes. A general concern is that the constructions of this program do appear to modify a local description of spacetime, but only at distances comparable to the Planck length, $\sim 1/M_D$. But, as I have just argued, the unitarity crisis really seems to require modifications of locality that can be relevant at very long distance scales, in appropriate circumstances.

String theory initially seems more promising. It has a mechanism for nonlocality: the extendedness of the string, which can become large at high energies. String theory has also furnished us with perturbative calculations of the S-matrix that avoid the infinities I described in discussing nonrenormalizability. And, there are candidate non-perturbative descriptions of string theory; the most promising have been argued to be via dualities such as AdS/CFT, to more complete theories such as matrix theories or $\mathcal{N} = 4$ super Yang-Mills theory. Finally, it has been argued that these dualities realize the idea of *holography*, that a D-dimensional gravity theory in a region has an equivalent description in terms of a $D - 1$ -dimensional theory on the boundary of that region – a notion that, if true, is clearly nonlocal.

In physics one must pose correct, sharp questions, and so we should carefully investigate what string theory actually says. We will only overview some of the basic arguments; a more complete treatment would involve another set of lectures.

First, consider extendedness. This was conjectured to either interfere with black hole formation [44] or to provide a mechanism for information to escape a black hole [31, 45]. A way to test this is to explore the behavior of strings gravitationally scattering in a high-energy collision. In such a collision, a string can become excited, and thus extended. This was initially described as "diffractive excitation" in [6]. There is a clear intuitive picture for the phenomenon [46], in terms of one string scattering off the Aichelburg-Sexl metric (3.26) of the other: the metric produces tidal forces, which excite the string. There is a threshhold

---

¶The black hole/inflation parallel has also been in particular pursued by [39, 40], who also argue for a breakdown at times $\sim RS$, but based on apparently different considerations.

impact parameter below which this effect occurs, given by [6, 46]

$$b_t \approx \frac{1}{M_D} \left(\sqrt{\alpha'} E \right)^{2/(D-2)} , \qquad (6.1)$$

where the tidal impulse is sufficient to excite the string – this is represented by the dotted line in the eikonal region of fig. 3. A similar effect is expected in scattering of other composite objects, *e.g.* a hydrogen atom or a proton, with different threshholds given in terms of the excitation energy.

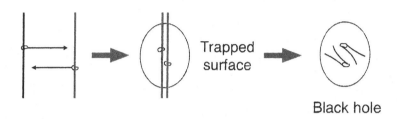

Figure 10: Pictured is the scattering of two ultra-high energy strings; it is argued in the text that a trapped surface (hence black hole) forms before the strings spread out in response to the tidal impulse they receive from each others gravitational shock.

String dynamics appears to respect notions of causality very similar to field theory. In particular, if tidal excitation is what excites the colliding strings, one would expect that only to happen only after one string had met the Aichelburg-Sexl shockwave of the other [46]. This can be tested [47], by explicit quantization of a string propagating in such a metric, where one indeed finds that the string spreads out after the collision, in response to the tidal impulse. On the other hand, the trapped surface, thus black hole horizon, forms *before* the two shockwaves meet. Thus, as illustrated in fig. 10, one ultimately excites the strings inside a black hole. Then, there so far doesn't appear to be any good reason that the excited string can escape the black hole – it is expected to behave like any extended object inside a black hole, and evolve locally/causally enough that it doesn't escape. There is thus no clear reason to believe that extendedness provides a mechanism for information to get out.

String theory is also famous for the improved properties of its perturbation theory, and in fact arguments for order-by-order finiteness in the loop expansion. However, as I have described, this expansion appears to break down precisely in the circumstances where we encounter our unitarity crisis, at $b \sim R(E)$. This statement, as explained at the end of section 3, appears generic to gravity, and there is no good indication that string theory escapes it; indeed, momentum fractionation indicates that the problem occurs independent of the short-distance degrees of freedom of the theory. Similarly, conjectured perturbative finiteness of supergravity [19] encounters the same limitation. Again, it appears that unitarity is a more profound problem than short-distance issues such as renormalizability.

But, if we need a non-perturbative description of the theory to confront these issues, AdS/CFT and related dualities have been claimed to furnish one. So, we should see if these solve our difficulties. In order to isolate the effects of the curvature of anti de Sitter space from the strong-gravity dynamics of black holes, we would like to take the AdS radius to be large, $R \gg 1/M_D$, and see whether we can extract, in this regime, an approximation to the flat-space S-matrix. Even if we can find a unitary S-matrix describing black hole formation and evaporation, we might not be satisfied that we have addressed the actual "paradox," since it results from the conflict I have described between the description of localized degrees of freedom inside a black hole and a unitary description. So far, in AdS/CFT, we do not have a description of localized observables, which we might expect to fit into our discussion of protolocal observables in the preceding section, so a local description of black hole interiors is lacking.[||] Nonetheless, we might justifiably feel that AdS/CFT providing such an S-matrix strongly indicates we're on the right track.

There has been significant discussion of the problem of defining such an S-matrix purely from CFT data, without additional input from the bulk theory – this is what is needed if the CFT is to give a non-perturbative definition of the theory. Relevant work includes [48–55].

In particular, a boundary CFT definition of the bulk theory would be achieved by a unitary equivalence between the CFT Hilbert space $\mathcal{H}$ and a Hilbert space of bulk states. If we consider a large AdS radius, say $R \sim 10^{10}$ lightyears, then this bulk Hilbert space should include the kinds of scattering states that we are familiar with, e.g. two particles in gaussian wavepackets, separated by 10 km, and prepared such that they will collide in a small region, resulting in outgoing states with similar asymptotic identification.

In the case where the bulk string coupling constant $g_s \sim g_{YM}^2$ vanishes, also implying $G_D = 0$, this can be achieved; free particle states of the bulk can be identified with states created by boundary operators, as reviewed in [56]. One might expect this holds also for small coupling. However, we can think of AdS as providing a gravitational "box" of size R, and the states of the free-particle Hilbert space correspond to particles that ricochet back and forth across this box for infinite time. Thus, while we would like to isolate the S-matrix corresponding to a single scattering event, such states undergo infinitely many interactions. In this sense the coupling does not necessarily have a small effect, and it is not clear we can extract from the interacting boundary Hilbert space the approximate S-matrix corresponding to the type of situation described in the preceding paragraph. One proposal for deriving the corresponding reduced transition matrix elements appears in [54], but it is also not clear that this approach in fact overcomes this "multiple-collision" problem.

In order to avoid this problem, Polchinski and Susskind [48,49] proposed that one should instead consider states incident from the boundary of AdS, as pictured in fig. 11. Here there are subtleties as well. First, a boundary source generically produces a non-normalizable bulk state, that is thus not in the Hilbert space, and does not have a clear scattering interpretation. A way around this is to prepare states using boundary sources with compact support, so that outside of this support the bulk state is normalizable.

[||]One idea is to seek an appropriate notion of observables that are relational in the matrix space.

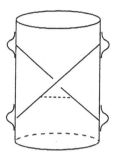

Figure 11: Wavepackets created on the boundary of AdS scatter in the bulk.

Indeed, [51, 52] explored such "boundary-compact" wavepackets, and their scattering. In particular, [51] investigated the plane-wave limit of such wavepackets, and found some necessary conditions on the CFT if it is to produce familiar properties of the bulk S-matrix. Specifically, the presence of a momentum-conserving delta function requires a certain kind of singularity in correlators of the boundary theory, which had not been previously investigated, and behavior $\sim s^2/t$ in the Born regime requires a specific subleading structure to this singularity. These necessary conditions appear non-trivial from the perspective of the boundary theory, although [53] provided some plausibility arguments that they are satisfied in certain types of CFTs, specifically those with large hierarchies of anomalous dimensions.

However, [52] pointed out another issue: the boundary-compactness condition places limitations on the types of bulk wavepackets that can be obtained. In careful treatments of scattering (see *e.g.* [57]) one uses wavepackets that fall at long distance faster than a power-law – such as gaussians. While boundary-compact wavepackets can be arranged to have characteristic widths, outside these widths they have tails that fall off as a power of the distance, $1/r^\Delta$, where Δ is the conformal dimension of the corresponding boundary operator. This is an obstacle to approximating the usual space of scattering states. Such tails can matter, when one is *e.g.* trying to resolve very small matrix elements. For example, the matrix elements to individual quantum states in the black hole evaporation problem are expected to be of size $\exp\{-S(E)/2\}$, and these are thus expected to be obscured by such tails.

A logical alternative to a complete equivalence between bulk and boundary theories, that allows one to derive all desired features of the bulk theory from the boundary, is that the boundary theory is a sort of effective theory for the bulk, that is, captures certain coarse-grained features, but not all the fine-grained detail of the bulk dynamics [50, 52]. Certainly the powerful principles of universality and symmetry are at work, and provide at least part of the explanation for relations between bulk and boundary.

Complementarity is an idea that has been widely discussed, in association with holography, to provide a picture of how information escapes a black hole in Hawking radiation. In the form advocated in [58], this states that observations of observers who stay away from the black hole, and of those who fall in, are complementary in an analogous fashion to complementarity of observations of x and p in quantum mechanics, and thus should not simultaneously enter a physical description. Picturesquely [59], an astronaut (named "Steve") falling into a black hole may feel healthy all the way to the vicinity of the singularity, but the external observer describes the astronaut as burning up at the horizon, in the process imprinting his bits of information on the Hawking radiation. This seems to be a radical departure from local quantum field theory. So far it has been hard to give a detailed account of this idea. It may be that a significantly less radical notion of locality is all that is needed to resolve the problem [32]. One possible distinguishing feature of different proposals is the thermalization or mixing time, which is the time scale on which information falling into the black hole undergoes quantum mixing with the degrees of freedom of the Hawking radiation. Advocates of the complementarity picture [59–61] suggest this time scale is of size $R \log R$. However, [32] suggests this time scale could be much longer; an upper bound is the time scale (4.4) found by Page. Such a longer time scale for mixing would represent a less radical departure from the semiclassical prediction [25] of infinite mixing time.

Another proposed story for black hole evolution, arising from string theory and related to the preceding ideas, is the fuzzball proposal [62]. This suggests that a macroscopic black hole is more fundamentally described in terms of an ensemble of "microstates" each of which significantly differs from the black hole geometry and from each other *outside* the horizon. With such large variations in the geometry of the states of the ensemble, it is very hard to see how an observer falling into such a configuration would not be uncomfortably disrupted when falling through the would-be horizon. Thus, such a proposal seems to fit into the general category of massive remnants [63], where the black hole evolves into a configuration that is massive but no longer has a meaningful horizon – it is more like a kind of exotic star. An essential question with such a proposal is to understand how such a configuration would dynamically evolve from a matter configuration collapsing to form a black hole, and to be consistent with our knowledge of black holes. In particular, as [63] outlined, after horizon formation this would require some nonlocal propogation of information – of a specific type, for the fuzzball proposal.

In short, the jury is still out on whether AdS/CFT, or other dualities, which offer a similar approach, can supply a definition of non-perturbative string S-matrix, sufficient to investigate our puzzles of gravitational scattering. If it does, it is very mysterious how it does so, that is, how *all* the features of *approximately* local D-dimensional gravitational dynamics are encoded in a lower-dimensional theory. In such a hypothesis, one would like to understand the mechanisms and principles both for how the boundary theory mimics this higher-dimensional physics of local QFT at a fine-grained level, and also for how it modifies the local QFT picture, particularly in describing the dynamics of black hole formation and evaporation. Perhaps, via such dualities, string theory accomplishes such miracles – or maybe we need to reach beyond. Either way, it is important to understand the mechanisms and principles in operation.

7 Beyond local quantum field theory

In reaching for the principles of a more complete theory, one can seek inspiration from what seems to have been a similar crisis encountered in physics: that of the classical atom. A classical treatment of the hydrogen atom yields singular evolution, as the electron spirals into the charge center at $r = 0$, producing an "atomic stability crisis." Before knowing quantum mechanics (and in the absence of experimental data), one theoretical approach would be to modify the laws of physics near $r = 0$ to smooth out this singular evolution. But, the ultimate resolution was quite different; we discovered that there is a new scale, the Bohr radius – indicated by experiment – and that the laws of classical physics simply do not furnish a good description of the atomic electron at shorter distances. Instead, they must be replaced by a new set of principles and a new framework in this domain; in particular, one encounters the uncertainty principle, and in the end describes the physics via the wave mechanics of quantum theory.

Our present situation seems similar. We have found a breakdown of the classical laws at the $r = 0$ singularity of a black hole. We might expect that there is some short-distance modification of the physics, say on the Planck length scale, $\sim 1/M_D$, that smooths out the singular evolution; perhaps it is provided by the short-distance regulation of the physics by loop quantum gravity or string theory. But, the (experimental) fact that black holes exist together with the black hole information problem strongly indicate that such a short distance solution is not sufficient. Instead, it appears that some new physical principles and framework beyond local QFT are needed, and that those should significantly modify the physical description on distance scales extending at least to the horizon radius, which can be a macroscopic length. Thus, the information problem could be an important guide, just as was the stability problem of the atom.

If we want to understand the physics involved, there is a basic set of starting questions. First, we would like to more sharply characterize the circumstances in which local QFT fails. This is important, because it tells us where the *correspondence boundary* lies, where familiar constructs must match onto those of a more basic theory. In the case of quantum mechanics, the correspondence boundary is described by the uncertainty principle and related statements, and this is a very important clue to the underlying physics. A second question is what mechanisms are used to describe the dynamics – for example what picture do we have of flow of information, relation to spacetime, *etc.* A third, related, question is what physical and mathematical framework replaces QFT, and, how do the familiar properties of local QFT emerge from it in domains away from the correspondence boundary? In trying to answer these questions, we might expect that unitarity in black hole evolution could be a key guide, along with the need to provide approximately local and quantum-mechanical descriptions for dynamical spacetimes such as cosmologies or black hole interiors.

We have specifically seen that there appear to be very good reasons to question locality, as described with respect to semiclassical geometry, at macroscopic length scales, under appropriate circumstances. One reason for this is that in such a semiclassical description of the black hole geometry, in order to restore unitary evolution, it appears that information must be relayed from deep inside the black hole, near the singularity, to outside the horizon, over spacelike separations that can be large, $\sim R(E) >>> 1/M_D$.** Also, we see a gravitational growth in the effective size of objects, through scattering. This is indicated, for example, by the scattering angle

$$\theta \sim \left[\frac{R(E)}{b}\right]^{D-3} \tag{7.1}$$

for a particle scattering from another particle of energy E, and becomes large at impact parameters $b \sim R(E)$ growing with the energy, as I have described. Finally, there is the lack of local observables; specifically, the construction of approximately local observables outlined in section 5 fails [1] at increasing distances when the energies involved are increased.

We in particular would like to more sharply describe the correspondence boundary where local QFT ceases to be a good description. There have been a number of such previous proposals. For example, it has been widely believed that a breakdown occurs when spacetime curvatures reach the Planck scale. Or in string theory, a "string uncertainty principle" has been formulated [64, 65], augmenting the usual uncertainty principle by a term accounting for the growth of strings with energy,

$$\Delta X \geq \frac{1}{\Delta p} + \alpha' \Delta p \ . \tag{7.2}$$

An alternative viewpoint, related to modified dispersion relations, is that validity of familiar QFT requires small momenta, $p < M_D$. These are all statements relevant to the description of a single-particle state. Holographic ideas have been encoded in a statement about multiparticle states, that local QFT is only valid in cases where one restricts the information content of a region, bounded by area A, as

$$I \leq A/4G_D \ ; \tag{7.3}$$

this finds its most sophisticated realization in the covariant entropy bound of [66, 67].

But, our discussion of scattering suggests a different boundary. In quantum mechanics, the uncertainty principle represents a limitation on the domain of validity of the classical dynamical description, given by phase space. If we consider QFT in semiclassical spacetime, the dynamical description is given in terms of Fock space states in the background geometry. In the example of a flat background, it is not clear how to formulate a single-particle bound, such as (7.2), respecting Lorentz invariance; a shortest distance is not a Lorentz-invariant concept. (We also have a "good" classical geometry, the Aichelburg-Sexl metric, describing a highly-boosted particle.) But, consider a two-particle state, with, e.g, minimum uncertainty wavepackets, with central positions x, x' and momenta $p = \hbar k$, $p' = \hbar k'$:

$$\phi_{x,k}\phi_{x',k'}|0\rangle \ . \tag{7.4}$$

**This suggests some basic inaccuracy of this semiclassical picture.

In local QFT, we allow states with arbitrary x, x' and k, k'. But, in light of our discussion, we recognize that gravity becomes strong, and that we cannot trust such a Fock-space description, when the inequality

$$|x - x'|^{D-3} > G|k + k'| \tag{7.5}$$

is violated, with $G \sim \hbar G_D$. This is the proposed *locality bound* of [69]. It also has generalizations for N-particle states [46], and within the context of de Sitter space [38]. This kind of bound thus might play a role similar to that of the uncertainty principle, and serve as an important principle giving clues to the more fundamental dynamics.

In seeking the mechanisms, principles, and mathematical framework of a more fundamental "nonlocal mechanics" of gravity, there are other important guides. In particular, nonlocality in a framework with a QFT approximation generically causes serious difficulty, and indeed inconsistency. The reason is that, with Lorentz invariance, locality and causality are linked, so nonlocality (*e.g.* signaling outside the light cone) implies acausality (propagation backward in time), and acausality in turn leads to paradoxes such as the "grandmother paradox," where one sends a message back in time instructing an evil accomplice to kill ones grandmother before ones mother is born. Moreover, locality is observed to hold to an excellent approximation! So very important questions are how usual locality can be absent as a fundamental property of the theory, yet emerge in an approximate sense, and how this can happen without associated inconsistencies – nonlocal needs to be nearly local, in the appropriate contexts. If this line of investigation is correct, these should be important guides, and are very interesting and challenging problems. One can begin to think about these by considering how locality is sharply described, in the QFT approximation. A standard formulation, termed "microcausality" is that local observables must commute outside the lightcone,

$$[O(x), O(y)] = 0 \quad \text{for} \quad (x - y)^2 > 0 . \tag{7.6}$$

As we have indicated, though, such observables are expected to only emerge in an approximation from protolocal observables of the more fundamental theory. In that case, we would like to understand what features of that theory lead to approximate validity of (7.6).

This story requires deeper understanding of relational observables, which would involve another set of lectures. But, there is another way to probe locality in QFT, that returns us to a main topic of these lectures: in terms of properties of the S-matrix. An important question is to understand how to distinguish S-matrices from local theories, and one important characteristic appears to be *polynomial* behavior at large momenta. A rough idea for why this is so is that interactions involving powers of derivatives are local, but nonpolynomial interactions, for example involving $\exp\{\partial^n\}$, are not.

8 The gravitational S-matrix: exploring general properties

I have argued that the S-matrix (or related inclusive probabilities, in $D = 4$) should be a well-defined object in a quantum gravity theory with a Minkowski-space solution, and should be a good approximate concept if there are solutions closely approximating Minkowski space. One possible source of clues to the principles of a theory can be sought in the general structure of this S-matrix. Indeed, as was realized in earlier studies of the S-matrix, *e.g.* in the context of strong interactions, the combined principles of unitarity, analyticity, and crossing are very powerful constraints. Combining these with other properties of a theory can yield nontrivial information; for example, as we've noted, string theory was *discovered* by Veneziano finding an expression for the S-matrix with duality between s and t channels. We might anticipate that we could gain important knowledge about quantum gravity by combining general properties of gravity with this powerful framework. Given our preceding discussion, we would particularly like to understand how locality and causality enter the story. This section will briefly summarize some preliminary investigation into these questions; for more discussion, see [15, 70].

8.1 Unitarity, analyticity, crossing, . . .

In the S-matrix approach, unitarity, $S^\dagger S = 1$, is a basic assumption. Analyticity and crossing are somewhat more subtle in a theory of gravity, as compared to theories without massless particles. Let us make the "maximal analyticity hypothesis," that the only singularities are those dictated by unitarity. For purposes of illustration, as above we will think about a theory with massive matter, *e.g.* a scalar, coupled to gravity. But, the masslessness of the graviton leads to various important new behavior.

To illustrate, in a theory with a mass gap, the commutativity of local observables (7.6) implies [71, 72] that amplitudes are polynomially bounded, for example at fixed t and large complex s. This, in turn, allows one to derive a bound on the total cross section known as the *Froissart bound*,

$$\sigma < c(\log E)^{D-2} , \tag{8.1}$$

for some constant c. Gravity manifestly violates this bound; for example, simply from strong gravity/"black hole formation" we find a cross-section

$$\sigma_{BH} \sim E^{(D-2)/(D-3)} , \tag{8.2}$$

and we find an even larger cross section from (3.20), describing eikonal scattering, $\sigma_E \sim E^{2(D-2)/(D-4)}$. We would like to understand what other properties necessarily change in a theory with gravity.[tt]

[tt]Ref. [73] suggests similar properties, particularly growth of cross sections, in other massless, nonrenormalizable theories.

124

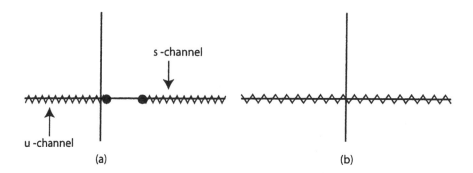

(a)

(b)

Figure 12: (a) The complex s-plane in the case $m \neq 0$. (b) The complex s-plane in the case $m = 0$.

In fact, even crossing is more subtle in a theory without a gap. Consider $2 \rightarrow 2$ scattering, and investigate the amplitude as a function of complex s at fixed $t < 0$. In a massive theory we have cuts shown as in fig. 12(a); the s-channel amplitude arises when we approach the right cut from above, but we can also analytically continue between the cuts and find the u-channel amplitude by approaching the left cut from below. But, in a massless theory, the cuts merge, and our picture is instead fig. 12(b). Thus, we would seem to lose the ability to analytically continue between channels, which is an important property for constraining amplitudes.

So, clearly some properties familiar from massive theories are lost. Without a theory that calculates the full gravitational S-matrix, we do not know which properties still hold. One approach to further investigation is to explore which properties – such as the preceding crossing path – we are forced to give up, and which can be maintained: in particular, we might assume that a familiar property from the massive case holds, until confronted with contrary information. In this way, one can try to find the outline of a consistent story. One set of checks on some of this discussion, which should be performed, comes from the fact that there is now significant explicit knowledge about supergravity to two and three loops, and beyond (see Bern's lectures at this school). So, we might check which properties hold at least to a low, but non-trivial, loop order.

For example, there is a different analytic continuation exhibiting crossing, found by Bros, Epstein, and Glaser [74]. Begin by keeping $u < 0$ fixed, and, at large $|s|$, follow the path $s \rightarrow e^{i\pi} s$. Then, maintaining $s < 0$ fixed, follow the path $t \rightarrow e^{-i\pi} t$. This is an alternate path between the s and u channels. (One also must check that the endpoints of the path can be connected.)

Exercise: *Check that the described path does indeed connect the s and u channels. What assumptions are needed to connect the two path segments?*

Exercise: *Check whether this describes a valid crossing path at one, two, and three loops in*

N = 8 supergravity, by working with the explicit amplitudes of that theory.

Crossing symmetry can be combined with other properties to provide important constraints. For example, another standard S-matrix property is hermitian analyticity,

$$T(s^*, t^*) = T(s, t)^* . \tag{8.3}$$

Exercise: *Determine whether the one, two, and three loop amplitudes in N = 8 supergravity are hermitian analytic.*

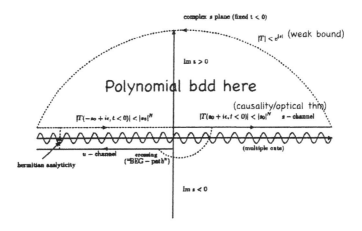

Figure 13: Causality (a forward dispersion relation), crossing, and hermitian analyticity indicate a polynomially-bounded amplitude in the upper-half s-plane.

Crossing, hermitian analyticity, and causality potentially give information about polynomiality, which I have indicated is also connected with locality. I will give a flavor of this, depending on certain assumptions that should be more carefully investigated. Specifically, consider fig. 13. A basic statement of causality is that forward scattering ($t = 0$) is polynomially bounded. With an additional assumption about the amplitudes being well behaved (this is an assumption of sufficient smoothness, bounding the real part of the amplitude, discussed in [70]) this can be converted into a statement that the amplitude is polynomially bounded as $s \to \infty$ along the upper side of the right cut. Crossing ensures the same behavior on the lower side of the left cut. Then, hermitian analyticity implies the same bound on the upper side of the left cut. Finally, if the amplitude satisfies a rather mild bound, $|T| < e^{|s|}$ in the upper-half s-plane, the Phragmen-Lindelof theorem of complex analysis implies that in fact T is *polynomially bounded* in the upper half plane [70]. In this sense, causality plus apparently mild and plausibly true analyticity properties imply a polynomial bound in the physical region $t < 0$ real, and $Im(s) > 0$.

However, this does not mean that the amplitude is *polynomial* – it could behave non-polynomially, but with nonpolynomial *growth* restricted to some other region besides the upper-half s plane. Indeed, basic features of gravitational scattering suggest that this occurs, in association with large cross sections such as (8.2).

8.2 Partial waves, strong gravity, and a "Black-hole ansatz"

One way to investigate the gravitational S-matrix is via the partial-wave expansion [15,70], which in D dimensions takes the general form

$$iT(s,t) = 2^{4\lambda+2}\pi^\lambda\Gamma(\lambda)E^{4-D}\sum_{l=0}^{\infty}(l+\lambda)C_l^\lambda(\cos\theta)\left[e^{2i\delta_l(s)-2\beta_l(s)} - 1\right] , \qquad (8.4)$$

where $\lambda = (D-3)/2$ and C_l^λ is a Gegenbauer polynomial. Here $\delta_l(s)$, $\beta_l(s)$ are the real and imaginary (absorptive) parts of the phase shift for the lth partial wave, and the full $2 \to 2$ amplitude is encoded by them. Basic features of gravity suggest specific behavior for these functions.

For example, consider scattering in the strong gravity/"black hole" regime. This corresponds to

$$l \lesssim ER(E) \sim S(E) . \qquad (8.5)$$

We might expect that a quantum version of black hole states enter like resonances in two-body scattering. Specifically, if such a black hole forms, it takes a time $\sim R(M)$ to decay to a different black hole state plus a Hawking particle, and thus has a width $\Gamma \sim 1/R(M)$. Thus, such a state is very narrow:

$$\frac{\Gamma}{M} \sim \frac{1}{R(M)M} \sim \frac{1}{S(M)} \ll 1 . \qquad (8.6)$$

For two-particle collisions with CM energy $E \sim M$ and angular momentum l, we expect a density of states of order

$$\rho_{acc}(M,l) \sim R(M) \qquad (8.7)$$

to be accessible directly via scattering [70]. Their properties suggest properties of δ_l, β_l.[‡‡]

For example, if we collide two particles at energy E to form a black hole, and if Hawking evaporation at least gives a good coarse-grained picture, we expect there to be $\sim \exp\{S(E)\}$ outgoing multi-particle states. (Here for simplicity we neglect the fact that $E \neq M$, which can be accounted for in a more careful treatment.) Moreover, the amplitude for an outgoing

---

[‡‡]One might ask about the full $\Delta\mathcal{N} \sim \exp\{S(M)\}R(M)\Delta M$ states in the interval ΔM, expected from the Bekenstein-Hawking entropy $S(M)$. Ref. [70] argues that this larger set of states can be accessed by forming a black hole with mass $> M$, which then is allowed to evaporate to M. This produces configurations where the $\Delta\mathcal{N}$ black hole states are entangled with the Hawking radiation emitted in reaching M. For further discussion of properties of the spectrum and the connection with scattering, see [70].

state with just two particles is of size $\sim \exp\{-S(E)/2\}$. The tininess of this corresponds to very strong absorption in the $2 \to 2$ channel, and in partial waves can be parameterized by

$$\beta_l \approx \frac{S(E,l)}{4} \quad , \quad \text{for} \quad l \lesssim ER(E) \,. \tag{8.8}$$

This is a first part of a "Black-hole ansatz" for the $2 \to 2$ scattering [15].

We also expect features of the spectrum of states and scattering to be reflected in δ_l. In particular, in the case of a resonance with a single decay channel back to the initial state, Levinson's theorem tells us that increasing the energy through that of the resonance increases the phase shift by $i\pi$. This suggests that accessible black hole states would lead to a formula $\delta_l(E) \sim \pi S(E,l)$. However, as discussed in [70], the multiple decay channels lead to a more complicated story; we parameterize the second part of our ansatz as

$$\delta_l(E) = \pi k(E,l) S(E,l) \tag{8.9}$$

where we expect $k(E,l) \sim \mathcal{O}(1)$ and $k(E,l) > 0$, the latter condition corresponding to time *delay* in scattering.

So, we have good indications of both strong absorption (β_l) and scattering (δ_l) to large impact parameters, $l \sim ER(E)$, corresponding to the growing range of the strong gravity region, out to impact parameters $b \sim R(E)$. This naïvely yields nonpolynomial $T(s,t)$, by (8.4) producing exponential behavior $\sim \exp\{E^p\}$.

A more complete story requires matching the physics onto the longer-distance region. This produces an even larger cross section and range, as indicated *e.g.* by (3.20). We also find corresponding nonpolynomial behavior in the eikonal amplitude (3.19), as is seen from its saddlepoint approximation, which using (3.16), is

$$T_{\text{eik}} \sim \exp\left\{i[s(-t)^{(D-4)/2}]^{1/(D-3)}\right\} \,. \tag{8.10}$$

This expression also exhibits nonpolynomial behavior.

Exercise: *Derive* (8.10) *from the saddlepoint of* (3.19).

8.3 Locality vs. causality

An interesting question is whether quantum gravity can have an appropriate sense of causality (which I have argued is linked with consistency, at least in the semiclassical approximation), yet not be local. The preceding discussion suggests a possible realization of these ideas, in terms of properties of the S-matrix.

I have outlined in section 8.1 an argument that causality implies a polynomial bound in the upper-half s plane, for $t < 0$, given certain assumptions. This is naturally associated with scattering producing a time *delay* (causal) rather than a time *advance* (acausal), as is most simply seen by considering $0 + 1$ - dimensional scattering, with initial and final amplitudes related by

$$\psi_f(t) = \int_{-\infty}^{\infty} dt' S(t-t') \psi_i(t') \,. \tag{8.11}$$

128

By causality, if the source ψ_i vanishes for $t' < 0$, the response ψ_f should as well. This is achieved if the Fourier transform $S(E)$ is polynomially bounded in the upper half E plane. **Exercise:** *Check this last statement*

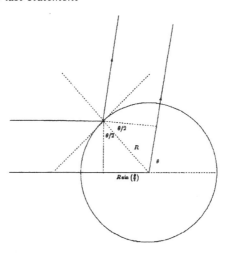

Figure 14: Scattering from a repulsive potential with range R produces a time advance proportional to R.

In higher-dimensional scattering, there is a simple argument that repulsive scattering with range growing as a power of the energy produces nonpolynomially-bounded behavior, associated with a time advance. Specifically, consider fig. 14. For scattering at angle θ, the path that scatters off the potential at $R(E) \propto E^p$ is shortened compared to that through the origin, corresponding to no scattering. Simple geometry gives the time advance, and its corresponding phase shift, yielding

$$ S \sim e^{-i\sqrt{-t}R(E)} . \tag{8.12} $$

which violates polynomial boundedness in the upper-half s plane.

While I have stressed the long-range nature of gravity, gravity is also *attractive*, indicating a different story. For example, consider instead a picture of scattering of a particle from an Aichelberg-Sexl shock, shown in fig. 15. The attractive behavior produces a time delay, and we now have polynomially-bounded behavior,

$$ S \sim e^{i\sqrt{-t}R(E)} . \tag{8.13} $$

Nonetheless, behavior such as (8.13) is nonpolynomial – its non-polynomial growth is just in a different region than $\text{Im}(s) > 0$, $t < 0$. This causal yet nonpolynomial behavior is

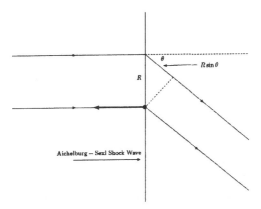

Figure 15: A schematic picture of (attractive) scattering from an Aichelburg-Sexl shock, producing a lengthened path, therefore a time delay.

the kind suggested by our arguments in the preceding subsection. So, if we think of locality as characterized in terms of nonpolynomiality of the S-matrix, these arguments appear to hang together, and suggest a sense in which scattering in gravity could walk a fine line of *nonlocal* and *causal* behavior, thus offering a possible outline for a consistent story. I have really only given an outline of how such arguments could be made. Further discussion of how such ideas could be part of the story of black hole evaporation is given in [32].

Finally, there is a slightly different way of describing locality/causality in the S-matrix framework; see *e.g.* Coleman's 1969 lectures [68] in this school. Specifically, one considers two subsets of scattered particles, in the limit as the separation between the subsets is taken large. *Cluster decomposition* is the statement that the leading term in the limit of large separation is that where the two sets of particles scatter independently. *Macrocausality* is the statement that the next most important terms arise from a single particle from one of the subsets scattering with the other subset. If we consider gravity, we find that such statements are expected to be true, but in a more limited region of validity than in other field theories.

Specifically, this arises from an order of limits question. If we fix a large separation $|x-x'|$ between the two clusters of particles, we expect cluster decomposition and macrocausality to break down when the CM energy of the two clusters, $|k + k'|$, becomes large enough to violate the locality bound (7.5). In fact, as we saw in section 3, we expect other important contributions even for

$$|x - x'|^{D-4} \sim G_D |k + k'|^2 \ . \tag{8.14}$$

There is still however a more limited sense in which clustering and macrocausality hold; for fixed CM energy, both of these statements are expected to be true in gravity for sufficiently large separation $|x - x'|$. So, part of the question is how to achieve this degree of locality, yet modify microcausality (7.6), and plausibly the very existence of a description in terms of quantum field states, in contexts where the locality bound is violated.

9 Conclusions and directions

There are many facets of the story of quantum gravity; some future directions are indicated by incompleteness, unproven statements, and exercises in these notes. More broadly, there are several important and general themes for investigation.

While there has historically been a lot of focus on the nonrenormalizability and singularities of gravity, I have argued that *unitarity* is a more profound problem in formulating a complete theory. This is particularly revealed in gedanken experiments involving ultrahigh-energy scattering. These notes have given sharpened statements that this unitarity crisis is a *long-distance* issue, and there is no clear path to its resolution in short-distance alterations of the theory. This appears to be a critically important and central aspect in the problem of quantizing gravity, suggesting that these considerations are an important guide towards the shape of the ultimate theory.

In investigating these questions, one seeks sharp physical quantities that a theory of gravity should be able to compute. One possibility is quantum mechanical observables that approximate local observables in certain states and circumstances. Another, in cases where, as we appear to observe, there is a good flat space limit, is the gravitational S-matrix.

In particular, investigation of ultrahigh-energy scattering, specifically in the black-hole regime, reveals deep conflict between the "local" approach to describing physics and a more "asymptotic" or scattering approach. This appears to indicate a failure or modification of the "local" approach even at macroscopic distances, under the appropriate circumstances, which does not appear to be curable by short-distance modifications of the theory. A profound question is how to reconcile such failure with the use and validity of local quantum field theory under ordinary circumstances.

While specific frameworks for quantum gravity have been proposed, they do not yet satisfactorily resolve these problems. Loop quantum gravity is still grappling with the problem of approximating flat space and producing an S-matrix. Despite initial promise, string theory has not yet advanced to the stage where it directly addresses the tension between the asymptotic and local approaches, or is able to compute a unitary S-matrix in the relevant strong gravity regime. Because of the long-distance and non-perturbative nature of the problem, it is also not clear how it would be addressed if other problems of quantum gravity were resolved, for example if supergravity indeed yields perturbatively finite amplitudes [19].

I have argued that the needed theory should have certain non-local features, yet be nearly-local, to avoid consistency problems associated with acausality, and also since we expect clustering and macrocausality to hold in certain limits.

One way to seek further information is to explore the interplay between the local and asymptotic descriptions, and the limitations of the former. Cosmology, particularly inflationary cosmology, provides another testing ground for this, whose discussion would require another set of lectures. In particular, one can explore the formulation and role of relational and protolocal observables [1,38], and apparent limitations to a perturbative approach [37,41,42]. One can look for other clues by investigating the nature of the nonperturbative dynamics, and the structure of a quantum framework [75], needed to give a more complete treatment of cosmology or of other dynamical spacetimes such as black holes.

Further clues can also be sought directly through the S-matrix. We know that such a matrix capable of describing gravity must have some rather special properties, in addition to general properties such as unitarity and analyticity. Particularly, we have seen that crossing becomes more subtle, and that there are interesting questions surrounding asymptotic behavior. Unitarization of the S-matrix in the strong gravity regime is a key question. It has been suggested that one can have such an S-matrix consistent with causality, but with modifications of the usual indicators of locality.

The bigger question is what mechanisms, principles, and framework underly the theory. Particularly, we seem to need to relax or modify locality such that it is not a fundamental property, but emerges approximately in appropriate circumstances, and moreover to avoid inconsistencies usually associated with acausality – this seems a critical tension. This "locality without locality" is plausibly a powerful constraint on the ultimate shape of such a "nearly-local" mechanics.

Acknowledgements I would like to thank the organizers for the invitation to present these lectures at Erice. I have greatly benefited from discussions with my collaborators J. Andersen, M. Gary, D. Gross, J. Hartle, M. Lippert, A. Maharana, D. Marolf, J. Pendeones, R. Porto, M. Schmidt-Sommerfeld, M. Sloth, M. Srednicki, as well as with N. Arkani-Hamed, B. Grinstein, and E. Witten; however, none of them should be blamed for misconceptions, which are my own. This work is supported in part by the U.S. Dept. of Energy under Contract DE-FG02-91ER40618, and by grant FQXi-RFP3-1008 from the Foundational Questions Institute (FQXi)/Silicon Valley Community Foundation.

References

[1] S. B. Giddings, D. Marolf and J. B. Hartle, "Observables in effective gravity," Phys. Rev. D **74**, 064018 (2006) [arXiv:hep-th/0512200].

[2] M. Gary, S. B. Giddings, "Relational observables in 2-D quantum gravity," Phys. Rev. **D75**, 104007 (2007). [hep-th/0612191].

[3] G. 't Hooft, "Graviton Dominance in Ultrahigh-Energy Scattering," Phys. Lett. **B198**, 61-63 (1987).

[4] I. J. Muzinich, M. Soldate, "High-Energy Unitarity of Gravitation and Strings," Phys. Rev. **D37**, 359 (1988).

[5] D. Amati, M. Ciafaloni, G. Veneziano, "Superstring Collisions at Planckian Energies," Phys. Lett. **B197**, 81 (1987);

"Can Space-Time Be Probed Below the String Size?," Phys. Lett. **B216**, 41 (1989); "Higher Order Gravitational Deflection And Soft Bremsstrahlung In Planckian Energy Superstring Collisions," Nucl. Phys. **B347**, 550-580 (1990).

[6] D. Amati, M. Ciafaloni, G. Veneziano, "Classical and Quantum Gravity Effects from Planckian Energy Superstring Collisions," Int. J. Mod. Phys. **A3**, 1615-1661 (1988).

[7] H. L. Verlinde, E. P. Verlinde, "Scattering at Planckian energies," Nucl. Phys. **B371**, 246-268 (1992). [hep-th/9110017].

[8] T. Banks, W. Fischler, "A Model for high-energy scattering in quantum gravity," [hep-th/9906038].

[9] G. Veneziano, "Construction of a crossing - symmetric, Regge behaved amplitude for linearly rising trajectories," Nuovo Cim. **A57**, 190-197 (1968).

[10] S. B. Giddings, S. D. Thomas, "High-energy colliders as black hole factories: The End of short distance physics," Phys. Rev. **D65**, 056010 (2002). [hep-ph/0106219].

[11] S. Dimopoulos, G. L. Landsberg, "Black holes at the LHC," Phys. Rev. Lett. **87**, 161602 (2001). [hep-ph/0106295].

[12] S. B. Giddings, "High-energy black hole production," AIP Conf. Proc. **957**, 69-78 (2007). [arXiv:0709.1107 [hep-ph]].

[13] K. Nakamura *et al.* [Particle Data Group Collaboration], "Review of particle physics," J. Phys. G **G37**, 075021 (2010).

[14] S. Weinberg, "Infrared photons and gravitons," Phys. Rev. **140**, B516-B524 (1965).

[15] S. B. Giddings, M. Srednicki, "High-energy gravitational scattering and black hole resonances," Phys. Rev. **D77**, 085025 (2008). [arXiv:0711.5012 [hep-th]].

[16] D. M. Eardley, S. B. Giddings, "Classical black hole production in high-energy collisions," Phys. Rev. **D66**, 044011 (2002). [gr-qc/0201034].

[17] P. C. Aichelburg, R. U. Sexl, "On the Gravitational field of a massless particle," Gen. Rel. Grav. **2**, 303-312 (1971).

[18] S. B. Giddings, M. Schmidt-Sommerfeld, J. R. Andersen, "High energy scattering in gravity and supergravity," Phys. Rev. **D82**, 104022 (2010). [arXiv:1005.5408 [hep-th]].

[19] Z. Bern, L. J. Dixon, R. Roiban, "Is N = 8 supergravity ultraviolet finite?," Phys. Lett. **B644**, 265-271 (2007). [hep-th/0611086].

[20] Z. Bern, L. J. Dixon, D. C. Dunbar, M. Perelstein, J. S. Rozowsky, "On the relationship between Yang-Mills theory and gravity and its implication for ultraviolet divergences," Nucl. Phys. **B530**, 401-456 (1998). [hep-th/9802162].

[21] S. Weinberg, Ultraviolet Divergences In Quantum Theories Of Gravitation, in *General Relativity: An Einstein centenary survey*, eds. S. W. Hawking and W. Israel, Cambridge University Press (1979).

[22] M. Niedermaier, M. Reuter, "The Asymptotic Safety Scenario in Quantum Gravity," Living Rev. Rel. **9**, 5 (2006).

[23] G. Dvali, S. Folkerts, C. Germani, "Physics of Trans-Planckian Gravity," [arXiv:1006.0984 [hep-th]].

[24] M. J. Duff, "Covariant gauges and point sources in general relativity," Annals Phys. **79**, 261-275 (1973); "Quantum Tree Graphs and the Schwarzschild Solution," Phys. Rev. **D7**, 2317-2326 (1973).

[25] S. W. Hawking, "Particle Creation by Black Holes," Commun. Math. Phys. **43**, 199-220 (1975).

[26] S. W. Hawking, "Breakdown of Predictability in Gravitational Collapse," Phys. Rev. **D14**, 2460-2473 (1976).

[27] A. Strominger, "Les Houches lectures on black holes," [hep-th/9501071].

[28] S. B. Giddings, Quantum mechanics of black holes, arXiv:hep-th/9412138; The black hole information paradox, arXiv:hep-th/9508151.

[29] C. G. Callan, Jr., S. B. Giddings, J. A. Harvey, A. Strominger, "Evanescent black holes," Phys. Rev. **D45**, 1005-1009 (1992). [hep-th/9111056].

[30] S. B. Giddings, W. M. Nelson, "Quantum emission from two-dimensional black holes," Phys. Rev. **D46**, 2486-2496 (1992). [hep-th/9204072].

[31] D. A. Lowe, J. Polchinski, L. Susskind, L. Thorlacius, J. Uglum, "Black hole complementarity versus locality," Phys. Rev. **D52**, 6997-7010 (1995). [hep-th/9506138].

[32] S. B. Giddings, "Nonlocality versus complementarity: A Conservative approach to the information problem," Class. Quant. Grav. **28**, 025002 (2011). [arXiv:0911.3395 [hep-th]].

[33] T. Banks, L. Susskind, M. E. Peskin, "Difficulties for the Evolution of Pure States Into Mixed States," Nucl. Phys. **B244**, 125 (1984).

[34] L. Susskind, "Trouble for remnants," [hep-th/9501106].

[35] S. B. Giddings, "Why aren't black holes infinitely produced?," Phys. Rev. **D51**, 6860-6869 (1995). [hep-th/9412159].

[36] D. N. Page, "Information in black hole radiation," Phys. Rev. Lett. **71**, 3743-3746 (1993). [hep-th/9306083].

[37] S. B. Giddings, "Quantization in black hole backgrounds," Phys. Rev. **D76**, 064027 (2007). [hep-th/0703116 [HEP-TH]].

[38] S. B. Giddings, D. Marolf, "A Global picture of quantum de Sitter space," Phys. Rev. **D76**, 064023 (2007). [arXiv:0705.1178 [hep-th]].

[39] N. Arkani-Hamed, talk at the KITP conference *String phenomenology 2006*.

[40] N. Arkani-Hamed, S. Dubovsky, A. Nicolis, E. Trincherini, G. Villadoro, "A Measure of de Sitter entropy and eternal inflation," JHEP **0705**, 055 (2007). [arXiv:0704.1814 [hep-th]].

[41] S. B. Giddings, M. S. Sloth, "Semiclassical relations and IR effects in de Sitter and slow-roll space-times," JCAP **1101**, 023 (2011). [arXiv:1005.1056 [hep-th]].

[42] S. B. Giddings, M. S. Sloth, "Cosmological observables, IR growth of fluctuations, and scale-dependent anisotropies," [arXiv:1104.0002 [hep-th]].

[43] C. P. Burgess, R. Holman, L. Leblond, S. Shandera, "Breakdown of Semiclassical Methods in de Sitter Space," JCAP **1010**, 017 (2010). [arXiv:1005.3551 [hep-th]].

[44] Private conversations with multiple prominent string theorists.

[45] G. Veneziano, "String-theoretic unitary S-matrix at the threshold of black-hole production," JHEP **0411**, 001 (2004). [hep-th/0410166].

[46] S. B. Giddings, "Locality in quantum gravity and string theory," Phys. Rev. **D74**, 106006 (2006). [hep-th/0604072].

[47] S. B. Giddings, D. J. Gross, A. Maharana, "Gravitational effects in ultrahigh-energy string scattering," Phys. Rev. **D77**, 046001 (2008). [arXiv:0705.1816 [hep-th]].

[48] J. Polchinski, "S matrices from AdS space-time," [hep-th/9901076].

[49] L. Susskind, "Holography in the flat space limit," [hep-th/9901079].

[50] S. B. Giddings, "Flat space scattering and bulk locality in the AdS / CFT correspondence," Phys. Rev. **D61**, 106008 (2000). [hep-th/9907129].

[51] M. Gary, S. B. Giddings, J. Penedones, "Local bulk S-matrix elements and CFT singularities," Phys. Rev. **D80**, 085005 (2009). [arXiv:0903.4437 [hep-th]].

[52] M. Gary, S. B. Giddings, "The Flat space S-matrix from the AdS/CFT correspondence?," Phys. Rev. **D80**, 046008 (2009). [arXiv:0904.3544 [hep-th]].

[53] I. Heemskerk, J. Penedones, J. Polchinski, J. Sully, "Holography from Conformal Field Theory," JHEP **0910**, 079 (2009). [arXiv:0907.0151 [hep-th]].

[54] A. L. Fitzpatrick, E. Katz, D. Poland, D. Simmons-Duffin, "Effective Conformal Theory and the Flat-Space Limit of AdS," [arXiv:1007.2412 [hep-th]].

[55] S. B. Giddings, "Is string theory a theory of quantum gravity?," invited contribution to *Forty Years of String Theory: Reflecting on the Foundations*, a special issue of *Found. Phys.*, to appear.

[56] O. Aharony, S. S. Gubser, J. M. Maldacena, H. Ooguri, Y. Oz, "Large N field theories, string theory and gravity," Phys. Rept. **323**, 183-386 (2000). [hep-th/9905111].

[57] M. Reed and B. Simon, *Methods Of Mathematical Physics. Vol. 3: Scattering The- ory,* New York, USA: Academic (1979) 463p.

[58] L. Susskind, L. Thorlacius, J. Uglum, "The Stretched horizon and black hole complementarity," Phys. Rev. **D48**, 3743-3761 (1993). [hep-th/9306069].

[59] L. Susskind and J. S. Lindesay, *Black Holes, Information, and the String Theory Revolution: The Holographic Universe* (Singapore, World Scientific, 2005).

[60] P. Hayden, J. Preskill, "Black holes as mirrors: Quantum information in random subsystems," JHEP **0709**, 120 (2007). [arXiv:0708.4025 [hep-th]].

[61] Y. Sekino, L. Susskind, "Fast Scramblers," JHEP **0810**, 065 (2008). [arXiv:0808.2096 [hep-th]].

[62] S. D. Mathur, "Fuzzballs and the information paradox: A Summary and conjectures," [arXiv:0810.4525 [hep-th]].

[63] S. B. Giddings, "Black holes and massive remnants," Phys. Rev. **D46**, 1347-1352 (1992). [hep-th/9203059].

[64] G. Veneziano, "A Stringy Nature Needs Just Two Constants," Europhys. Lett. **2**, 199 (1986).

[65] D. J. Gross, "Superstrings And Unification," PUPT-1108 *Plenary Session talk given at 24th Int. Conf. on High Energy Physics, Munich, West Germany, Aug 4-10, 1988.*

[66] W. Fischler, L. Susskind, "Holography and cosmology," [hep-th/9806039].

[67] R. Bousso, "A Covariant entropy conjecture," JHEP **9907**, 004 (1999). [hep-th/9905177].

[68] S. Coleman, "Acausality," In *Erice 1969, Ettore Majorana School On Subnuclear Phenomena*, New York 1970, 282-327.

[69] S. B. Giddings and M. Lippert, "Precursors, black holes, and a locality bound," Phys. Rev. D **65**, 024006 (2002) [arXiv:hep-th/0103231];
"The information paradox and the locality bound," Phys. Rev. D **69**, 124019 (2004) [arXiv:hep-th/0402073];
S. B. Giddings, "Black hole information, unitarity, and nonlocality," Phys. Rev. D **74**, 106005 (2006) [arXiv:hep-th/0605196];
"(Non)perturbative gravity, nonlocality, and nice slices," Phys. Rev. D **74**, 106009 (2006) [arXiv:hep-th/0606146].

[70] S. B. Giddings, R. A. Porto, "The Gravitational S-matrix," Phys. Rev. **D81**, 025002 (2010). [arXiv:0908.0004 [hep-th]].

[71] M. Gell-Mann, M. L. Goldberger, W. E. Thirring, "Use of causality conditions in quantum theory," Phys. Rev. **95**, 1612-1627 (1954).

[72] A. Martin, "Extension of the axiomatic analyticity domain of scattering amplitudes by unitarity. 1.," Nuovo Cim. **A42**, 930-953 (1965).

[73] G. Dvali, G. F. Giudice, C. Gomez, A. Kehagias, "UV-Completion by Classicalization," [arXiv:1010.1415 [hep-ph]].

[74] J. Bros, H. Epstein, and V. Glaser, "A proof of the crossing property for two-particle amplitudes in general quantum field theory," Comm. Math. Phys. **1**, 240 (1965).

[75] S. B. Giddings, "Universal quantum mechanics," Phys. Rev. **D78**, 084004 (2008). [arXiv:0711.0757 [quant-ph]].

DISCUSSION I

- *Burda:*

I have a question regarding the information loss paradox. Suppose that we have two objects, say toasters, with some non-trivial structure. Assume they collide in such a way that we know a black hole forms. You said the black hole will evaporate with a thermal spectrum and, therefore, information will be lost. But if we have two big toasters with some structure they will not form a spherical black hole. The radiation will then not be thermal, but more complicated, e.g. there will be different harmonics. So the information about the toasters will be encoded in this complicated radiation. At much later times, there might be a spherical black hole, but by this time a lot has happened so I don't think there is a paradox.

- *Giddings:*

People have thought about whether the information could come out in, say, the "balding" radiation emitted as the black hole loses the multipole moments resulting from the detailed configuration of the incoming state. In fact, there is an argument, which I outlined, based on the principles of local quantum field theory, that this is not sufficient to get the information out. And if you make a black hole from very simple objects, say, even via spherical collapse, it turns out that you end up with a lot of missing information according to the naive Hawking argument, just from missing correlations between outgoing Hawking radiation and the correlated excitations that fall into the singularity. So you really do seem to have a serious problem of missing information, if you analyze things in the naive traditional fashion.

- *Alba*:

You said you would like to calculate the S-matrix for the ultraplanckian regime and that in some part of the phase diagram you can just use general relativity coupled to scalars. You computed some loop diagrams. But it seems that if the energies are large, the main effect should maybe come from some non-perturbative effects and so your calculation is inappropriate.

- *Giddings:*

Let's look at the relevant slide. We should all be interested in the regime of very high energies. Of course, at high energies and short distances gravity gets strong. But we actually know that there is a situation where gravity is quite weak and in fact very classical, that is also a very high energy situation, and that is in the collision of the earth and the moon. The energy of the earth and the moon are very very very large in Planck units. Of course, they are also separated by macroscopic distances, so that gravity is relatively weak. So that is the basic idea to think about. More precisely, you can see that even in the high energy limit, for large enough impact parameters, the leading contribution comes from the Born approximation and this is all you really need to describe gravity. As you decrease the impact parameter and go into the next region in the energy-impact parameter plane, you need some of the loop contributions and in essence this is just building up the classical weak field. Earth-moon scattering is in this eikonal regime rather than the Born regime, but this is not really strong gravity -- certainly graviton loop contributions are not important. It is tree-level exchange of multiple gravitons between the very high energy sources. So in fact this is not a regime where gravity is becoming strongly non-perturbative. In a sense it is non-perturbative, but it is really iterated perturbative single graviton exchange.

- Alba:

You said that every theory of quantum gravity should give us an expression for the S-matrix. How can we check that the expression is correct, because in principle we can invent some bad theory of quantum gravity, which isn't correct. What shall we do if we have two theories of quantum gravity that give us different answers for the S matrix. How can we check the answer?

- Giddings:

That is a very good question. The statement that a theory of quantum gravity should give the S-matrix is based on very general principles and observations: there are particle states in nature, and we can create them at very large separations. It is also based on the assumption of Lorentz invariance, at least if you are interested in the S-matrix at very high energies. You can boost things differently at very distant points and thus arrange a very high energy collision. You might then say can we experimentally test or are there other checks if we have two theories that give us an S-matrix, but that is not really the problem. The problem is that we do not have one theory that gives us a fully consistent S-matrix and in fact when we contemplate what would be involved in a theory giving us a consistent S-matrix, particularly in this very high energy regime, we really run into some profound difficulties, so even one would be a good start and there are strong constraints on what kind of theory could possibly do that.

- Alba:

I have the impression we would like to compute the S-matrix and then from its analytical properties we would like to invent our theory of quantum gravity.

- Giddings:

Well, there is a question of how to go about quantizing gravity and it is a very thorny subject. But part of the question is what are physical quantities in quantum gravity and for example the expectation value of the metric isn't really a diffeomorphism (hence gauge) invariant quantity, so one of the motivations for thinking about the S-matrix is that it should be a sharply defined object and then we can try to study it and its properties. If the S-matrix is describing some of the basic features of gravity and is unitary that should give us some important clues about the ultimate structure of the theory. We should also think about other approaches, for example you can try to investigate how you would get out more familiar types of observables like the approximately local observables that we need to describe cosmology. We need various clues, I think, in trying to figure out the quantum structure of gravity. It is not obvious to me how we are going to do that, but another parallel is that I think we are in a situation a little bit like that when classical physics was collapsing and we hadn't yet invented quantum mechanics. There one had data, which we are a little scarce on, but on the other hand we have basic consistency principles to help guide us. Also, one might say, well, don't we already have that theory, it is string theory? That is a very good question; string theory might answer these questions, but in this high energy regime there are some profound puzzles about whether string theory is actually giving us answers.

- Dunin-Barkowski:

The argument with the black hole suggests that we cannot probe spacetime at distances shorter than certain distances. Are there any theories of gravity that use some new geometry of spacetime that have the inherent property that there cannot be distances shorter than something?

- Giddings:

This is a very interesting question and there have been such proposals; for example non-commutative geometry suggests some kind of minimal length, and there have been other attempts to formulate statements about there being a minimal length in nature. That suggests that there really is some limit on the impact parameter, that you can't go to shorter distances than, say, the Planck scale. One would like to somehow parameterize that statement. One of the things we run into, however, is that there are very strong constraints from Lorentz invariance. If you think about the statement that you have a shortest distance, and consider a boost, the shortest distance becomes shorter. So how do we reconcile the statement that there is some limitation on the resolution with Lorentz invariance? I believe that a correct way of formulating this is probably not as a limitation on, say, the configuration space of a single particle like in quantum mechanics, but that the limitations are on two- or multi-particle states. One attempt to parameterize such a limitation is what I call a locality bound and the idea is that if you have a two-particle state that has a high enough center of mass energy and at the same time you are trying to collide the two particles at a small enough impact parameter, then you encounter fundamental limitations. I should say that there is a question of how to implement this in a consistent mathematical formalism and this is a really good and profound question.

- Schmidt-Sommerfeld:

Maybe you can tell us about your point of view on the following: Does it make a difference whether we scatter the earth and the moon, which are big objects where we expect classical physics to apply as opposed to electrons boosted to very high energies where we expect quantum effects to be important?

- Giddings:

Let us first consider one intermediate situation and that is suppose the earth and the moon were much smaller, i.e. very massive particles, and there I think the scattering problem is essentially the same. Then you can ask whether the relativistic situation, where you have very highly boosted particles is somehow fundamentally different from the non-relativistic situation, where you just have big masses. We are still investigating all the features of this kind of scattering and it looks so far that the ultrarelativistic situation is a very close parallel to the non-relativistic situation where you have heavy masses. There are other issues like radiation and so on which are somewhat detail-dependent, but my coarse-grained answer would be that these are the same kind of problem.

- Alba:

You said it is in principle possible to use non-commutative geometry. Non-commutative field theories are in general highly non-local, but it seems that gravity is quite local, so what should we do with non-locality, which we obtain when we use non-commutative geometry.

- Giddings:

I mentioned non-commutative geometry because that is what some people have tried and proposed. I am not advocating that because I see other issues and difficulties with non-commutative geometry. On the other hand, this statement about non-locality is an important thing to consider. The difficulty we get into in trying to describe unitary evolution that corresponds to formation and evaporation of black holes strongly suggest that a) we have to modify some basic principles of physics, we really are in a situation like the downfall of classical mechanics where something has to change and that b) the easiest way out, i.e. the softest principle that we might suspect should be gotten rid of or modified is that of locality. In fact, in quantum gravity there is no precise formulation of locality, for example we don't have local gauge-invariant observables and there are various other reasons based on what is going on with the black hole information paradox that suggest that we have to abandon locality. We better not abandon locality completely, because locality is intimately tied up with causality and with consistency, so there is a very critical tension: how can the theory ultimately have some degree of non-locality, but still all the good properties, in particular match onto local quantum field theory in familiar regimes.

- Arnowitt:

In Yang-Mills, the vacuum expectation value of the vector field A is also not gauge invariant, yet with much effort one can prove the S-matrix is gauge invariant. Is the difference with general relativity the fact of black hole formation or is there some other pathology?

- Giddings:

It is not necessarily the question of black holes. For example in Yang-Mills there are gauge invariant local observables, say the trace of the field strength tensor, so it is a different kind of issue. In fact, you might think that in gravity you should form something like the trace of the field strength tensor, and try to write something diffeomorphism invariant, and when you do so you apparently have to write down things that are not local objects.

Scientific Secretaries: A. Khmelnitskiy, M. Schmidt-Sommerfeld

DISCUSSION II

- *Mulhearn:*
Could you explain the nice slice diagram?

- *Giddings:*
First of all it's a Penrose diagram that is supposed to be representing the geometry of an evaporating black hole. I haven't drawn collapsing matter, but here are the singularity, the horizon and future null infinity. The Hawking radiation is produced in the vicinity of the horizon. It evaporates from the black hole and ultimately reaches null infinity, one quantum at a time, roughly speaking. The basis of the argument for missing information involves drawing space-like slices through the Hawking radiation. When you actually describe not just the state of the Hawking radiation but the state on this entire slice, you find that this radiation is not alone. If you start with the vacuum on some spatial slice down here and evolve that, it turns into something where your outgoing Hawking quantum is paired with excitations going into the singularity, in a semi-classical picture. So that is what this state represents. I have written it in a particular basis and it is a little bit schematic -- one can be more detailed. It involves excitation inside and outside the black hole. If we want a description of Hawking radiation we form a density matrix, which we get by taking the product of two copies of the nice slice state traced over the internal degrees of freedom. And this density matrix describes a mixed state that has missing information corresponding to that basically carried by the internal degrees of freedom. A characterization of this missing information is the entropy of the density matrix, which is Von Neumann entropy: $\mathrm{Tr}(\rho \ln \rho)$

- *Burda:*
Could you say a few words about non-local observables, which, you said, are proper observables of Quantum Gravity?

- *Giddings:*
First of all, local observables cannot be gauge invariant in gravity, where the gauge symmetry is diffeomorphisms. Local observables sit at a point, and diffeomorphisms move points around, roughly speaking. The question is: "what should you do?" There is a possible analogue in gauge theory, where you can take a trace over the colour indices. Doing an integral is a little bit like taking this trace over colour indices. So this is a proposal of how you can write down gauge invariant objects that are however non-local because you have integrated over space-time. The idea is that if you have an operator acting on a background field, you can prepare a certain background state so that this operator is peaked in a certain region -- with a state that is diffeomorphism invariant. Technically say, it satisfies the Wheeler-De Witt equation. But it essentially

determines a feature, that is, tells you that there is a special location in your space-time. And if you have such a state and this operator that is sensitive to this state and you do this integral with another operator, then you are going to pick out the value of the second operator where the feature in the state is. Of course implementing this on the quantum level is very difficult and ultimately it is only a semi-classical or effective field theory description. It is what people really do in inflationary cosmology, where there is the same issue of gauge (non)invariance and you refer things to the reheating time when you exit inflation. Implementing it in the quantum theory of gravity involves knowing more about the underlying theory. We are trying to bootstrap up and get hints about how you can write down any kind of sensible observables. There is also a toy-model where you can explicitly study such things and see some properties that I have just described implemented in detail. If you think about living on world sheet of the string, with two-dimensional gravity, that is a system people in string theory know quite well how to treat mathematically. This provides an interesting example of these basic statements and that is something I worked out with my student Michael Gary. We have shown how it serves as an example of some of these basic features, where you can prepare certain states so you can recover, at least in an approximation, local observables. Developing this story further is an open question; it is one of the things to be done.

- Danish Azmi:

Does the thermal radiation from black holes contain information about their inner structure as suggested by gauge/gravity duality?

- Giddings:

I think that is a big mystery: how does the Hawking radiation end up carrying information about the interior of the black hole. In the semi-classical calculation of Hawking, as we see in the nice slice argument, there is no possible way that outgoing radiation could contain the information. The idea of AdS/CFT duality suggests that the information in fact can be imprinted in Hawking radiation. It does not give an explicit implementation of these ideas explaining how it happens. That is one of the reasons why I have been trying to make these questions and answers sharper in the AdS/CFT context and in particular tried to understand the problem how to extract the gravitational S-matrix in the AdS/CFT context where you would like to describe that incoming matter forms a black hole and produces outgoing Hawking radiation. And if AdS/CFT gives you the unitary description of that process, then you would like to be able to take that apart and understand what it is telling you about how the information gets imprinted into Hawking radiation, what is wrong with the initial Hawking argument. But before that you have to understand how to get the S-matrix from AdS/CFT, and there are non-trivial obstacles to that. And it is not clear that those obstacles will be surmounted. For those who believe that AdS/CFT does furnish the answers to these questions this is a very very important problem, to understand how it does so. So we do not have a theory that describes in detail how the information gets out, but I think there is a pretty good reason to believe it does. The question is what physics is responsible for that. And that is a big mystery and may involve, as I was explaining, new principles that possibly go beyond anything we know now. Another question one may ask is what we are supposed to make of this in terms of practical suggestions on how to proceed. We really do seem to be at a juncture in physics that is analogous to the quandary that we were in with classical physics with the catastrophe of the collapse of the atom. We apparently really need a significant jump and implementation of new principles to understand what is going on here. The fact that we could be in that situation is very exciting.

- Haidt:

You have argued about locality. Shouldn't you at the same time argue about the meaning of the "event"?

- Giddings:

Well I did not include events in my discussion, and it is definitely not clear that in a theory where you have no fundamental notion of locality, you have the notion of an event as a fundamental concept. You may have some quantum description of the theory, whatever it is, such that events are not part of it in a precise fashion -- that could well be.

- Korthals-Altes:

Does the back-reaction of the Hawking radiation play any role in the "information crisis"? After all the black hole gets lighter by radiating so the back-reaction has to kick in.

- Giddings:

That has been calculated on a semi-classical, averaged level and there is a very precise implementation of that, which I was involved in many years ago, in a two-dimensional model, the so-called CGHS model. There you can be quite systematic in writing down a quantum stress tensor describing the Hawking radiation. That quantum stress tensor gives you corrections to the Einstein's equations, basically, so you can see the back-reaction. A lot of people have looked at the evolution of that. In practice it is difficult in, say, four dimensions, or in higher dimensions. But in principle I do not see any obstacle there to similarly having the Einstein tensor sourced by the quantum stress tensor of Hawking radiation. There is one subtle issue which is not settled down until now. If you are really trying to compute the quantum state of the nice slice after it has evolved for a very long time then the back-reaction of the radiation on the nice slice can become significant depending on what was emitted earlier. I've given arguments based on this that the perturbative expansion is breaking down when you try to compute the state on the nice slice.

- 't Hooft:

This is a comment rather than a question. Once you assume the existence of a unitary S-matrix relating all in-going things to all out-going things, to see that Hawking radiation has imprinted in it the information of ingoing matter making the black hole is rather straightforward. You must start with the Ansatz that all amplitudes form a unitary matrix. *Then*, making any change in the ingoing matter, like creating or annihilating a particle changes the paths of the Hawking particles and you can calculate that. The effect, particularly the effect of the gravitational force is *big*. The only problem left is that the Hilbert space of all states you get is too big; we want only one piece of information per

Planck length squared.

- Giddings:

I think we have different intuitions and we are taking our guidelines from different places. Let me give you at least a counterpoint. This is based on calculations done in quantum field theory in the black hole background. Actually at one point I thought that there was such an important effect based on these calculations, but then I convinced myself that I was wrong. First of all the basic picture of the Hawking radiation if you think about quantum field theory in a curved space is that you have something that looks, for, say, an infalling observer, like vacuum and that evolves over time into this correlated pair of quanta, that ultimately separate from the horizon when their wavelengths become comparable with the horizon size. You have an outgoing Hawking quantum emitted to infinity and an excitation that falls into the black hole. If you come in earlier, before they separate, then you trace back this Hawking quantum to a very high-energy excitation. But it is completely mixed up with its partner and if you look at the scattering off both of these excitations it looks like scattering off the vacuum, which has no effect. On the other hand if you wait until these quanta have separated, these quanta reach the wavelength of order of the radius of the black hole. And then you do not have big scattering because it is no longer a high energy situation. That is the basic picture I have for why you do not get an important imprint from the infalling quanta changing the state.

- 't Hooft:

That is because you assumed that Hawking quanta originate at those points, but for the S-matrix argument to work you have to follow the Hawking particles' wave functions back to beyond the moment when the black hole is actually formed. And then they are actually very thick, they carry very high momenta. Maybe it just does not look very physical, but their gravitational interaction is very strong.

- Giddings:

Let me just give you one more statement about the calculation. I do not know if it helps to settle this or not. What is the leading interaction between the infalling particle and any state that sits here? Say "stress tensor – graviton exchange – stress tensor". So in order to compute the interaction we need the stress tensor of the state describing the Hawking particles. And if you look at this stress tensor for the outgoing Hawking particle it looks like it is huge. But there are cancellations. When you combine the outgoing Hawking particle and its partner inside their contributions cancel.

- 't Hooft:

The partner is not really there, the partner cannot be seen outside of the horizon and it plays no role. If you have the other particle, all the problems will arise.

- Giddings:

But I think there is a sensible picture when you keep it in.

- *'t Hooft:*

Then we are in trouble.

- *Giddings:*

We know that we are in trouble. The question is how to get out.

- *Galakhov:*

Consider a charged black hole. It looks like it can't evaporate to nothing due to charge conservation. Suppose we throw some information into the black hole; it is stored inside until the horizon disappears. Maybe you should just regularize with a small charge instead of treating it as a standard black hole?

- *Giddings:*

I am not sure I understood the actual question. I could try to describe what happens with charged black holes. First of all, if you have electrically charged black holes, they typically discharge, at least the small ones, because the electron is very light and has a large charge to mass ratio. So actually, the small ones neutralize. A way around that is to think about magnetically charged black holes and those might, in ordinary QED or grand unified theories, be stable. So then one can ask what happens if you form a black hole with magnetic charge. It is going to evaporate down to extremality and stop. That is a little bit like a remnant. Then you can ask what happens when you throw more stuff in and it becomes non-extremal again. It will evaporate back down. If information is actually returned in the Hawking radiation, you would expect that the information comes out when it radiates and you always come back to a state with a certain fixed amount of information or a fixed number of internal states, roughly speaking. On the other hand if you have something like a remnant scenario in the neutral sector, then in the charged sector you end up with an infinite number of internal states and that is actually where you see most precisely these infinite pair production issues.

- *Galakhov:*

Well, it looks like if you consider such black holes you will not have a problem with loss of information.

- *Giddings:*

Because the information is stored inside the black hole? Well, there the problem with Hawking's argument is that you say the information does not get out. You can throw in more information many times and it does not come out. If you keep doing that, you can put an arbitrarily large amount of information into the black hole and that means it has an infinite number of internal states, which implies for pair production in a macroscopic magnetic field, say via the Schwinger mechanism, that the rate is a tiny number times that degeneracy factor, the number of internal states, which is infinite. So that is where things go wrong if you assume that there are remnants.

- *Azmi :*

Can you please explain what non-renormalisability of gravity is?

- Giddings:

Because of the dimensionful coupling constant, you find that you need to introduce counterterms with higher and higher numbers of the fields and derivatives, and there is no a priori way of fixing the coefficients of those counterterms - they can be anything. That is the essence of the problem and it means that if you are trying to make predictions for what happens in scattering near the Planck scale, you need to know the coefficients of all those counterterms, but there is no principle that fixes them - barring some magic like string theory or supergravity. So you have no way to make predictions.

- Borsten:

I wondered what your thoughts on the fuzzball proposal as an approach to the information loss paradox are.

- Giddings:

First of all, you have to ask what the fuzzball picture would really be saying in the context of something like a Schwarzschild black hole. I think that a fuzzball is a realization of a proposal, that came many years ago, of a massive remnant that has the property that if you try to explore the region near the horizon, it looks like something very different from a black hole. It seems inevitable to me that if you really have an implementation of that proposal for macroscopic Schwarzschild black holes the proposed sum over microstates would produce something that, if you fell into it, would be painful to you. So it does not look like a nice ordinary horizon. That is something we might consider, and it has been considered before. There are various questions about how that actually gets implemented, how that object forms from something that initially looks like a Schwarzschild black hole. I think it has not been sorted out at any level of detail how such a radical picture would work. It is perhaps in some sense more radical than we need, but who knows.

- Alba:

You told us that we should introduce non-locality in gravity and some non-local observables. So does that mean that other interactions have to be non-local, too? E.g. high energy limits of the standard model. Do we have experimental constraints?

- Giddings:

Non-locality in general is a very dangerous thing. In many ways it seems like a very foolish thing. Most implementations of non-locality immediately lead to inconsistencies due to acausality and so on. So you might ask why one would even consider such silly thoughts. But it turns out that this seems to be - barring someone proposing something else - the most conservative way out of this dilemma that we are in. Where should it manifest itself? It seems that it better not manifest itself in most usual circumstances. Effective field theory, e.g. that of the standard model, is very much based on locality, and locality is manifest in all ordinary circumstances and throwing it out gets you into trouble. So if this is the right answer - it seems to me the most plausible one - to this set of puzzles, it has to be a very subtle kind of non-locality implemented in the basic principles of gravity. Ultimately, you can't isolate gravity from everything else, but if you could turn off gravity, you would not necessarily expect any modification to the framework of local quantum field theory.

- Alba:

Do you know some framework in which we can force non-locality in some theory to disappear by introducing something else?

- Giddings:

One naive way of getting something that looks non-local is to integrate out fields corresponding to excitations in extra dimensions. There are various ideas along these lines. I think what we are after is probably something different.

- Alba:

Do we have some toy model where we can describe some non-local thing in terms of local theory.

- Giddings:

If you have a theory with some fields coupled to one another, one of which is massless, and you integrate it out, you get a non-local theory. That is a trivial example, but you get something that looks non-local and that is because you integrated out the massless field. But I don't think that is what we should take as an example for what we are after here.

- Alba:

Can you consider radiation in even dimensions?

- Giddings:

You are thinking about the tails? I think that is a standard story, but I am not sure how it relates to the current story of what is the issue with black holes.

- Alba:

But it is also non-local.

- Giddings:

Not in the same sense. There is a precise implementation of locality in quantum field theory and the precise sharp statement is commutativity of local observables outside the light cone. And that can be made in any dimension. But that is something that we don't know how to implement in gravity and that is another clue I think.

Y. DECLAIS

"Direct Evidence of Oscillation from II to III Family Neutrinos"

The Slides of the Lecture can be found at:

http://www.ccsem.infn.it/issp2010/

Scientific Secretaries: A. Vinogradov, C. Pagnutti

DISCUSSION

- Preghenella:

Are there any future plans for neutrino physics with the CERN beam in Italy?

- Declais:

It's a complicated story. The plan now for neutrino physics would be the search for the cross-mixing matrix for the measurement of theta-13. To measure this effect you need a high intensity beam. And unfortunately at the time being the CERN machines in place are not designed in order to provide a high intensity beam. That has been some plan in the U.S., and we will probably hear Pier Oddone's discussion about the Project X. There is now an experiment running in Japan with Super-K and a new high power accelerator in JPARC designed in order to perform those measurements. So at the time being there have been some proposals to reduce the beam energy in order to be at the first peak, as I said this morning. In order to get a good optimization for Nu_tau detection it was needed to go to higher energy, but with the distance between CERN and LNGS that was not at the maximum. But, unfortunately, I would say that it's not really competitive with respect to the facility at SUPER-K and the coming facility at Fermilab. With the CNGS I'm afraid that probably – this is my personal point of view – for sure there has been lots of discussions, there has been many proposals, but I didn't see any interesting proposal which can compete with SUPER-K or NOvA with new high intensity machines.

- Grelli:

I'm not really familiar with neutrino experiment. During your presentation this morning, at a certain point, you talked about charm, and then you showed a plot. The title of the plot was charm decay event. How are they detected, the charm events? And what does it mean in this case, charm?

- Declais:

It's a CC production. It means, as it is shown in charged current, or in neutral current, a pair of c,c-bar. And the topology for those events are discerned since you have multi-pronged events, and the p_t is not the same. So it's quite easy to distinguish between tau decay and a charm. But what is interesting with charm, the cross section is rather well known, there has been a lot of measurement, and especially at CERN with the WANF neutrino beam. So this is in fact a perfect sample for us in order to measure the efficiency of secondary decay detection in our experiment. This is why it is interesting in our case to look at charm. Charm events can only be a background if you do not see the muon in a charged current event, or if you miss one of the charms in a c,c-bar charm event in neutral current. This is also why it is so important to have a good signature of particles, and to get a good signature even at low energy for muons. The main component now for the background is coming from muons at the edge of the detector, which then escapes the detector, and does not allow us to measure free tracks and to get the muon ID.

- Haidt:

You have talked about the charm background. I could think right away of the conventional background coming from hadron production at relative low energies, it's easy to produce a nice kink.

- Declais:

In our case, for the detection of the tau decaying into hadrons, the hadron re-interaction process is the most important background. But the main cut to throw away hadron re-interaction as well as the kaon decay is to use p_t cut, because the kinematics of tau decay allows much larger p_t than for the hadron re-interactions or the kaon decays. This is the main point. What is critical, what we have to prove experimentally is our control of the selection criteria for hadron re-interaction. It's a little bit premature at the time being since we have not measured enough hadrons in our detector in order to measure this probability. But it will be done. We cannot rely only on Monte Carlo for this kind of selection.

- Haidt:

Another thing would be, for example, you get a proton out of the primary interaction. This proton interacts and produces a neutron and a pion. This again would give you a nice kink. It should be very frequent, however observable. You could identify such cases.

-Declais:

Again, we should make some calculations, but the p_t of such interactions would be negligible. The key point is that the kinematics of the tau decay provides a signal with very high p_t.

-Haidt:

The initial proton can have quite a high energy, and if you don't observe the neutron, then you can get kink angles.

-Declais:

I don't think that this can be a background that we have to worry about.

- Preghenella:

Suppose that the MINOS confirms that the antineutrino parameters are different from the neutrino ones. Is there a way to save the CPT then?

- Declais:

I don't know. There have been many papers on this. But it is really a preliminary result. Those experiments are very difficult. You cannot rely on 2-3 sigma effects. We are not able to get 4-sigma. At the time being at super-k they didn't see any difference between the neutrino and the antineutrino. In the case of super-k you can find the charge of the muon: they didn't see any effect. But the statistics isn't very large. This is the same case in [other experiments]. It's a small effect, and we need bigger effects. I hope there is not such a difficulty with ongoing experiments with theta-13, which also appears in experiments with a small signal. The control of the background is very difficult. But MINOS is a two position experiment. You measure in the near detector and the far detector. So we'll see. But, for example, the first results from MINOS for theta-13, two years ago, were positive at 1.5 sigma.

- Mao:

Can you explain how you do particle identification like pion, kaon, proton in your experiment? I am wondering whether they are different from the experiments at the LHC.

- Declais:

The particle ID is made by the dE/dx. The number of grains for a track is related. So you can measure dE/dx. And this is one possibility to get particle ID. The second point is made by the Multiple Coulomb Scattering (MCS), and we can a distinction between muons and hadrons. And we can also access the momentum of the tracks in using this method. What is important in our case is to be able to take particle ID below 10 GeV. What is important is the muon ID. The distinction between pion, kaon is not really important in our case. It's a coarse-grain detector. You have one millimeter of lead. For example, what we plan to do for complicated events, in order to be sure that we cannot miss low energy muons, is to measure the dE/dx before the stopping point of the muons. So we can follow the muon into our detector, into the brick, and measure dE/dx. We performed tests to validate the method, and it works. So we can get muon ID with respect to pion in that case. But it is not as good as with a TPC or something like that.

ITP-UU-10/30
SPIN-10/25
arXiv:1009.0669v2 [gr-qc]

Probing the small distance structure of canonical quantum gravity using the conformal group

Gerard 't Hooft

Institute for Theoretical Physics
Utrecht University
and

Spinoza Institute
Postbox 80.195
3508 TD Utrecht, the Netherlands
e-mail: g.thooft@uu.nl
internet: http://www.phys.uu.nl/~thooft/

Abstract

In canonical quantum gravity, the formal functional integral includes an integration over the local conformal factor, and we propose to perform the functional integral over this factor before doing any of the other functional integrals. By construction, the resulting effective theory would be expected to be conformally invariant and therefore finite. However, also the conformal integral itself diverges, and the effects of a renormalization counter term are considered. It generates problems such as unitarity violation, due to a Landau-like ghost, and conformal anomalies. Adding (massive or massless) matter fields does not change the picture. Various alternative ideas are offered, including a more daring speculation, which is that no counter term should be allowed for at all. This has far-reaching and important consequences, which we discuss. A surprising picture emerges of quantized elementary particles interacting with a gravitational field, in particular gravitons, which are "partly classical". This approach was inspired by a search towards the reconciliation of Hawking radiation with unitarity and locality, and it offers basic new insights there.

1. Introduction: splitting the functional integral

The Einstein-Hilbert action of the generally covariant theory of gravity reads

$$S^{\text{total}} = \int \mathrm{d}^4 x \sqrt{-g} \left(\frac{1}{16\pi G_N} R + \mathcal{L}^{\text{mat}} \right) , \tag{1.1}$$

where the matter Lagrangian $\mathcal{L}^{\text{mat}}$ is written in a generally covariant manner using the space-time metric $g_{\mu\nu}(x)$, and G_N is Newton's constant. In this paper, we begin studying the case where $\mathcal{L}^{\text{mat}}$ is conformally symmetric, which means that under a space-time transformation

$$x^{\mu\prime} = C \frac{x^\mu - a^\mu}{(x-a)^2} + b^\mu, \tag{1.2}$$

we have a transformation law for the matter fields such that

$$g_{\mu\nu}(x') = \lambda(x)^2 g_{\mu\nu}(x) ; \qquad \sqrt{-g(x')} \mathcal{L}^{\text{mat}\prime}(x') = \sqrt{-g(x)} \mathcal{L}^{\text{mat}}(x) ; \tag{1.3}$$
$$S^{\text{mat}\prime} = S^{\text{mat}} ,$$

so that in n dimensions, $\qquad \mathcal{L}^{\text{mat}\prime}(x') = \lambda^n \mathcal{L}^{\text{mat}}(x) . \tag{1.4}$

For the conformal transformation (1.2) we have $\lambda(x) = C/(x^\mu - a^\mu)^2$, which leaves flat spacetime flat, but for curved background space-times, where we drop the condition of flatness, $\lambda(x)$ may be any function of x^μ. There are several examples of such conformally invariant matter systems such as $\mathcal{N} = 4$ super-Yang-Mills theory in $n = 4$ space-time dimensions. We will concentrate on $n = 4$.

We begin by temporarily assuming conformal invariance of the matter fields, only for convenience; later we will see that allowing matter fields to be more general will only slightly modify the picture.

In canonical gravity, the quantum amplitudes are obtained by functionally integrating the exponent of the entire action over all components of the metric tensor at all space-time points x^μ:

$$\Gamma = \int \mathcal{D}g_{\mu\nu}(x) \, \mathcal{D}\varphi^{\text{mat}}(x) \, e^{iS^{\text{total}}} . \tag{1.5}$$

Although one usually imposes a gauge constraint so as to reduce the size of function space, this is not necessary formally. In particular, one has to integrate over the common factor $\omega(x)$ of the metric tensor $g_{\mu\nu}(x)$, when we write

$$g_{\mu\nu}(x) \overset{\text{def}}{=} \omega^2(x) \, \hat{g}_{\mu\nu}(x) , \tag{1.6}$$

where $\hat{g}_{\mu\nu}(x)$ may be subject to some arbitrary constraint concerning its overall factor. For instance, in any coordinate frame one may impose

$$\det(\hat{g}) = -1 , \tag{1.7}$$

besides imposing a gauge condition for each of the $n = 4$ coordinates. The quantity $\hat{g}_{\mu\nu}(x)$ in Eq. (1.7) does not transform as an ordinary tensor but as what could be called a "pseudo"tensor, meaning that it scales unconventionally under coordinate transformations with a non-trivial Jacobian. $\omega(x)$ is then a "pseudo"scalar.

The prefix "pseudo" was put in quotation marks here because, usually, 'pseudo' means that the object receives an extra minus sign under a parity transformation; it is therefore preferred to use another phrase. For this reason, we replace "pseudo" by 'meta', using the words 'metatensor' and 'metascalar' to indicate fields that transform as tensors or scalars, but with prefactors containing unconventional powers of the Jacobian of the coordinate transformation.

Rewriting

$$\int \mathcal{D}g_{\mu\nu}(x) = \int \mathcal{D}\omega(x) \int \mathcal{D}\hat{g}_{\mu\nu}(x) \ , \tag{1.8}$$

while imposing a gauge constraint[1] that *only* depends on $\hat{g}_{\mu\nu}$, not on ω, we now propose *first* to integrate over $\omega(x)$ and then over $\hat{g}_{\mu\nu}(x)$ and $\varphi^{\mathrm{mat}}(x)$. This has peculiar consequences, as we will see.

In the standard perturbation expansion, the integration order does not matter. Also, if dimensional regularization is employed, the choice of the functional metric in the space of all fields $\omega(x)$ and $\hat{g}_{\mu\nu}(x)$ is unambiguous, as its effects are canceled against all other quartic divergences in the amplitudes (any ambiguity is represented by integrals of the form $\int \mathrm{d}^n k\, \mathrm{Pol}(k)$ which vanish when dimensionally renormalized). $\omega(x)$ acts as a Lagrange multiplier. Again, perturbation expansion tells us how to handle this integral: in general, $\omega(x)$ has to be chosen to lie on a complex contour. The momentum integrations may be carried out in Euclidean (Wick rotated) space-time, but even then, $\omega(x)$ must be integrated along a complex contour, which will later (see Section 2) be determined to be

$$\omega(x) = 1 + i\alpha(x) \ , \qquad \alpha \text{ real.} \tag{1.9}$$

If $\omega(x)$ itself had been chosen real then the Wick rotated functional integral would diverge exponentially so that ω would no longer function properly as a Lagrange multiplier.

If there had been no further divergences, one would have expected the following scenario:

- The functional integrand $\omega(x)$ only occurs in the gravitational part of the action, since the matter field is conformally invariant (non-conformal matter does contribute to this integral, but these would be sub dominating corrections, see later).

- After integrating over all scale functions $\omega(x)$, but not yet over $\hat{g}_{\mu\nu}$, the resulting effective action in terms of $\hat{g}_{\mu\nu}$ should be expected to become scale-invariant, *i.e.*

---

[1] A fine choice would be, for instance, $\partial_\mu \hat{g}^{\mu\nu} = 0$. Of course, the usual Faddeev Popov quantization procedure is assumed.

if we would split $\hat{g}_{\mu\nu}$ again as in Eq. (1.6),

$$\hat{g}_{\mu\nu}(x) \overset{?}{=} \hat{\omega}^2(x)\hat{\hat{g}}_{\mu\nu} , \tag{1.10}$$

no further dependence on $\hat{\omega}(x)$ should be expected.

- Therefore, the effective action should now describe a conformally invariant theory, both for gravity and for matter. Because of this, the effective theory might be expected to be renormalizable, or even finite! If any infinities do remain, one might again employ dimensional renormalization to remove them.

However, this expectation is jeopardized by an apparent difficulty: the ω integration is indeed ultraviolet divergent.[1] In contrast with the usual procedures in perturbation theories, it is not associated with an infinitesimal multiplicative constant (such as the coupling constant in ordinary perturbation theories), and so a renormalization counter term would actually represent an infinite distortion of the canonical theory. Clearly, renormalization must be carried out with much more care. Later, in Section 5, we suggest various scenarios.

First, the main calculation will be carried out, in the next section. Then, the contributions from conformal matter fields are considered, and subsequently the effect of non conformal matter, by adding mass terms. Finally, we will be in a position to ask questions about renormalization (dimensional or otherwise). We end with conclusions, and an appendix displaying the details of the matter field calculations.

2. Calculating the divergent part of the scalar functional integral

Calculations related to the conformal term in gravity, and their associated anomalies, date back from the early 1970s and have been reviewed amnong others in a nice paper by Duff[2]. In particular, we here focus on footnote (4) in that paper.

First, we go to n space-time dimensions, in order later to be able to perform dimensional renormalization. For future convenience (see Eq. (2.2)), we choose to replace the parameter ω then by $\omega^{2/(n-2)}$, so that Eq. (1.6) becomes

$$g_{\mu\nu}(x) = \omega^{\frac{4}{n-2}} \hat{g}_{\mu\nu}(x) . \tag{2.1}$$

In terms of $\hat{g}_{\mu\nu}$ and ω, the Einstein-Hilbert action (1.1) now reads

$$S = \int \mathrm{d}^n x \sqrt{-\hat{g}} \left(\frac{1}{16\pi G_N} \left(\omega^2 \hat{R} + \frac{4(n-1)}{n-2} \hat{g}^{\mu\nu} \partial_\mu \omega \, \partial_\nu \omega \right) + \mathcal{L}^{\mathrm{mat}}(\hat{g}_{\mu\nu}) \right) . \tag{2.2}$$

This shows that the functional integral over the field $\omega(x)$ is a Gaussian one, which can be performed rigorously: it is a determinant. The conformally invariant matter Lagrangian is independent of ω, but at a later stage of the theory we shall consider mass terms for matter, which still would allow us to do the functional integral over ω, but as for now we wish to avoid the associated complications, assuming that, perhaps, at scales close to the Planck scale the ω dependence of matter might dwindle.

We use the caret (^) to indicate all expressions defined by the metatensor $\hat{g}_{\mu\nu}$, such as covariant derivatives, as if it were a true tensor.

Note that, in 'Euclidean gravity', the ω integrand has the wrong sign. This is why ω must be chosen to be on the contour (1.9). In practice, it is easiest to do the functional ω integration perturbatively, by writing

$$\hat{g}_{\mu\nu}(x) = \eta_{\mu\nu} + \kappa\, h_{\mu\nu}(x)\,, \qquad \eta_{\mu\nu} = \mathrm{diag}(-1,1,1,1)\,, \qquad \kappa = \sqrt{8\pi G_N}\,, \qquad (2.3)$$

and expanding in powers of κ (although later we will see that that expansion can sometimes be summed). A factor $\sqrt{-\hat{g}}\,8(n-1)/16\pi G_N(n-2)$ in Eq. (2.2) can be absorbed in the definition of ω.[2] This turns the action (2.2) into

$$S = \int \mathrm{d}^n x \sqrt{-\hat{g}} \left(\tfrac{1}{2}\hat{g}^{\mu\nu}\partial_\mu\omega\partial_\nu\omega + \tfrac{1}{2}\frac{n-2}{4(n-1)}\hat{R}\omega^2 + \mathcal{L}^{\mathrm{mat}}(\hat{g}_{\mu\nu}) \right)\,. \qquad (2.4)$$

Regardless the ω contour, the ω propagator can be read off from the action (2.4):

$$P^{(\omega)}(k) = -\frac{1}{k^2 - i\varepsilon}\,, \qquad (2.5)$$

where k_μ is the momentum. The $i\varepsilon$ prescription is the one that follows from the conventional perturbative theory. We see that there is a kinetic term (perturbed by a possible non-trivial space-time dependence of $\hat{g}_{\mu\nu}$), and a direct interaction, "mass" term proportional to the background scalar curvature $\hat{R}$:

$$\frac{n-2}{4(n-1)}\hat{R} \xrightarrow{n\to 4} \tfrac{1}{6}\hat{R}\,, \qquad (2.6)$$

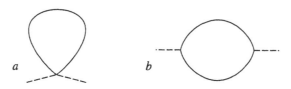

Figure 1: Feynman diagrams for the ω determinant

The most important diagrams contributing to the effective action for the remaining field $\hat{g}_{\mu\nu}$ are the ones indicated in Fig. 1, which include the terms up to $\mathcal{O}(\kappa^2)$. The

[2]Note that, therefore, Newton's constant disappears completely (its use in Eq. (2.3) is inessential). This a characteristic feature of this approach.

"tadpole", Fig. 1a, does not contribute if we apply dimensional regularization, since there is no mass term in the single propagator that we have, Eq. (2.5). So, in this approximation, we have to deal with the 2-point diagram only.

We can compute the integral

$$F(q) \stackrel{\text{def}}{=} \int_{\text{Eucl}} \frac{d^n k}{k^2 (k-q)^2} = \frac{\pi^{\frac{1}{2}n+\frac{3}{2}} 2^{3-n} (q^2)^{\frac{1}{2}n-2}}{\Gamma(\frac{1}{2}n - \frac{1}{2}) \sin \pi(2 - \frac{1}{2}n)} . \tag{2.7}$$

Now we will also need integrals containing extra factors k_μ in the numerator. Therefore, we define

$$\langle k \cdots k \rangle \stackrel{\text{def}}{=} \frac{1}{F(q)} \int_{\text{Eucl}} \frac{d^n k \; k \cdots k}{k^2 (k-q)^2} . \tag{2.8}$$

Then

$$\langle k_\mu \rangle = \tfrac{1}{2} q_\mu ; \tag{2.9}$$

$$\langle k_\mu k_\nu \rangle = \frac{1}{4(n-1)} \left(n q_\mu q_\nu - q^2 \delta_{\mu\nu} \right) ; \tag{2.10}$$

$$\langle k_\mu k_\nu k_\lambda \rangle = \frac{1}{8(n-1)} \left((n+2) q_\mu q_\nu q_\lambda - q^2 (\delta_{\mu\nu} q_\lambda + \delta_{\nu\lambda} q_\mu + \delta_{\lambda\mu} q_\nu) \right) ; \tag{2.11}$$

$$\langle k_\mu k_\nu k_\alpha k_\beta \rangle = \frac{1}{16(n-1)(n+1)} \Big((n+2)(n+4) \, q_\mu q_\nu q_\alpha q_\beta$$
$$- q^2 (n+2) (\delta_{\mu\nu} q_\alpha q_\beta + [5 \text{ terms}])$$
$$+ q^4 (\delta_{\mu\nu} \delta_{\alpha\beta} + \delta_{\mu\alpha} \delta_{\nu\beta} + \delta_{\mu\beta} \delta_{\nu\alpha}) \Big) , \tag{2.12}$$

where the 5 terms are simply the remaining 5 permutations of the previous term.

These expressions can now be used to compute all diagrams that contribute to the ω determinant, but the calculations are lengthy and not very illuminating. More important are those parts that diverge as $n \to 4$. The expression (2.7) for $F(q)$ diverges at $n \to 4$, so that all integrals in Eq. (2.8) diverge similarly. By using general covariance, one can deduce right away that the divergent terms must all combine in such a way that they only depend on the Riemann curvature. Dimensional arguments then suffice to conclude that the coefficients must be local expressions in the squares of the curvature.

The key calculations for the divergent parts have already been performed in 1973 [3]. There, it was found that a Lagrangian of the form

$$\mathcal{L} = \sqrt{-g} \left(-\tfrac{1}{2} g^{\mu\nu}(x) \, \partial_\mu \varphi \partial_\nu \varphi + \tfrac{1}{2} M(x) \, \varphi^2 \right) , \tag{2.13}$$

will generate an effective action, whose divergent part is of the form

$$S^{\text{div}} = \int d^n x \, \Gamma^{\text{div}}(x) , \quad \Gamma^{\text{div}} = \frac{\sqrt{-g}}{8\pi^2(4-n)} \left(\tfrac{1}{120}(R_{\mu\nu} R^{\mu\nu} - \tfrac{1}{3} R^2) + \tfrac{1}{4}(M + \tfrac{1}{6} R)^2 \right) \tag{2.14}$$

(we use here a slightly modified notation, implying, among others, a sign switch in the definition of the Ricci curvature, and a minus sign as ref. [3] calculated the Lagrangian $\mathcal{L} + \Delta\mathcal{L}$, $\Delta\mathcal{L} = -\Gamma^{\text{div}}$ needed to obtain a finite theory.)

In our case, we see that, in the Lagrangian (2.4) (with the dynamical part of ω imaginary),

$$M = -\tfrac{1}{6}\hat{R} , \qquad \Gamma^{\text{div}} = \frac{\sqrt{-\hat{g}}}{960\pi^2(4 - n)}(\hat{R}_{\mu\nu}\hat{R}^{\mu\nu} - \tfrac{1}{3}\hat{R}^2) , \qquad (2.15)$$

since the second term in (2.14) cancels out exactly. Indeed, it had to cancel out, as we will see shortly.

To see what the divergence here means, we use the fact that the mass dependence of a divergent integral typically takes the form

$$C(n)m^{n-4}\Gamma(2 - \tfrac{1}{2}n) \to \frac{C}{4 - n}\left(1 + (n - 4)\log\left(\frac{m}{\Lambda}\right)\right) \to C\left(\log\Lambda + \frac{1}{4 - n}\right) + \text{ finite} ,$$

$$(2.16)$$

where m stands for a mass or an external momentum k, and Λ is some reference mass, such as an ultraviolet cutoff. Thus, the divergent expression $1/(4 - n)$ generally plays the same role as the logarithm of an ultraviolet cutoff Λ.

3. Local scale invariance and the Weyl curvature

Assume for a moment that, after having dealt with the divergent expression (2.15), the functional integral over the conformal variable $\omega(x)$ could somehow be made to produce a finite and meaningful result. We would have a finite effective action Γ^{eff} that is completely conformally invariant, and we would expect to be left with $\hat{g}_{\mu\nu}(x)$ as our remaining dynamical variables.

It is this theory that could be used to handle the black hole complementarity issue. It was explained in Ref.[4] that black hole complementarity means that an observer on his way into a black hole may experience the surrounding space-time differently from what an outside observer sees. They disagree about the back reaction from Hawking radiation, and it was argued that this disagreement must include the metascalar field $\omega(x)$. A completely conformally invariant theory as a starting point could explain this situation; we return to this issue in Section 6.

Thus, we now consider an effective theory with not only general covariance,

$$\hat{g}_{\mu\nu} \to \hat{g}_{\mu\nu} + \hat{D}_\mu u_\nu + \hat{D}_\nu u_\mu , \qquad (3.1)$$

where $u_\mu(x)$ are the generators of infinitesimal coordinate transformations, and $\hat{D}_\mu$ is the covariant derivative with respect to $\hat{g}_{\mu\nu}$; but now we also have a new kind of gauge invariance, being local scale invariance, which we write in infinitesimal notation, for convenience:

$$\hat{g}_{\mu\nu} \to \hat{g}_{\mu\nu} + \lambda(x)\hat{g}_{\mu\nu} , \qquad (3.2)$$

and we demand invariance under that as well. Note that this transformation is quite distinct from scale transformations in the coordinate frame, which of course belongs to

(3.1) and as such is always an invariance of the usual theory. In short, we now have a theory with a 5 dimensional local gauge group. Theories of this sort have been studied in detail[5].

The Riemann tensor $\hat{R}^{\alpha}{}_{\beta\mu\nu}$ transforms as a decent tensor under the coordinate transformations (3.1), but it is not invariant (or even covariant) under the local scale transformation (3.2). Now, in four space time dimensions, we can split up the 20 independent components of the Riemann tensor into the 10 component Ricci tensor

$$\hat{R}_{\mu\nu} = \hat{R}^{\alpha}{}_{\mu\alpha\nu} \, , \tag{3.3}$$

and the components orthogonal to that, called the Weyl tensor,

$$\hat{W}_{\mu\nu\alpha\beta} = \hat{R}_{\mu\nu\alpha\beta} + \\ \tfrac{1}{2}(-g_{\mu\alpha}\hat{R}_{\nu\beta} + g_{\mu\beta}\hat{R}_{\nu\alpha} + g_{\nu\alpha}\hat{R}_{\mu\beta} - g_{\nu\beta}\hat{R}_{\mu\alpha}) + \tfrac{1}{6}(g_{\mu\alpha}g_{\nu\beta} - g_{\nu\alpha}g_{\mu\beta})\hat{R} \, , \tag{3.4}$$

which has the remaining 10 independent components.

The transformation rules under coordinate transformations (3.1) are as usual; all these curvature fields transform as tensors. To see how they transform under (3.2), first note how the connection fields transform:

$$\hat{\Gamma}_{\alpha\mu\nu} \to (1+\lambda)\hat{\Gamma}_{\alpha\mu\nu} + \tfrac{1}{2}(\hat{g}_{\alpha\nu}\partial_{\mu}\lambda + \hat{g}_{\alpha\mu}\partial_{\nu}\lambda - \hat{g}_{\mu\nu}\partial_{\alpha}\lambda) + \mathcal{O}(\lambda^2) \, , \tag{3.5}$$

from which we derive

$$\hat{R}_{\alpha\beta\mu\nu} \to (1+\lambda)\hat{R}_{\alpha\beta\mu\nu} + \tfrac{1}{2}(\hat{g}_{\alpha\nu}\hat{D}_{\beta}\partial_{\mu}\lambda - \hat{g}_{\alpha\mu}\hat{D}_{\beta}\partial_{\nu}\lambda - \hat{g}_{\beta\nu}\hat{D}_{\alpha}\partial_{\mu}\lambda + \hat{g}_{\beta\mu}\hat{D}_{\alpha}\partial_{\nu}\lambda) \, . \tag{3.6}$$

From this we find how the Ricci tensor transforms:

$$\hat{R}_{\mu\nu} \to \hat{R}_{\mu\nu} - \hat{D}_{\mu}\partial_{\nu}\lambda - \tfrac{1}{2}\hat{g}_{\mu\nu}\hat{D}^2\lambda \, , \qquad \hat{R} \to \hat{R}(1-\lambda) - 3\hat{D}^2\lambda \, . \tag{3.7}$$

The Weyl tensor (3.4), being the traceless part, is easily found to be invariant (apart from the canonical term):

$$\hat{W}_{\alpha\beta\mu\nu} \to (1+\lambda)\hat{W}_{\alpha\beta\mu\nu} \, . \tag{3.8}$$

Since the inverse, $\hat{g}^{\mu\nu}$, and the determinant, $\hat{g}$, of the metric transform as

$$\hat{g}^{\mu\nu} \to (1-\lambda)\hat{g}^{\mu\nu} \, ; \qquad \hat{g} \to (1+4\lambda)\hat{g} \, , \tag{3.9}$$

we establish that exactly the Weyl tensor squared yields an action that is totally invariant under local scale transformations in four space-time dimensions (remember that $\hat{g}^{\mu\nu}$ is used to connect the indices):

$$\mathcal{L} = C\sqrt{-\hat{g}}\,\hat{W}_{\alpha\beta\mu\nu}\hat{W}^{\alpha\beta\mu\nu} = C\sqrt{-\hat{g}}(\hat{R}_{\alpha\beta\mu\nu}\hat{R}^{\alpha\beta\mu\nu} - 2\hat{R}_{\mu\nu}\hat{R}^{\mu\nu} + \tfrac{1}{3}\hat{R}^2) \, , \tag{3.10}$$

which, due to the fact that the integral of

$$\hat{R}_{\alpha\beta\mu\nu}\hat{R}^{\alpha\beta\mu\nu} - 4\hat{R}_{\mu\nu}\hat{R}^{\mu\nu} + \hat{R}^2 \tag{3.11}$$

is a topological invariant, can be further reduced to

$$\mathcal{L} = 2C\sqrt{-\hat{g}}(\hat{R}_{\mu\nu}^2 - \tfrac{1}{3}\hat{R}^2) \ , \tag{3.12}$$

to serve as our locally scale invariant Lagrangian.

The constant C may be any dimensionless parameter. Note that, according to Eq. (3.7), neither the Ricci tensor nor the Ricci scalar are invariant; therefore, they are locally unobservable at this stage of the theory. Clearly, in view of Einstein's equation, *matter*, and in particular its stress-energy-momentum tensor, are locally unobservable in the same sense. This will have to be remedied at a later stage, where we must work on redefining what matter is at scales much larger than the Planck scale.

Thus we have verified that, indeed, the action (2.15) is the only expression that we could have expected there (apart from its overall constant) since we integrated out the scale component of the original metric $g_{\mu\nu}$. Demanding locality immediately leads to this expression.

In fact, gravity theories with this action as a starting point have been studied extensively [5], and there the suspicion was expressed that such theories might be unitary, in spite of the higher time derivatives in the action. Model calculations show[6] that unitarity can be regained if one modifies the hermiticity condition, which is equivalent to modifying the boundary conditions of functional amplitudes in the complex plane. Effectively then, the fields become complex. Before following such a route further, we would have to understand the underlying physics.

In Eq. (2.15), we arrived at the conformal action with an essentially infinite coefficient in front. Before deciding what to do with this infinity, and to obtain more insight in the underlying physics, let us study the classical equations that correspond to this action.

To this end, consider an infinitesimal variation $h_{\mu\nu}$ on the metric: $\hat{g}_{\mu\nu} \to \hat{g}_{\mu\nu} + \delta\hat{g}_{\mu\nu}$, $\delta\hat{g}_{\mu\nu} = h_{\mu\nu}$. The infinitesimal changes of the Ricci tensor and scalar are

$$\delta\hat{R}_{\mu\nu} = \tfrac{1}{2}(\hat{D}_\alpha\hat{D}_\mu h_\nu^\alpha + \hat{D}_\alpha\hat{D}_\nu h_\mu^\alpha - D^2 h_{\mu\nu} - \hat{D}_\mu\partial_\nu h_\alpha^\alpha) \ ; \tag{3.13}$$

$$\delta\hat{R} = -h^{\alpha\beta}\hat{R}_{\alpha\beta} + \hat{D}_\alpha\hat{D}_\beta h^{\alpha\beta} - \hat{D}^2 h_\alpha^\alpha \ . \tag{3.14}$$

Using the Bianchi identity

$$D_\mu R^\mu_{\ \nu} = \tfrac{1}{2}\partial_\nu R \ , \tag{3.15}$$

the variation of the Weyl action (3.10), (3.12) is then found to be

$$\delta\mathcal{L} = -2C \int \mathrm{d}^n x \sqrt{-\hat{g}} \, h^{\alpha\beta} \square_{\alpha\beta}^R \ , \qquad \text{with}$$

$$\square_{\alpha\beta}^R = \hat{D}^2\hat{R}_{\alpha\beta} - \tfrac{1}{3}\hat{D}_\alpha\hat{D}_\beta\hat{R} - \tfrac{1}{6}g_{\alpha\beta}\hat{D}^2\hat{R} - 2\hat{R}_\alpha^\mu\hat{R}_{\mu\beta} + 2\hat{R}^{\mu\nu}\hat{R}_{\alpha\mu\beta\nu} - \tfrac{2}{3}\hat{R}\hat{R}_{\alpha\beta} \ . \tag{3.16}$$

The classical equations of motion for the Ricci tensor as they follow from the Weyl action are therefore:

$$\square_{\alpha\beta}^R = 0 \ . \tag{3.17}$$

To see their most salient features, let us linearize in $\hat{R}_{\mu\nu}$ and ignore connection terms. We get

$$\hat{R}_{\mu\nu} - \tfrac{1}{6}\hat{R}\delta_{\mu\nu} \overset{\text{def}}{=} S_{\mu\nu} \; ; \qquad \partial_\mu S_{\mu\nu} = \partial_\nu S_{\alpha\alpha} \; ; \quad \partial^2 S_{\mu\nu} - \partial_\mu \partial_\nu S_{\alpha\alpha} = 0 \; . \qquad (3.18)$$

Defining $\lambda(x)$ by the equation

$$\partial^2 \lambda \overset{\text{def}}{=} -S_{\alpha\alpha} \; , \qquad (3.19)$$

we find that the solution $S_{\mu\nu}$ of Eq. (3.18) can be written as

$$S_{\mu\nu} = -\partial_\mu \partial_\nu \lambda + A_{\mu\nu} \; , \quad \text{with} \quad \partial^2 A_{\mu\nu} = 0 \; , \; A_{\alpha\alpha} = 0 \; , \; \partial_\mu A_{\mu\nu} = 0 \; . \qquad (3.20)$$

From Eq. (3.7) we notice that the free function $\lambda(x)$ corresponds to the local scale degree of freedom (3.2), while the equation for the remainder, $A_{\mu\nu}$, tells us that the Einstein tensor, after the scale transformation $\lambda(x)$, can always be made to obey the d'Alembert equation $\partial^2 G_{\mu\nu} = 0$, which is basically the field equation for the stress-energy-momentum tensor that corresponds to massless particles[3]. Thus, it is not true that the Weyl action gives equations that are equivalent to Einstein's equations, but rather that they lead to Einstein equations with only massless matter as their source.

4. Non conformal matter

To generalize to the case that our matter fields are not conformal, the easiest case to consider is a scalar field $\phi(x)$. Conformally invariant scalar fields are described by the action

$$\mathcal{L}^\phi_{\text{conf}} = -\tfrac{1}{2}\sqrt{-g}(g^{\mu\nu}\partial_\mu\phi\,\partial_\nu\phi + \tfrac{1}{6}R\phi^2) \; , \qquad (4.1)$$

where the second term is a well-known necessity for complete conformal invariance. Indeed, substituting the splitting (1.6) we find that the field $\phi(x)$ must be written as $\omega^{-1}\hat{\phi}(x)$, and then

$$\sqrt{-g}\,R \;=\; \omega^2\Big(\sqrt{-\hat{g}}\hat{R} \;-\; 6\partial_\mu(\sqrt{-\hat{g}}\,\hat{g}^{\mu\nu}\tfrac{1}{\omega}\partial_\nu\omega) \;+\; 6\sqrt{-\hat{g}}\,\hat{g}^{\mu\nu}\partial_\mu\omega\partial_\nu\omega\Big) \; , \qquad (4.2)$$

$$\sqrt{-g}\,g^{\mu\nu}\partial_\mu\phi\,\partial_\nu\phi \;=\; \sqrt{-\hat{g}}\,\hat{g}^{\mu\nu}(\partial_\mu\hat{\phi} - \frac{\partial_\mu\omega}{\omega}\phi)(\partial_\nu\phi - \frac{\partial_\nu\omega}{\omega}\phi) \; , \qquad (4.3)$$

$$\mathcal{L}^\phi_{\text{conf}} \;=\; -\tfrac{1}{2}\sqrt{-\hat{g}}(\hat{g}^{\mu\nu}\partial_\mu\hat{\phi}\partial_\nu\hat{\phi} + \tfrac{1}{6}\hat{R}\hat{\phi}^2) \; . \qquad (4.4)$$

The extra term with the Ricci scalar is in fact the same as the insertion (2.6) in Eq. (2.4). Inserting this as our matter Lagrangian leaves everything in the sections 2 and 3 unaltered.

Now, however, we introduce a mass term:

$$\mathcal{L}^{\phi,\text{mass}} = \mathcal{L}^\phi_{\text{conf}} - \tfrac{1}{2}\sqrt{-g}\,m^2\phi^2 \; . \qquad (4.5)$$

---

[3]Not quite, of course. The statement only holds when these particles form classical superpositions of plane waves such as an arbitrary function of $x - t$.

After the split (1.6), this turns into

$$\mathcal{L}^{\phi,\text{mass}} = \mathcal{L}^{\phi}_{\text{conf}} - \tfrac{1}{2}\sqrt{-\hat{g}}\, m^2 \omega^2 \hat{\phi}^2 \ . \tag{4.6}$$

Thus, an extra term proportional to ω^2 arises in Eq. (2.4). But, as it is merely quadratic in ω, we can still integrate this functional integral exactly.[4] At $n \to 4$, and remembering that we had scaled out a factor $6/\kappa^2$ in going from Eq. (2.2) to Eq. (2.4), the quantity M in Eq. (2.13) is now replaced by

$$M = -\tfrac{1}{6}\hat{R} + \tfrac{1}{6}\kappa^2 m^2 \hat{\phi}^2 \ , \tag{4.7}$$

and plugging it into the divergence equation (2.14) replaces Eq. (2.15) by

$$\Gamma^{\text{div}} = \frac{\sqrt{-\hat{g}}}{8\pi^2(4-n)}\left(\tfrac{1}{120}(\hat{R}_{\mu\nu}\hat{R}^{\mu\nu} - \tfrac{1}{3}\hat{R}^2) + \tfrac{1}{144}(\kappa^2 m^2 \hat{\phi}^2)^2\right) \ , \tag{4.8}$$

where $\kappa^2 = 8\pi G_N$. Indeed, the extra term is a quartic interaction term and as such again conformally invariant. $\kappa^2 m^2$ is a dimensionless parameter and, usually, it is quite small.

The two terms in Eq. (4.8) have to be treated in quite a different way. As was explained in Section 3, the first term would require a non canonical counter term, which we hesitate to add just like that, so it presents real problems that will have to be addressed.

This difficulty does not play any role for the second term. Its divergent part can be renormalized in the usual way by adding a counter term representing a quartic self interaction of the scalar field. There will be more subtle complications due to the fact that renormalization of these non gravitational interaction terms in turn often (but not always) destroys scale invariance. As for the matter fields, these complications will not be further considered here. Suffices to say that in some special cases, such as in supersymmetric theories, the problems simplify.

It is important to conclude from this section that non-conformal matter does not affect the formal conformal invariance of the effective action after integrating over the metascalar ω field. Also, the non conformal parts, such as the mass term, do not have any effect on the dangerously divergent term in this effective action.

5. The divergent effective conformal action

Let us finally address our real problem, the divergence of the effective action (2.15) as $n \to 4$. This really spoils the beautiful program we outlined at the beginning of Section 3. One can imagine five possible resolutions of this problem.[5]

A. Cancelation against divergences due to matter. Besides scalar matter fields, one may have Dirac spinors and/or gauge fields that also propagate in the conformal metric

---

[4]Note that a cosmological constant would add a term $C\Lambda\omega^4$ to the action, so that the ω integration can then no longer be done exactly. Thus, there is good reason to omit the cosmological constant, but it would be premature to speculate that this adds new views on the well-known cosmological constant problem.

[5]This section is the most important revision in version # 2 of this paper.

$\hat{g}_{\mu\nu}(\vec{x}, t)$. These also lead too divergences. Ignoring interactions between these matter fields, one indeed finds that these fields contribute to the divergence in the effective action (2.15) as well. In fact, all these divergences take the same form of the Weyl action (3.12), and they each just add to the overall coefficient. So, with a bit of luck, one might hope that all these coefficients added up might give zero. That would certainly solve our problem. It would be unlikely that also the finite parts of the effective action would completely cancel out, so we would end up with a perfectly conformally invariant effective theory.

A curious problem would have to be addressed, which is that the effective action scales as the fourth power of the momenta of the conformal $\hat{g}_{\mu\nu}$ fields, so that there should be considerable concern that unitarity is lost. One might hope that unitarity can be saved by observing that the theory is still based on a perfectly canonical theory where we started off with the action (1.1).

Unfortunately, this approach is ruled out for a very simple reason: the matter fields can never cancel out the divergence because they all contribute with the same sign! This is a rather elaborate calculation, of a kind already carried out in the early 1970s [7][8]. As we are only interested in the part due to the action of scalar, spinor and vector fields on a conformal background metric $\hat{g}_{\mu\nu}$, we repeated the calculation and summarize its result in the Appendix. It is found that, if the matter fields consist of N_0 elementary scalar fields, $N_{1/2}$ elementary Majorana spinor fields (or $\frac{1}{2}N_{1/2}$ complex Dirac fields) and N_1 real Maxwell or Yang-Mills fields (their mutual interctions are ignored), then the total coefficient C in front of the divergent effective action

$$S^{\text{eff}} = C \int \mathrm{d}^n x \frac{\sqrt{-\hat{g}}}{8\pi^2(4-n)} \left(\hat{R}^{\mu\nu}\hat{R}_{\mu\nu} - \tfrac{1}{3}\hat{R}^2 \right) , \tag{5.1}$$

is

$$C = \tfrac{1}{120}(1 + N_0) + \tfrac{1}{40}N_{1/2} + \tfrac{1}{10}N_1 . \tag{5.2}$$

Here, the first 1 is the effect of the metascalar component ω of gravity itself. All contributions clearly add up with the same sign. This, in fact, could have been expected from simple unitarity arguments, but as such arguments famously failed when the one-loop beta functions for different particle types were considered, it is preferred to do the calculation explicitly. In any case, option A is excluded. Although the coefficients are known from the literature [7], we reproduce the details of the calculation in the Appendix.

B. Make the integral finite with a local counter term, of the same form as Eq. (5.1), but with opposite sign. This is the option most physicists who are experienced in renormalization would certainly consider as the most reasonable one. However, a combination of two observations casts serious doubts on the viability of this option. First, in conventional theories where renormalization is carried out, this is happening in the context of a perturbation expansion. The expression that has to be subtracted has a coefficient in front that behaves as

$$\frac{g^\alpha}{(4-n)^\beta} , \tag{5.3}$$

where g is a coupling strength, and the power α is usually greater than the power β. If we agree to stick to the limit where *first* g is sent to zero and *then* n is sent to 4, the total coefficient will still be infinitesimal, and as such not cause any violation of unitarity, even if it does not have the canonical form. This is exactly the reason why the consideration of non-canonical renormalization terms is considered acceptable when perturbative gravity is considered, as long as the external momenta of in- and outgoing particles are kept much smaller than the Planck value.

Here, however, Newton's constant has been eliminated, so there is no coupling constant that makes our counter term small, and of course we consider all values of the momenta. The Weyl action is quadratic in the Riemann curvature $R^{\alpha}_{\beta\mu\nu}$ and therefore quartic in the momenta. As stated when we were considering option A, one might hope that the original expression we found can be made compatible with unitarity because it itself follows from the canonical action (1.1). The counter term itself cannot be reconciled with unitarity. In fact, it not only generates a propagator of the form $1/(k^2 - i\varepsilon)^2$, which at large values of k^2 is very similar to the difference of two propagators: $1/(k^2-i\varepsilon)-1/(k^2+m^2-i\varepsilon)$, where the second one would describe a particle with indefinite metric, but also, the combination with the total action would leave a remainder of the form

$$\frac{1}{4-n}\left((k^2)^{n/2} - \mu^{n-4}(k^2)^2\right) \;\rightarrow\; (k^2)^2 \log(k^2/\mu^2)\,, \tag{5.4}$$

where μ is a quantity with the dimension of a mass that defines the subtraction point. The effective propagator would take a form such as

$$\frac{1}{(k^2 + m^2 - i\varepsilon)^2 \log(k^2/\mu^2)}\,, \tag{5.5}$$

which develops yet another pole, at $k^2 \approx \mu^2$. This is a Landau ghost, describing something like a tachyonic particle, violating most of the principles that one would like to obey in quantizing gravity. All these objections against accepting a non canonical renormalization counter term are not totally exclusive[9], but they are sufficient reason to search for better resolutions. For sure, one would have to address the problems, and as yet, this seems to be beyond our capacities.

C. An observation not yet included in an earlier version of this paper, can put our argument in a very different light. If one follows what actually happens in conventional, perturbative gravity, one would be very much tempted to conclude that it is incomplete: one has to include the contribution of the $\hat{g}_{\mu\nu}$ field itself to the infinity! Only this way, one would obtain the complete renormalization group equations for the coefficient C in the action (5.1). What is more, in some supergravity theories, the *conformal anomaly* then indeed cancels out to zero.[10][6] However, arguing this way would not at all be in line with the entire approach advocated here: *first* integrate over the ω field and *only then* over the fields $\hat{g}_{\mu\nu}$. We here discuss only the integral over ω, with perhaps in addition the matter fields, and this should provide us with the effective action for $\hat{g}_{\mu\nu}$. If that is no longer conformally invariant, we have a problem. Treating C as a freely adjustable,

---

[6]I thank M. Duff for this observation.

running parameter, even if it turns out not to run anymore, would be a serious threat against unitarity, and would bring us back to perturbative gravity as a whole, with its well-known difficulties. In addition, an important point then comes up: how does the *measure* of the $\hat{g}_{\mu\nu}$ integral scale? This might not be reconcilable with conformal invariance either, but also difficult if at all possible to calculate: the measure is only well-defined if one fixes the gauge à la Faddeev-Popov, and this we wish to avoid, at this stage. We neither wish to integrate over $\hat{g}_{\mu\nu}$, nor fix the gauge there. In conclusion therefore, we dismiss option C as well.

Therefore, yet another option may have to be considered:

D. No counter term is added at all. We accept an infinite coefficient in front of the Weyl action. To see the consequences of such an assumption, just consider the case that the coefficient $K = C/(4 - n)$ is simply very large. In the standard formulation of the functional integral, this means that the quantum fluctuations of the fields are to be given coefficients going as $1/\sqrt{K}$. The *classical* field values can take larger values, but they would act as a background for the quantized fields, and not take part in the interactions themselves. In the limit $K \to \infty$, the quantum fluctuations would vanish and only the classical parts would remain. In short, this proposal would turn the $\hat{g}_{\mu\nu}$ components of the metric into classical fields!

There are important problems with this proposal as well: classical fields will not react upon the presence of the other, quantized, fields such as the matter fields. Therefore, there is no back reaction of the metric. This proposal then should be ruled out because it violates the action = reaction principle in physics. Furthermore, the reader may already have been wondering about gravitons. They are mainly described by the parts of $\hat{g}_{\mu\nu}$ that are spacelike, traceless and orthogonal to the momentum. If we would insist that $\hat{g}_{\mu\nu}$ is classical, does this mean that gravitons are classical? It is possible to construct a gedanken experiment with a device that rotates gravitons into photons; this device would contain a large stretch of very strong, transverse magnetic fields. Turning photons into gravitons and back, it would enable us to do quantum interference experiments with gravitons. This then would be a direct falsification of our theory. However, the classical behavior of gravitons that we suspect, comes about because of their interactions with the logarithmically divergent background fluctuations. If the usual renormalization counter term of the form (5.1) is denied to them, this interaction will be infinite. The magnetic fields in our graviton-photon transformer may exhibit fluctuations that are fundamentally impossible to control; gravitons might still undergo interference, but their typical quantum features, such as entanglement, might disappear.

Yet, there may be a different way to look at option D. In previous publications[11], the author has speculated about the necessity to view quantum mechanics as an *emergent* feature of Nature's dynamical laws. 'Primordial quantization' is the procedure where we start with classical mechanical equations for evolving physical variables, after which we attach basis elements of Hilbert space to each of the possible configurations of the classical variables. Subsequently, the evolution is re-expressed in terms of an effective Hamiltonian, and further transformations in this Hilbert space might lead to a description of the world as we know it. This idea is reason enough to investigate this last option further.

There is one big advantage from a technical point of view. Since $\hat{g}_{\mu\nu}$ is now considered to be classical, there is no unitarity problem. All other fields, both the metascalar field ω (the 'dilaton') and the matter fields are described by renormalizable Lagrangians, so that no obvious contradictions arise at this point.

How bad is it that the action = reaction principle appears to be violated? The metric metatensor does allow for a source in the form of an energy momentum tensor, as described in Eqs. (3.17)—(3.20). This, however, would be an unquantized source. We get a contradiction if sources are described that evolve quantum mechanically: the background metric cannot react. In practice, this would mean that we could just as well mandate that

$$\hat{g}_{\mu\nu} = \eta_{\mu\nu} \ , \tag{5.6}$$

in other words, we would live in a flat background where only the metascalar component of the metric evolves quantum mechanically.[7]

Could a non-trivial metric tensor $\hat{g}_{\mu\nu}$ be *emergent*? This means that it is taken either to be classical or totally flat beyond the Planck scale, but it gets renormalized by dilaton and matter fields at much lower scales. This may be the best compromise between the various options considered. Spacetime is demanded to be conformally flat at scales beyond the Planck scale, but virtual matter and dilaton fluctuations generate the $\hat{g}_{\mu\nu}$ as we experience it today. A problem with this argument, unfortunately, is that it is difficult to imagine how dilaton fluctuations could generate a non trivial effective metric. This is because, regardless the values chosen for $\omega(x)$, the light cones will be the ones determined by $\hat{g}_{\mu\nu}$ alone, so that there is no 'renormalization' of the speed of light at all. We therefore prefer the following view:

Consider a tunable choice for a renormalization counter term in the form of the Weyl action (5.1), described by a subtraction point μ. If μ were chosen to be at low frequencies, so at large distance scales, then the Landau ghost, Eq. (5.5), would be at low values of k^2 and therefore almost certainly ruin unitarity of the amplitudes. Only if μ would be chosen as far as possible in the ultraviolet, this ghost would stay invisible at most physical length scales, so the further away we push the subtraction point, the better, but perhaps the limit $\mu \to \infty$ must be taken with more caution.

The previous version of this paper was incomplete without the following alternative option. A more mainstream standpoint would be :

E. The action (1.1) no longer properly describes the situation at scales close to the Planck scale. At $|k| \approx M_{Pl}$, we no longer integrate over $\omega(k)$, which has two consequences: a natural cut-off at the Planck scale, and a breakdown of conformal invariance. Indeed, this would have given the badly needed scale dependence to obtain a standard interpretation of the amplitudes computed this way. Note that, in our effective action (4.8), all dependence on Newton's constant has been hidden in an effective quartic interaction term for the ϕ field. That could have been augmented with a 'natural' quartic interaction already present in the matter Lagrangian, so we would have lost all explicit references

---

[7]This idea goes back to, among others, Nordström[12].

to Newton's constant. Now, with the explicit breakdown of conformal invariance, we get Newton's constant back.

Adopting this standpoint, it is also easy to see how a subtraction point wandering to infinity, as described in option D, could lead to a classical theory for $\hat{g}_{\mu\nu}$. It simply corresponds to the classical limit. Letting the subtraction point go to infinity is tantamount to forcing M_{Pl} to infinity, in which limit, of course, gravity is classical. Only if we embrace option D fully, we would insist that the physical scale is not determined by M_{Pl} this way, but by adopting some gauge convention at a boundary at infinity. This is the procedure demanded by black hole complementarity.

The price paid for option E is, that we lost the fundamental advantages of exact conformal invariance, which are a calculable and practically renormalizable effective interaction, and a perfect starting point for a conformally invariant treatment of the black hole correspondence principle as was advocated in Ref. [4]. The idea advocated in this paper is *not* to follow option E representing what would presumably be one of the mainstream lines of thought. With option E, we would have ended up with just another parametrization of non-renormalizable, perturbative, quantum gravity. Instead, we are searching for an extension of the canonical action (1.1) that is such that the equivalent of the ω integration can be carried out completely.

6. Conclusions

Our research was inspired by recent ideas about black holes [4]. There, it was concluded that an effective theory of gravity should exist where the metascalar component either does not exist at all or is integrated out. This would enable us to understand the black hole complementarity principle, and indeed, turn black holes effectively indistinguishable from ordinary matter at tiny scales. A big advantage of such constructions would be that, due to the formal absence of black holes, we would be allowed to limit ourselves to topologically trivial, continuous spacetimes for a meaningful and accurate, nonperturbative description of all interactions. This is why we searched for a formalism where the metascalar ω is integrated out first.

Let us briefly summarize here how the present formulation can be used to resolve the issue of an apparent clash between unitarity and locality in an evaporating black hole. An observer going into the hole does not explicitly observe the Hawking particles going out. (S)he passes the event horizon at Schwarzschild time $t \to \infty$, and from his/her point of view, the black hole at that time is still there. For the external observer, however, the black hole has disappeared at $t \to \infty$. Due to the back reaction of the Hawking particles, energy (and possibly charge and angular momentum) has been drained out of the hole. Thus, the two observers appear to disagree about the total stress-energy-momentum tensor carried by the Hawking radiation. Now this stress-energy-momentum tensor was constructed in such a way that it had to be covariant under coordinate transformations, but this covariance only applies to *changes* made in the stress-energy-momentum when creation- and/or annihilation operators act on it. About these covariant changes, the two observers do not disagree. It is the *background subtraction* that is different, because the two observers do not agree about the vacuum state. This shift in the background's source of gravity can be neatly accommodated for by a change in the conformal factor $\omega(x)$ in the metric seen by the two observers.

This we see particularly clearly in Rindler space. Here, we can generate a modification of the background stress-energy-momentum by postulating an infinitesimal shift of the parameter $\lambda(x)$ in Eqs. (3.7) and (3.20). It implies a shift in the Einstein tensor $G_{\mu\nu}$ (and thus in the tensor $T_{\mu\nu}$) of the form

$$G_{\mu\nu} \rightarrow G_{\mu\nu} - D_\mu \partial_\nu \lambda + g_{\mu\nu} D^2 \lambda \ . \tag{6.1}$$

If now the transformation λ is chosen to depend only on the lightcone coordinate x^-, then

$$G_{--} \rightarrow G_{--} - \partial_-^2 \lambda \ , \tag{6.2}$$

while the other components do not shift. Thus we see how a modification only in the energy and momentum of the vacuum in the x^+ direction (obtained by integrating G_{--} over x^-) is realized by a scale modification $\lambda(x^-)$.

In a black hole, we choose to modify the pure Schwarzschild metric, as experienced by an ingoing observer, by multiplying the entire metric with a function $\omega^2(t)$ that decreases very slowly from 1 to 0 as Schwarzschild time t runs to infinity. This then gives the metric of a gradually shrinking black hole as seen by the distant observer. Where ω has a non vanishing time derivative, this metric generates a non vanishing Einstein tensor, hence a non vanishing background stress-energy-momentum. This is the stress-energy-momentum of the Hawking particles.

Calculating this stress-energy-momentum yields an apparently disturbing surprise: it does not vanish at apacelike infinity. The reason for this has not yet completely been worked out, but presumably lies in the fact that the two observers not only disagree about the particles emerging from the black hole, but also about the particles going in, and indeed an infinite cloud of thermal radiation filling the entire universe around the black hole.

All of this is a sufficient reason to suspect that the conformal (metascalar) factor $\omega(x)$ must be declared to be locally unobservable. It is fixed only if we know the global spacetime and after choosing our coordinate frame, with its associated vacuum state. If we would not specify that state, we would not have a specified ω. In 'ordinary' physics, quantum fields are usually described in a flat background. Then the choice for ω is unique. Curiously, it immediately fixes for us the sizes, masses and lifetimes of all elementary particles. This may sound mysterious, until we realize that sizes and lifetimes are measured by using light rays, and then it is always assumed that these light rays move in a flat background. When this background is not flat, because $\hat{g}_{\mu\nu}$ is non-trivial, then sizes and time stretches become ambiguous. We now believe that this ambiguity is a very deep and fundamental one in physics.

Although this could in principle lead to a beautiful theory, we do hit a real obstacle, which is, of course, that gravity is not renormalizable. This 'disease' still plagues our present approach, unless we turn to rather drastic assumptions. The usual idea that one should just add renormalization counter terms wherever needed, is found to be objectionable. So, we turn to ideas related to the 'primitive quantization' proposal of Ref. [11]. Indeed, this quantization procedure assumes a basically classical set of equations of motion as a starting point, so the idea would fit beautifully.

Of course, many other questions are left unanswered. Quite conceivably, further research might turn up more alternative options for a cure to our difficulties. One of these, of course, is superstring theory. Superstring theory often leads one to avoid certain questions to be asked at all, but eventually the black hole complementarity principle will have to be considered, just as the question of the structure of Nature's degrees of freedom at distance and energy scales beyond the Planck scale.

Acknowledgements

The author thanks S. Giddings, R. Bousso, C. Taubes , M. Duff and P. Mannheim for discussions, and P. van Nieuwenhuizen for his clarifications concerning the one-loop pole terms. He thanks R. Jackiw for pointing out an inaccuracy in the Introduction, which we corrected.

A. The calculation of the one-loop pole terms for scalars, spinors and vectors interacting with a background metric.

The general algorithm for collecting all divergent parts of one-loop quantum corrections in quantum field theories was formulated in Ref. [13], applied to a gravitational background metric in Ref. [3], and worked out much further in [8]. Here, we briefly summarize the calculations that lead to the coefficients in Eq. (5.2), see also Ref. [7].

Consider a quantized complex, possibly multi-component, scalar field $\phi(x)$: let its Lagrangian in a curved background be

$$\mathcal{L} \; = \; \sqrt{-g}(-g^{\mu\nu}D_\mu\phi^* D_\nu\phi) + \sqrt{g}\,\phi^*(2N^\mu D_\mu\phi + M\phi) \,, \qquad \text{then} \qquad (A.1)$$

where $g_{\mu\nu}$ is a 4 by 4 matrix (of course, $g_{\mu\nu}$ is expected to have an inverse, $g^{\mu\nu}$), and N^μ and M may be arbitrary, differentiable functions of the spacetime coordinates x^μ, as well as matrices in the internal indices of the ϕ field. The gradient D_μ may contain a background gauge field Z_μ:

$$D_\mu\phi = \partial_\mu\phi + Z_\mu\phi \,, \qquad\qquad (A.2)$$

where Z_μ may again be any function of space-time.[8]

---

[8]Actually, having both a gauge field Z_μ and an external field N^μ is redundant, but we keep them both for later convenience.

It was derived in Ref. [3] that the infinite component of the effective action (which we will call the 'pole term') is

$$\Delta\mathcal{L} = \frac{\sqrt{-g}}{\varepsilon}\mathrm{Tr}\left(\tfrac{1}{12}Y_{\mu\nu}Y^{\mu\nu} + \tfrac{1}{2}X^2 + \tfrac{1}{60}(R_{\mu\nu}^2 - \tfrac{1}{3}R^2)\right) , \tag{A.3}$$

where we slightly modified the notation:[9]

$$\varepsilon = 8\pi^2(4-n) ; \quad , \quad Z_{\mu\nu} \overset{\mathrm{def}}{=} \partial_\mu Z_\nu - \partial_\nu Z_\mu + [Z_\mu, Z_\nu] , \tag{A.4}$$
$$X = M - N_\mu N^\mu - D_\mu N^\mu + \tfrac{1}{6}R , \quad Y_{\mu\nu} = Z_{\mu\nu} + D_\mu N_\nu - D_\nu N_\mu + [N_\mu, N_\nu] ,$$

and 'Tr' stands for the trace in the internal ϕ indices. The Lorentz indices are assumed to me moved up and down, and summed over, using the metric $g_{\mu\nu}$ in the usual way. Naturally, the covariant derivative of the background function N^μ is defined to be

$$D_\mu N^\alpha = \partial_\mu N^\alpha + \Gamma^\alpha_{\mu\nu}Z^\nu + [Z_\mu, N^\alpha] , \tag{A.5}$$

For the metascalar field ω in Section 2, we have $M = -\tfrac{1}{6}R$, while for the scalar matter field in Section 4 we have $M = m - \tfrac{1}{6}R$, and in both these cases there is no further gauge field or N^μ field, so $Y_{\mu\nu} = 0$. Since both ω and ϕ (Eq. (4.1) were chosen to have only one single, real component, the resulting pole term has to be divided by 2. That gives Eqs. (2.14), (2.15), and Eq. (4.8), leading to the first coefficient, $\tfrac{1}{120}$ in Eq. (5.2).

Next, Eq. (A.3) can be used as a stating point to compute the pole term for Dirac and for vector fields. First, let us consider a quantized Maxwell field $B_\mu(x)$.

We add to the Maxwell Lagrangian the gauge fixing term $\mathcal{L}_g = -\tfrac{1}{2}\sqrt{-g}\,(D_\mu B^\mu)^2$, which, for convenience, was chosen to be covariant for general coordinate transformations. Because of this choice, the Faddeev Popov ghost fields η, $\bar{\eta}$ now couple to the background metric. The total Lagrangian thus becomes

$$\mathcal{L} = \sqrt{-g}\left(-\tfrac{1}{2}(D_\mu B_\nu D^\mu B^\nu - \tfrac{1}{2}B^\mu(D_\nu D_\mu - D_\mu D_\nu)B^\nu + \bar{\eta}D^2\eta\right) =$$
$$= \sqrt{-g}\left(-\tfrac{1}{2}(D_\mu B_\nu)^2 - \tfrac{1}{2}B_\mu R^{\mu\nu}B_\nu + \bar{\eta}D^2\eta\right) , \tag{A.6}$$

where indices are moved up and down using the background metric $g_{\mu\nu}$, using the fact that the metric commutes with the covariant derivative D_μ.

We can now use Eq. (A.3) as a master equation, provided that the Lorentz indices $\mu, \nu, \cdots$ of the Maxwell field B are replaced by internal Lorentz indices $a, b, \cdots$, using the Vierbein field e^a_μ obeying

$$g_{\mu\nu} = e^a_\mu e^a_\nu , \qquad e^a_\mu e^{a\nu} = \delta^\nu_\mu , \text{ etc,} \tag{A.7}$$

where the summation over the internal Lorentz index $a, b, \cdots$ is assumed to have the sign convention $(-, +, +, +)$ in the usual way. The covariant derivative of the Maxwell field

---

[9]Including an overall sign switch, since in Refs. [3], [7] and [8], the *counter* term was computed.

now contains the Lorentz connection field A_μ^{ab} as a gauge field, whose curvature coincides with the Riemann tensor:

$$D_\mu B^a = \partial_\mu B^a + A_\mu^{ab} B^b \;;\qquad F_{\mu\nu}^{ab} = \partial_\mu A_\nu^{ab} - \partial_\nu A_\mu^{ab} + [A_\mu, A_\nu]^{ab} = R_{\mu\nu}^{ab} \;. \qquad \text{(A.8)}$$

Inserting the variable B^a in (A.1), and remembering that now it has 4 real components, we have

$$Z_\mu^{ab} = A_\mu^{ab} \;,\qquad X_{ab} = -R_{ab} + \tfrac{1}{6}R\delta^{ab} \;,\qquad Y_{\mu\nu}^{ab} = F_{\mu\nu}^{ab} \;,$$
$$\Delta\mathcal{L}^B = \tfrac{\sqrt{-g}}{\varepsilon}\left(\tfrac{1}{24}Y_{\mu\nu}^{ab}Y_{\mu\nu}^{ba} + \tfrac{1}{4}(R^{ab} - \tfrac{1}{6}R\delta^{ab})^2 + \tfrac{4}{120}(R_{\mu\nu}^2 - \tfrac{1}{3}R^2)\right) =$$
$$= \tfrac{\sqrt{-g}}{\varepsilon}\left(\tfrac{7}{60}R_{\mu\nu}^2 - \tfrac{1}{40}R^2\right) \;, \qquad \text{(A.9)}$$

where use was made of the fact that the combination (3.11) is a pure derivative and so can be put equal to zero.

The ghost contribution, including its sign switch, is

$$\Delta\mathcal{L}_g = \frac{\sqrt{-g}}{\varepsilon}\left(-\tfrac{1}{60}(R_{\mu\nu}^2 - \tfrac{1}{3}R^2) - \tfrac{1}{72}R^2\right) \;, \qquad \text{(A.10)}$$

and the result, when added up,

$$\Delta\mathcal{L}_{\text{Maxwell}} = \frac{\sqrt{-g}}{\varepsilon}\left(\tfrac{1}{10}R_{\mu\nu}^2 - \tfrac{1}{30}R^2\right) \;, \qquad \text{(A.11)}$$

gives the last coefficient $\tfrac{1}{10}$ in Eq. (5.2).

For a derivation of the pole term coming from the Dirac fields, we can also use the master formula (A.3). Here, the use of the Vierbein field will be seen to be crucial. Let γ^a, $a = 1,2,3,4$, be the four Dirac γ matrices. We write $\gamma_\mu = e_\mu^a \gamma^a$; $\gamma_\mu \gamma_\nu = g_{\mu\nu} + \sigma_{\mu\nu}$, $D_\mu \gamma_\nu = 0$, and as the Lagrangian for a complex Dirac field we use

$$\mathcal{L} = -\sqrt{-g}\,\overline{\psi}(\gamma^\mu D_\mu + M)\psi \;,\qquad D_\mu\psi \overset{\text{def}}{=} (\partial_\mu + B_\mu + \tfrac{1}{4}\sigma^{ab}A_\mu^{ab})\psi \;. \qquad \text{(A.12)}$$

The mass term and the external gauge field B_μ will actually not be used in this paper, but it is convenient to keep them for later use, and for checking the correctness of the formalism.

Now this is a first order Lagrangian, while Eq. (A.1) is second order. So, instead of Eq. (A.12), we take a squared Lagrangian:

$$\mathcal{L} = \sqrt{-g}\,\overline{\psi}(\gamma D - m_1)(\gamma D + m_2)\psi = \sqrt{-g}\left(-g^{\mu\nu}D_\mu\overline{\psi}D_\nu\psi + \overline{\psi}(2N^\mu D_\mu + M)\psi\right) \;,$$
$$N_\mu = \tfrac{1}{2}(m_2 - m_1)\gamma^\mu \;,\qquad M = \tfrac{1}{2}\sigma^{\mu\nu}G_{\mu\nu} + \tfrac{1}{8}\sigma^{\mu\nu}\sigma^{ab}F_{\mu\nu}^{ab} + \gamma\partial m_2 - m_1 m_2 \;,$$
$$X = m_1 m_2 - m_1^2 - m_2^2 + \tfrac{1}{2}\gamma\partial(m_1 + m_2) + \tfrac{1}{6}R + \tfrac{1}{8}\sigma^{\mu\nu}\sigma^{ab}R_{\mu\nu ab} + \tfrac{1}{2}\sigma^{\mu\nu}G_{\mu\nu} \;,$$
$$Y_{\mu\nu} = G_{\mu\nu} + \tfrac{1}{4}\sigma^{ab}F_{\mu\nu}^{ab} + \tfrac{1}{2}(\gamma^\mu\partial_\nu - \gamma^\nu\partial_\mu)(m_1 - m_2) + \tfrac{1}{2}(m_2 - m_1)^2\sigma_{\mu\nu} \;. \qquad \text{(A.13)}$$

Here, $G_{\mu\nu}$ is the covariant curl of the external B field.

Next, assuming that we have 4 complex spinor components, derive

$$\frac{1}{4}\mathrm{Tr}\,(\gamma_\mu\gamma_\nu\gamma_\alpha\gamma_\beta\,R_{\mu\nu\alpha\beta})^2 = 4R^2\,,$$
$$\frac{1}{4}\mathrm{Tr}\,(\gamma_\alpha\gamma_\beta R_{\mu\nu ab})^2 = -8R_{\mu\nu}^2 + 2R^2\,,$$
$$\frac{1}{4}\mathrm{Tr}\,\gamma^\mu\gamma^\nu\gamma^\alpha\gamma^\beta R_{\mu\nu\alpha\beta} = -2R\,,$$
$$\frac{1}{4}\mathrm{Tr}\,(\sigma^{\mu\nu}G_{\mu\nu})^2 = -2G_{\mu\nu}^2\,,$$
$$\frac{1}{4}\mathrm{Tr}\,(\sigma_{\mu\nu})^2 = -12\,. \tag{A.14}$$

One finds

$$\frac{1}{4}\mathrm{Tr}\,X^2 = m_1^4 + m_2^4 + 3m_1^2 m_2^2 - 2m_1 m_2^3 - 2m_1^3 m_2 + \frac{1}{4}(\partial(m_1+m_2))^2 - \frac{1}{2}G_{\mu\nu}^2$$
$$+ R^2(\tfrac{1}{36} + \tfrac{1}{16} - \tfrac{1}{12}) + (m_1 m_2 - m_1^2 - m_2^2)(\tfrac{1}{3}R - \tfrac{1}{2}R)\,,$$
$$\frac{1}{4}\mathrm{Tr}\,Y_{\mu\nu}Y_{\mu\nu} = G_{\mu\nu}^2 - \frac{1}{2}R_{\mu\nu}^2 + \frac{1}{8}R^2 + \frac{3}{2}(\partial(m_1-m_2))^2 - 3(m_1^2 + m_2^2 - 2m_1 m_2)^2$$
$$- \frac{1}{2}R(m_1 - m_2)^2\,.$$
$$\frac{1}{60}\mathrm{Tr}\,(R_{\mu\nu}^2 - \tfrac{1}{3}R^2) = \frac{1}{15}R_{\mu\nu}^2 - \frac{1}{45}R^2\,. \tag{A.15}$$

This, inserted into Eq. (A.3) for 4 complex fields, adding the Fermionic minus sign, leads to[10]

$$-\Delta\mathcal{L} = \frac{\sqrt{-g}}{\varepsilon}\left(m_1^4 + m_2^4 + \partial m_1^2 + \partial m_2^2 - \frac{2}{3}G_{\mu\nu}^2 + \frac{1}{6}R(m_1^2 + m_2^2) - \frac{1}{10}(R_{\mu\nu}^2 - \tfrac{1}{3}R^2)\right)\,. \tag{A.16}$$

Notice that all cross terms containing products such as $m_1 m_2^3$ cancel out, as they must, because what was computed here is the combined effect of two fermion species, with masses m_1 and m_2. One concludes that the pole term produced by a single fermion of mass M is given by

$$\Delta\mathcal{L} = \frac{\sqrt{-g}}{\varepsilon}\mathrm{Tr}\left(-M^4 - g^{\mu\nu}\partial_\mu M\partial_\nu M - \frac{1}{6}RM^2 + \frac{1}{3}G_{\mu\nu}^2 + \frac{1}{20}(R_{\mu\nu}^2 - \tfrac{1}{3}R^2)\right)\,. \tag{A.17}$$

A Majorana spinor counts as half a Dirac spinor, so this is how we derived the coefficient $\frac{1}{40}$ in Eq. (5.2).

We observe that conformal invariance is obeyed throughout. If the mass terms are treated as metascalars, as they should, we see that the Lagrangians we start off with are totally conformally invariant, and so are the pole terms that we found. Not only does the Riemann curvature only appear in the Weyl combination, $R_{\mu\nu}^2 - \frac{1}{3}R^2$, but we also see that the conformal combination $(\partial M)^2 + \frac{1}{6}RM^2$ emerges in the Dirac pole term (A.17).

References

[1] D.M. Capper and M.J. Duff, *Conformal Anomalies and the Renormalizability Problem in Quantum Gravity*, Phys. Lett. **53A** 361 (1975).

---

[10]In the derivation, it was assumed that m_1 and m_2 were commuting matrices, but one easily checks that the result (A.16) continues to hold when they do not commute.

[2] M.J. Duff, *Twenty Years of the Weyl Anomaly*, Talk given at the Salamfest, ICTP, Trieste, March 1993, arXiv:hep-th/9308075.

[3] G. 't Hooft and M. Veltman, *One Loop Divergences in the Theory of Gravitation*, Ann. Inst. Henri Poincaré, **20** (1974) 69.

[4] G. 't Hooft, *Quantum Gravity without Space-time Singularities or Horizons*, Erice School of Subnuclear Physics 2009, to be publ.; arXiv:0909.3426

[5] P. D. Mannheim and D. Kazanas, Astrophys. J. **342**, 635 (1989); D. Kazanas and P. D. Mannheim, Astrophys. J. Suppl. **76**, 431 (1991); P. D. Mannheim, Prog. Part. Nucl. Phys. **56**, 340 (2006), astro-ph/0505266; P. D. Mannheim, *Intrinsically Quantum-Mechanical Gravity and the Cosmological Constant Problem*, arXiv:1005.5108 [hep-th].
G.U. Varieschi, *A Kinematical Approach to Conformal Cosmology*, Gen. Rel. Grav. **42** 929 (2010), arXiv:0809.4729

[6] C.M. Bender and P.D. Mannheim, *No-ghost theorem for the fourth-order derivative Pais-Uhlenbeck oscillator model*, Physical Review Letters **100**, 110402 (2008), arXiv:0706.0207 [hep-th]; id., *Exactly solvable PT-symmetric Hamiltonian having no Hermitian counterpart*, Phys. Rev. **D 78**, 025022 (2008), arXiv:0804.4190 [hep-th].

[7] S. Deser and P. van Nieuwenhuizen, *One-loop divergences of quantized Einstein-Maxwell fields*, Phys. Rev. **D10** 401 (1974);

[8] S. Deser and P. van Nieuwenhuizen, *Nonrenormalizability of the Quantized Einstein-Maxwell System*, Phys. Rev. Lett. **32**, no 5, 245 (1973); id., *Nonrenormalizability of the quantized Dirac-Einstein system*, Phys. Rev. **D10** 411 (1974); P. van Nieuwenhuiozen and J.A.M.Vermaseren, *One loop divergences in the quantum theory of supergravity*, Phys. Lett. **65 B** 263 (1976).

[9] B. Hasslacher and E. Mottola, *Asymptotically free quantum gravity and black holes*, Phys. Letters **B 99**, 221 (1981).

[10] E.S. Fradkin and A.A. Tseytlin, *Conformal Anomaly in Weyl Theory and Anomaly Free Superconformal Theories*, Phys. Lett. **134B** 187 (1984).

[11] G. 't Hooft, *Entangled quantum states in a local deterministic theory*, 2nd Vienna Symposium on the Foundations of Modern Physics (June 2009), ITP-UU-09/77, SPIN-09/30; arXiv:0908.3408.

[12] G. Nordström, Phys. Zeit. **13**, 1126 (1912); F. Ravndal, *Scalar Gravitation and Extra Dimensions*, Invited talk at The Gunnar Nordström Symposium on Theoretical Physics, Helsinki, August 27 - 30, 2003, arXiv:gr-qc/0405030.

[13] G. 't Hooft, *An algorithm for the poles at dimension 4 in the dimensional regularization procedure.* Nucl. Phys. **B62** 444 (1973) .

Scientific Secretaries: L. Alberte, G. Inguglia

DISCUSSION

- Inguglia:

I was wondering if the Hawking radiation does not violate energy conservation? In the sense that it seems that a black hole changes continuously virtual particles coming from the vacuum into real particles, one of them goes to increase the mass of the black hole, the other escapes somewhere in the universe.

- 't Hooft:

This is indeed an important question. It is an important problem for the picture that Hawking came up with when he first derived this phenomenon. Hawking did the derivation using totally standard quantum field theory, so I am not saying that he did anything wrong; he did notice the problem himself. His derivation gives a black hole behaving like a light bulb; what is derived is the probability that a particle is emitted, but it does not give quantum amplitudes, whereas to describe the energy you need oscillating waves including their amplitudes. Consequently, it seems nearly impossible to interpret this calculation in such a way that at the very end energy is conserved. The black hole is like a battle ship radiating very, very soft particles. You would not notice the energy conservation violation, because the thing itself is so heavy. Since this is a classical limit calculation, you do not notice the tiny amount of energy taken away by the Hawking radiation. But if you try to do the calculation precisely then indeed you hit this serious problem. It is one of the reasons why many people, including Steve Giddings, Lenny Susskind, and many other particle physicists, said that we have to do it differently. Probably, the Hawking radiation calculation was a very good classical approximation, but not infinitely precise. We want a precise theory. How do we do that? I think my approach is one possible answer. At all stages I keep exact quantum amplitudes; the black hole complementarity principle says that what happens inside the black hole is information that does not get lost, but it is transformed, translated into amplitudes emerging outside the black hole. The Hawking radiation is not a probabilistic thermal effect anymore, but it becomes quantum mechanically pure in a formal sense. Note that in principle a light bulb is also quantum mechanically pure but in all practical calculations you treat it as a thermal thing. It then seems as if the light bulb also does not exactly conserve the energy. Of course we know that the light bulbs do conserve energy, so that we know exactly what we have to pay for the electricity there. The same holds for the black hole: formally it should preserve energy exactly. It is my aim to have such a theory, but also, I should add, that also my theory at present is not infinitely precise. We have to work at all these details. I think the ultimate answer has to wait. If the theory is successful I am sure that it will conserve energy. You can compare a question that was answered by Niels Bohr, whether the electrons going around the atom actually conserve energy exactly. His guess was: Perhaps not? Who knows? So a valid answer for me would be: Perhaps not. Who knows? Remember that we enter a new field of physics. Maybe our familiar laws are no longer valid? It so happened that Niels Bohr did not have to worry at his time, and also in this case, my expectation is that the energy is conserved.

- Alberte:

I have a question about the conformal transformation which you proposed. You said: let us call the new $\hat{g}_{\mu\nu}$ now $\dfrac{g_{\mu\nu}}{(\det(-g))^{1/4}}$. You said that it corresponds to inversion from the conformal group of transformations?

- 't Hooft:

What I intended to say was that, if you modify the factor in front of $g_{\mu\nu}$, then this could be called a conformal transformation, but we must remember that the curvature of spacetime gets modified, and the Riemann tensor is modified in a complicated way. There is a special subgroup of this set of transformations that is such that flat Minkowski space remains flat Minkowski space, and this is what is usually called the conformal group. This group is obtained if you add to the Poincaré group the pure inversions: x to 1/x , or more precisely, x^μ going to $\lambda x^\mu / x^2$. That is the only extra thing. Complete that with all Poincaré transformations and you get the complete conformal group. Now that is how it is if you wish to transform flat spacetime to flat spacetime. Now add Einstein's curvature. Then you might transform a flat spacetime into any curved spacetime. This gives you much more options and now you can choose the conformal factor in front of the metric to be any spacetime dependent function. So this version of the conformal group is a very large local group. Very different from what we had in flat space. So you have to distinguish these two kinds of conformal transformations.

- Alberte:

But did you use somewhere explicitly the expression for g hat? Because in the rest of your calculations you integrate out the omega. Did you consider somewhere explicitly the omega you proposed?

- 't Hooft:

No. Let's call $\hat{g}_{\mu\nu}$ a pseudo-tensor — or I would rather call it a meta-tensor, because pseudo really means something else in physics. This meta-tensor means that it is an object which is just like the metric tensor, but it transforms in a strange way under coordinate transformations. It transforms, but it has an anomalous component in the transformation rule concerning the Jacobian of the transformation. If you go from a small volume to a large volume in your general coordinate transformation then normally speaking $g_{\mu\nu}$ goes into other structures; the conformal factor in front is determined by the coordinate transformation. Now I am saying that I am going to modify the conformal factor such that the determinant of $g_{\mu\nu}$ is now an invariant, which it was not before. I can postulate that, whenever I make a coordinate transformation, I put an extra factor in front such that the determinant remains the same. This is only a rule about the way that it transforms; it does not affect any of its other properties in principle. So you can use that metric exactly the way you use the original $g_{\mu\nu}$. However, since the transformation law is now different, you have to reconsider what the coordinate transformation is. In practice it is a transformation to which you add a conformal transformation as well. Although the transformation rules are different the dynamics looks very much the same. Except that if you want to see what is invariant under these special transformations you can only allow conformal theories. This leads not to ordinary gravity, but Weyl gravity where you have the square of the Weyl curvature in the action. And all the particles must be coupled conformally invariantly, so no masses or anything like that are allowed. Such a theory is completely conformally invariant. What I derived in my lecture was that I can get such a theory from canonical gravity where the action is the Einstein-Hilbert action, which is not conformally invariant. I have ordinary particles with masses, also not conformally invariant. But integrating over the conformal factor leads to a theory which is completely conformally invariant, except for the conformal anomaly which is now playing tricks. Does this answer the question?

- Alberte:

Almost. So you say that the effective action, after you have integrated out the conformal factor, is a completely conformally invariant theory?

- 't Hooft:

That is what you would have expected if you did not know about the conformal anomalies. Because of the conformal anomalies the problem is more complex. In principle, I want to use a conformally invariant version of gravity. If I get that as an intermediate result using formal manipulations, that would be extremely welcome, because that is what I need for understanding black holes. This way I can put the complementarity principle in a different scheme, and now things can be done which otherwise would be very difficult.

- Alberte:

At the end what you got were modified equations of motion on the slide 35, where you said that $G_{\mu\nu} = G_{\mu\nu}^{(1)} + G_{\mu\nu}^{(2)}$. Could you explain something more about what you meant?

- 't Hooft:

I was talking about the Einstein's tensor $G_{\mu\nu}$. Consider the Weyl action, which is the Weyl curvature squared rather than the curvature. After partial integrations, this amounts to $\sqrt{-g}\left(R_{\mu\nu}^2 - \frac{1}{3}R^2\right)$. That action is quadratic in the Ricci tensor, so if you now compute the variation with respect to the metric tensor, then the Ricci tensor gets a second derivative. Since the Ricci tensor in turn is the second derivative of the metric, all together you get a fourth derivative equation for the metric tensor; you can also say that it comes out as a second derivative equation for the Einstein tensor. Now I simplify things a little bit: I linearize in $G_{\mu\nu}$ and the solutions then basically behave as follows: the solution, $G_{\mu\nu}(x)$, comes in two parts. The first one just corresponds to the conformal transformation, you should have expected that all along. The second part, however, is a slight deviation from the Einstein equations. Einstein's equations would set the $G_{\mu\nu}^{(2)}$ (actually $G_{\mu\nu}^{(1)}$ as well) to be zero. Whereas now it is not itself zero, but its second derivative is zero. This is just the second derivative of $G_{\mu\nu}$ which is zero, ignoring some non leading terms that have higher powers of the curvature.

- Alberte:

Why do you call it "massless" matter?

- 't Hooft:

That is a good question because I've actually been cheating a bit. I called it "massless" because it would be the equation for the massless field if $G_{\mu\nu}$ would be linear in the fields. But of course it is a quadratic expression, so that is not really legal. They are certainly solutions which move with the speed of light so if there are particles generating that energy momentum tensor, those particles better also move with the speed of light. So I think, quite generally this looks like the energy momentum tensor of massless matter, but I put this in quotation marks and do not intend to delete those out.

- Alba:

Can one write our action in Vierbein formalism and preserve conformal invariance? I would like to do it in order to be able to add some spinors to the theory.

- 't Hooft:

I could not quite hear the question. Could you rephrase?

Added in proof: Yes, of course, one has to rewrite the metric in terms of the Vierbein if one wishes to include fermions. There are no insurmountable problems with that.

- Alba:

You are presenting the action in the usual Einstein form, with R tensor, with metric. And also you are presenting a conformal transformation with respect to which this action is invariant.

- 't Hooft:

Yes, I started, as I explained already, with the Einstein-Hilbert action, totally standard action for gravity with approximately standard matter, that is, particles which may or may not have a mass. I coupled those to the gravitational field, and now the non standard thing is to first integrate over the conformal factor, not the metric and then keep what you are left over with. And that is an effective action at that scale. You do such things more often in a field theory. Think, for instance, of QCD, where you integrate over the gluons, you get the effective action for the quarks, which are confined, making mesons. You might think of an effective field theory in terms of mesons and baryons rather than the original fields such as quarks and the gluons. So that is a stage in the calculation. I do something similar here. I have integrated over a part of the fields and not all of them. This gives an effective field theory which unfortunately has a divergent action, and that is the new feature of this whole thing.

- Alba:

You told that it is possible to make such a transformation that changes in and out parts of a black hole. Is it important what kind of black hole is it? Just a usual black hole, or with angular momentum, or with charge?

- 't Hooft:

This picture was designed for the simplest case. Even simpler than the simplest case usually considered. The Schwarzschild black hole formed by a single shell of imploding matter. Normally when you have a Schwarzschild black hole, it is formed by an imploding star which is itself rather complex. Or two black holes or two objects meeting to form a black hole. That is a rather complex physical event. This I simplified as much as possible. Take a spherically symmetric arrangement of ingoing matter in a single shell. It collapses just like that and the advantage of this is that you can write down the exact metric for this configuration, when it collapses. And, if the Hawking radiation could also be described in this simple way, you also have an exact metric for the outgoing situation. Now I have an exact metric which I can map on the other exact metric. Here, I make the observation that the inside invites me to do a conformal transformation of the kind I mentioned where the inversion point is basically the tip of these two triangles. So the tip of one triangle is the inversion point which goes to infinity in the other triangle, and the converse.

- Alba:

But what will you obtain if you replace the Schwarzschild metric with Reissner-Nordström metric?

- 't Hooft:

Well, I think that is a hard and very good question because the Maxwell theory itself is conformally invariant. In other words, it should be not such a problem to put electro magnetic fields in, because it so happens that they are conformally invariant as well. So I would have a charge on the tip there somewhere or a source which is charged. Or perhaps if I have a charge in the ingoing and outgoing fields then the inside is the Faraday cage and nothing changes because the charge sits at the outside of the imploding shell. However, I could also put a charged particle in, I could put it here and then it would also be a charged particle in the out-frame, because Maxwell's theory happens to be conformally invariant. If you would make a Reissner-Nordström metric by having a charged shell of matter then the inside would not change at all because it is a Faraday cage.

- Alba:

What if you would add some angular momentum?

- 't Hooft:

I was worried that you'd ask this question. In that case I do not know because then the inside would not be the Faraday cage anymore and you would have to describe how such a black hole is produced out of a rotating collapsing dust shell. That is much more complicated. The rotating black holes, as you know, are no longer spherically symmetric; they are only cylindrically symmetric. That is a very tough calculation. So I would not really have an answer to that.

- Alba:

You have told us that there is some special kind of quantization and you have explained us in the question which was asked right after the lecture that in this quantization we consider the limit when $\hbar$ goes to zero. This is actually a classical limit when $\hbar$ is equal to zero. So I was confused – why do we still call it quantization?

- 't Hooft:

A system where $\hbar$ is equal to zero is a classical system; you can treat it as a very heavy particle. Think of the heavy ion collisions at LHC, for instance. These ions are practically classical objects, because they are so big. But you can still say: I am working in a quantum theory. So, yes, $\hbar$ is flying around everywhere, but under some circumstances I can ignore it, consider it to be very small. If you make $\hbar$ very, very small but not zero, as you may well know, you get a theory, which at first sight looks just classical: the commutators vanish. But if you look more precisely, you see that a lot of things are happening. For instance the wave packets of the objects now have to be replaced by the waves in WKB approximation which only in a very weak sense look classical, while, if you look more precisely it is quite a bit more complicated. Then I would say: Well, I have nearly but not quite taken a classical limit. I have replaced all waves with a very, very dense WKB approximation of waves. Hence, this is not quite a classical theory. In the same sense I would say that this is what I would prefer to do; the theory is sort of pseudo classical but not quite because I am still talking about WKB approximations.

- Alba:

Do I understand correctly that $\hbar$ is very, very small but not exactly equal to zero. Because, equally, I can see Schrödinger equation and regard some special ansatz when $\psi = e^S$, then I will get just the Jacobi equation and full classical equation?

- 't Hooft:

You have to remember that the $\hbar$ in the physical world that we know is not a tunable number. It is a definite number, you cannot change it. I am not saying that $\hbar$ is different for gravity than it is for any other matter. No, it is the number it is, but these fields that I am talking about essentially behave as if they were very, very massive. So rather than saying that I let $\hbar$ go to zero, I could instead better say that the effective mass of these fields goes to infinity. These fields are like classical fields. They are very strong fields. Quantum effects in gravity are very difficult to detect because the action describing even a tiny amount of curvature in the space-time metric is very large. You need a planet as heavy as the Earth to cause any noticeable curvature in space-time. In every practical sense $\hbar$ is very small for the gravitational field.

- Schmidt-Sommerfeld:

Time reversal invariance was sort of one of your driving principles, if I understand it correctly?

- 't Hooft:

Yes.

- Schmidt-Sommerfeld:

So in the naïve traditional picture the formation of the black hole proceeds rather quickly whereas the evaporation takes very long time. Does that get changed in your picture?

- 't Hooft:

No, I do want a theory of black holes which approaches the standard theory as well as possible, in particular large black holes should be effectively correctly described the way we usually do. I do not want to upset everything in physics. I only want to consider the changes which become noticeable at or near the Planck length. At scales that can be studied experimentally today, we believe that the gravitational field is correctly described by Einstein. Hawking did his calculation using what we all think is correct quantum field theory, and we all think it is a good theory for a black hole. So his calculation cannot be entirely wrong. It has to be basically correct. Only in the way we interpret things quantum mechanically there must be some very subtle and very important changes. That is the same thing as saying that although high energy physics and atomic physics are described by quantum mechanics, if we take a bottle of hydrogen this big it behaves classically. In the same sense our theory should approach ordinary physics in a sensible way when I look at ordinary circumstances.Time reversal invariance is in the amplitudes, but not in the average physical effect. You know that a light bulb is very well approximated by the Standard Model. The Standard Model is time reversal invariant, but the light bulb is not. A glass of water is described by a PCT invariant model, the Standard Model. But if I pour a glass of water then trying to do time reversal is very, very hard. In the same sense a large black hole presents itself to us as if it is not at all invariant under time reversal but we suspect that the microphysics is. My aim is a theory of microphysics that is time reversal invariant.

- Dunin-Barkowski:

Let us consider the spherically symmetric formation of a black hole. For example, if we have some spherically symmetric body which collapses and turns into a Schwarzschild black hole. If we are distant observers we would see that the black hole formation takes infinite amount of time. What if the formation of the black hole were asymmetric? Can it happen that we could see a black hole being formed in a finite amount of time?

- 't Hooft:

No, as I explained earlier, the theory which I want to construct should coincide with the standard theory for large objects, certainly for astronomical black holes, and to some extent even for small black holes. It should be the standard theory. There, the formation happens apparently very quickly; in fact, a black hole this big is formed in the amount of time that light needs to go about this distance. Then we get an exponential approach to the black hole. It is exponential, so that in principle it takes infinite amount of time for it to become an exact copy of the Schwarzschild metric, but we note that the coefficient in the exponential is what takes the light to travel this distance. So for all practical purposes, after the light has traveled this distance, one has a very, very good approximation of a Schwarzschild metric. And as soon as that happens, we know that the evaporation, the temperature of the radiation coming out of the black hole is such that the radiation has this average wavelength, which is actually quite large so that we are talking about a very low temperature. This means that it takes ages for a black hole to evaporate all the way! A black hole this big has the mass of the Earth, so how long does it take for the mass of the Earth to be radiated away with thermal wavelengths like that? It takes basically forever. That should not affect what I am talking about: a quantum theory which reproduces this behavior that we think we understand quite well for large black holes. It is the amplitude I am talking about. And of course the whole problem becomes extremely nontrivial when the black holes are not this big but as small as the Planck length. Then everything happens very fast and then in the exact theory it is very important to understand what is going on.

The QGCW Project - Technological Challenges to Study the New World
Horst Wenninger, CERN, Geneva, CH

During his lecture at the 2006 Erice School (ISSP 2006) Professor A. Zichichi[1] presented the **QGCW Project** to study the properties of - what he called - the **" new world "** which should be a source of totally unexpected phenomena, produced in a collision between heavy nuclei ($_{208}Pb^{82+}$) at extreme energy soon available with the LHC heavy ion collisions at CERN.

The " new world " is the Quark-Gluon-Coloured-World (**QGCW**). We avoid calling it " quark-gluon-plasma " since in the extremely high energy collisions between heavy ions, many QCD open-colour-states should be produced, far higher in number than the baryons and mesons known so-far, since these baryons and mesons have to obey the condition of being QCD-colourless.

A. Zichichi also concluded then, that we must be prepared with the most advanced accelerator and detector technology for the discovery of totally unexpected events. A recent outline of the QGCW Project can be found in the document:

CONFERENCE IN HONOUR OF
MURRAY GELL-MANN'S 80TH BIRTHDAY CELEBRATIONS

MURRAY GELL-MANN
AND THE LAST FRONTIER
OF LHC PHYSICS:
THE QGCW PROJECT

A. Zichichi
INFN and University of Bologna, Italy
CERN, Geneva, Switzerland
Enrico Fermi Centre, Rome, Italy
Ettore Majorana Foundation and Centre for Scientific Culture, Erice, Italy

---

24–26 February 2010
Nanyang Executive Centre, Nanyang Technological University, Singapore

A number of questions have been addressed in the above document, of which I cite below the most important ones:

(i) In the QGCW there are all states allowed by the $SU(3)_C$ colour group. As mentioned above, the number of possible states is by far more numerous than the number of colourless baryons and mesons, which have so far been built in all Labs, since the colourless condition is not needed. What are the consequences on the properties of the QGCW?

(ii) Light quarks versus heavy quarks: are the coloured quark masses **the same** as the values we derive from the fact that baryons and mesons need to be in a colourless state? It could be that all six quark flavours are associated with nearly 'mass-less' states like those of the 1st family (u, d). In other words the reason why the 'top' quark appears to be so heavy ($\simeq 10^2$ GeV) could be due to the fact that it must satisfy some, so far unknown, condition related to the fact that the final state must be QCD-'colourless'. We know that confinement produces masses of the order of a GeV.

Therefore, according to our present understanding, the QCD 'colourless' condition could not explain the heavy quark mass, but since the origin of the quark masses is still not known, it cannot be excluded that in a QCD coloured world, the six quarks are all nearly mass-less. If this was the case, the masses we measure are heavier than the effective coloured quark masses. In this case all possible states generated by 'heavy' quarks would be produced in the **QGCW** at much less temperature than the one needed in our world made with baryons and mesons, i.e. QCD colourless states. Here again we should try to see if with masses totally different from those expected, on the basis of what we know about colourless baryons and mesons, new **effects** could be detected due to the existence of all six flavours at relatively low temperature in the QGCW world.

(iii) We need to search for effects on the thermodynamic properties of the QGCW.

(iv) We need to derive the equivalent Stefan-Boltzmann Radiation Law for the QGCW.

The relation between energy density at emission **U** and Temperature **T** of the source is in classical Thermodynamics.

$$U = c \cdot T^4$$

In the QGCW the correspondence should be $U = p_\perp$ (transverse momentum); T = average energy $\langle E \rangle$ in the cm-system. In the QGCW the production of 'heavy' flavours should be studied versus $\langle p_\perp \rangle$ and versus $\langle E \rangle$. The expectation is

$$\langle p_\perp \rangle = C \cdot \langle E \rangle^4$$

and any deviation would be extremely important!

How to study the new world: QGCW?

Beams of known particles (p, n, π, K, μ, e, γ, ν) bombard the QGCW and a special set of detectors measures the properties of the out coming particles (*see schematic drawing from the lecture by Prof. A. Zichichi (ISSP 2006)*).

Totally unexpected effects should show up.

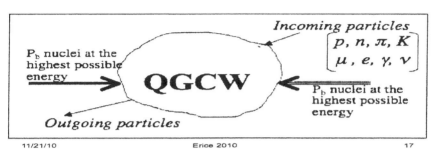

Before entering into the discussion and question of the **Technological Challenges to Study the New World** we will review evidence and characteristic signatures for the formation of the quarks gluon plasma in nuclear collisions.

It started in 1969 when the SLAC scattering experiment showed that the proton is a composite object made of partons. With first indications from the Gargamelle experiment at CERN in 1973 and the experimental evidence in 1974 of the charm quark in BNL / SLAC experiments (called the November 1974 Revolution) the QCD theory of strong interactions, developed during that period, had its breakthrough

An intensive experimental search for the predicted fractional charged quarks started

Experimental search for fractional charged quarks

A search for quarks in the CERN SPS neutrino beam
Basile, M ; Cara Romeo, G ; Cifarelli, Luisa ; Contin, A ; D'Ali, G ; Giusti, P ; Massam, Thomas ; Palmonari, F ; Sartorelli, G ; Valenti, G et al. A. Zichichi
CERN 1978 - Published in : Nuovo Cimento A: 46 (1978) , pp. 281-299 Fulltext

Results of a search for free quarks with a large avalanche chamber in high-energy neutrino interactions
/ Basile, M ; Berbiers, Julien ; Cara Romeo, G ; Castelvetri, A ; Cifarelli, Luisa ; Contin, A ; D'Ali, G ; Giusti, P ; Itoh, H ; Laakso, I et al   A. Zichichi
CERN 1991 - Published in : Nuovo Cimento A: 104 (1991) , pp. 405-436

until it became evident that one cannot get free quarks out of hadrons, in other words quarks are confined (**colour confinement**).

First mention of a possible " quark liberation" can already be found in a theoretical paper by Cabibbo et al. in 1975[2], proposing that the observed exponentially increasing hadronic spectrum is connected to the existence of a different phase of the vacuum in which quarks are not confined.

188

Such a phase transitions is observed in relativistic heavy ion collisions above a critical point when a quark gluon plasma is developing with "free" quarks.

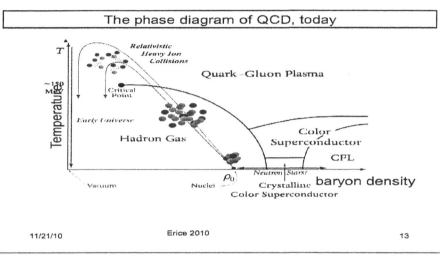

The phase diagram of QCD, today

11/21/10   Erice 2010   13

Courtesy: Summer Student Lectures 2010 CERN by Carlos Lourenço (3). Copyright CERN

With the start of the LHC supercollider, the highest energy density of sub-nuclear matter will become real physics, when lead-lead nuclei will collide at the maximum possible energy of 10^{15} eV. A quark-gluon plasma is formed and within a short moment a tiny interaction volume contains the quark gluon colored world "QGCW". Therefore, if we bombard this QGCW–volume with photons and charged leptons they should go more or less through. But if we bombard the same QGCW–volume with protons, pions or any other hadron, the hadrons should not go through.

What did we learn from RHIC about the formation of the quark-gluon-plasma ?
Paul Sorensen presented in his lecture during the Erice ISSP 2008 a schematic layout of the different phases of heavy ion hard collisions illustrating dimensions and timings inherent in the collision process.

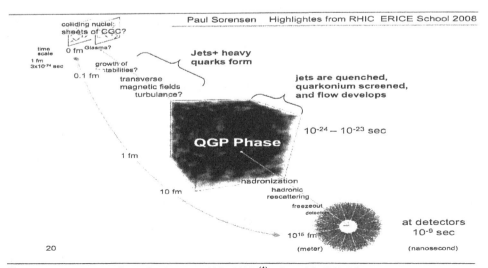

Courtesy: ERICE lecture by Prof P. Sorensen (ISSP 2008) [4]. Copyright ISSP Erice

At first the two colliding nuclei penetrate one another – the quarks and gluons collide and transfer large amount of energy from projectile to the vacuum, energy later observed as new particles (time 3 x 10^{-24} sec). From the very energetic "hard" collisions additional gluons and quarks are produced. These new gluons and quarks together with those present initially undergo a cascade of further collisions, which slow down and stop the nuclei. The plasma phase in local equilibrium lasts 3 times the penetration phase – about $10 - 30$ x 10^{-23} sec. The hadronization transition from plasma to cooled expansion is $10 - 30$ x 10^{-24} sec.

Following the understanding of the detailed collision process and learning from RHIC we need to analyze the best signatures of quark-gluon-plasma formation in high energy heavy-ion collisions. **We need to identify good probes of QCD matter and calibrate them thus finding new physics related to the QGCW.**

Leptonic probes, γ, e+e–, μ+μ– carry information about the spectrum of electromagnetic current fluctuations in the QGP state;

the abundance of quarkonia Ψ, Ψ', Y, Y' (also observed via l+l–) carry information about the chromoelectric field fluctuations in the QGP ;

the arsenal of hadronic probes, π, K, p, p$^-$, Λ, Ξ, Ω, φ, ρ, . . . provide information on the quark flavor chemistry and baryon number transport.

Theory suggests that with decays such as $\rho \rightarrow$ e+e– the properties of the hadronization and chiral symmetry breaking can be indirectly studied.

The central problem with the above probes is that they are all indirect messengers. The "probes" must be synchronized with the system they probe.

Considering the time scales and evolution of the QCD matter produced in heavy ion collisions that is of order

$$\boxed{10^{-22} - 10^{-24}\ seconds}$$

good "probes" are produced together with the system they probe! They must also be created very early in the collision evolution, so that they exist *before* the QGP might be formed. This suggests that indeed *Jets* and *heavy quarkonia* might be good QCD matter probes!

We also must have probes, not affected by the dense QCD matter, to serve as reference: photons, Drell-Yan di-muons. We also must have reference collision systems, to understand how the probes are affected in the absence of " new physics ": pp, p-nucleus, and light ion collisions.

With all the accumulated knowledge learned at SPS and RHIC experiments, we have to understand the expected changes from RHIC to LHC energies as are shown below:

In the table shown below, we compare the machine parameters of SPS, RHIC , LHC and expect to learn more about the QGCW at LHC energies.

	SPS	RHIC	LHC	
from SPS to RHIC to LHC				
$\sqrt{s_{NN}}$ (GeV)	17	200	5500 [2750 in 2010]	
dN_{ch}/dy	500	850	1500–4000	
τ^0_{QGP} (fm/c)	1	0.2	0.1	
T/T_c	1.1	1.9	3–4	Hotter
ε (GeV/fm^3)	3	5	15–60	Denser
τ_{QGP} (fm/c)	≤2	2–4	≥10	Longer
τ_f (fm/c)	~10	20–30	30–40	
V_f (fm^3)	few 10^3	few 10^4	few 10^5	Bigger

40/22/10   Erice 2010

The LHC is operational and heavy ion runs have meanwhile started with impressive pictures from Alice, the dedicated heavy ion experiment at the LHC.

ACCELERATOR TECHNOLOGY

The LHC physics program foresees lead-lead collisions with a design luminosity of 10^{27} cm^{-2} sec^{-1}. For this to be achieved an upgrade of the ion injector chain comprising Linac3, LEIR, PS and SPS machine was needed. Each LHC ring will be filled in 10 min by almost 600 bunches, each of 10^7 lead ions. Central to the scheme is the Low Energy Ion Ring (LEIR), which transforms long pulses from Linac3 into high brilliance bunches by multi-turn injection, electron cooling and accumulation. The total collision energy between heavy ions, $_{208}$Pb^{82+} (fully stripped), is 1150 TeV.

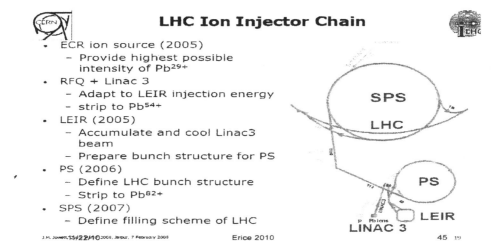

Courtesy: J. Jowett [5]. Copyright CERN

The first step in the accelerator technology to study the QGCW refers to the simultaneous availability of a proton beam able to bombard the QGCW produced in the lead-lead collisions. These protons could be provided by a beam-line from the SP2 to the Alice pit at LHC to be included into the future injector chain

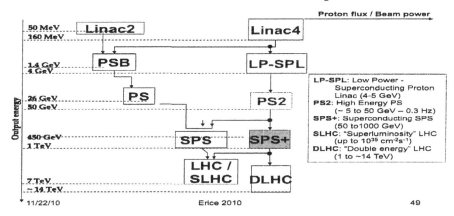

Courtesy: J. Jowett Copyright CERN

A design of a beam extraction from PS2 to a fixed target underground area is under study. A prolongation of the tunnel / proton beam line to pit 2 of LHC would be required.

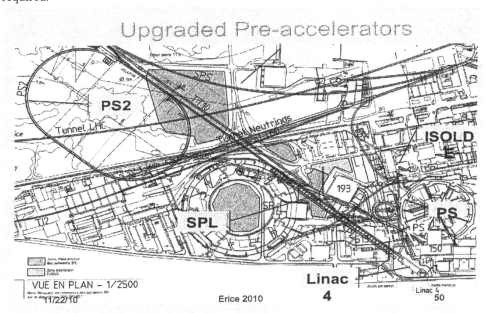

R&D work for the QGCW project

In the following part we describe the proposed first steps of R&D work in preparation for the QGCW Project:

Development work proposed at 3 levels to prepare the future of the QGCW project

(1) **Beams** to inject into pit 2 of LHC (sc- magnet R&D)

(2) **Timing** and beam synchronization (optical fibers)

(3) **Detectors** (TOF and electronics)

(1) To get a proton beam as described above we follow the R&D initiative for the high luminosity LHC, which needs new fast cycling superconducting dipole magnets. A joint proposal by CERN and FAIR is under preparation to be co-funded by the EU.

The DISCORAP Project, Genoa, Italy (courtesy: Lucio Rossi[6] , CERN)
The principal aim of this R&D activity is the design and construction of a cold mass model of the short dipole, with target performances reported in Table I.

Nominal central field	(T)	4.5
Nominal field ramp-rate	(T/s)	1
Maximum operating (coolant) temperature	(K)	4.7
Maximum allowable AC loss	(W/m)	10
Coil aperture	(mm)	100
Magnetic length	(m)	3.9
Beam curvature radius	(m)	66.67
Field quality	(unit @ 35 mm)	1

Table I – target performances of the DISCORAP model magnet

The model magnet represents the first representative test in view of the construction of the curved fast-cycled superconducting dipole magnets for SIS300 at FAIR. After a preliminary test in a vertical cryostat, the cold mass will be integrated into a horizontal cryostat for a test at GSI. The test will provide a first validation of the technical solutions adopted and of the construction technology. The next step should then be the realization of prototype magnets, to lead into the series production.

(2) Once the lead-lead collision is available, the problem is to synchronize the "proton" beam with the QGCW produced. This problem is at present under study.

Precise timing and synchronization between the colliding heavy ions, forming in their collision the quark gluon colored world, and the bombarding particle, injected simultaneously to probe the QGCW, and the detectors, triggering on the emerging particles is the challenge of the proposed experiment.

A summary of the timing topic is given below. R&D started on optical fibers.

REVIEW OF ACCELERATOR TIMING SYSTEMS
T. Korhonen, Paul Scherrer Institut, Switzerland

Citation of this review: *" The required time resolution depends mainly on RF frequency of accelerators. The control of one particle bunch or one RF bucket is usually the smallest required unit of accuracy."*

Ongoing R&D on timing and synchronization systems have currently started for the FAIR project at GSI, aiming at synchronizing with maximum precision and stability a series of complex heavy ion accelerators ring, experiments, and a high power laser. Synchronization and timings at **sub-nanosecond scale** are discussed.

Interesting new results on optical synchronization of a Free-Electron Laser FLASH with precision at **femtosecond scale** are reported fro DESY.

11/22/10   Erice 2010   59

At CERN the present timing systems all rely on the performance of a 35 km optical fiber network:

Types of CERN timing systems

- **General Machine Timing (GMT)**
  - Based on UTC-synchronous 40.000 MHz.
  - 500 kbit/s over fiber and twisted pair (RS-422).
  - Granularity: 1 ms.
  - Jitter < 1 ns.
- **Beam Synchronous Timing (BST)**
  - Based on TTC technology (see below).
  - Encodes messages in TTC data channel using bunch crossing frequency for LHC (40.079 MHz). Fiber-based.
  - Granularity: 1 LHC revolution (89 μs).
  - Jitter < 1ns.
- **Timing Trigger and Control (TTC)**
  - Technology to multiplex Revolution tick and data in a single stream.
  - Experiments use it without data to have better clock recovery.

11/22/10   Erice 2010   60

CERN's 35,000 km of optical fibers are used to synchronize the accelerators, take measurements of the beams and to send controls to the LHC. Over the past 7 years, special optical fibers have been developed at CERN that can resist the radiation levels of the LHC. The qualification testing for these fibers is now being finalized. Over 2,500 km of special radiation-resistant optical fibers have been installed in the LHC, and their success has garnered the interest of other institutes dealing with radiation. "

Advanced R&D has also started for FAIR at GSI. The "Bunch-phase Timing System (BuTiS)" for FAIR concentrates on the research of thermal stability properties of optical fibers. The work is done by the optical communications institute of TU Darmstadt under contract with GSI. The term "bunch phase" shall emphasize the kind of timing precision that is dealt with in the system.

(3) Apart from R&D for the accelerator technological challenges related to the QGCW project, there is also a need to improve on the detectors and resume intense R&D works.

One example is the further improvement of time resolution for the Alice Time-of-Flight (TOF) detector. This detector is based on the technology of "Multigap Resistive Plate Chambers (MRPC) ", developed in particular in the framework of the LAA project at CERN, a long-range R&D project, which is lead by Prof. Zichichi [7].

Excellent time resolution has already been obtained (see below) allowing for particle identification in the Alice heavy ion experiment, but improvement are possible and required to meet the goals of the QGCW project.

THE DETECTOR TECHNOLOGY
The detector technology is also under intense R&D since the synchronization needed is at a very high level of precision.
The basic point is that our instruments must allow us to measure, as precisely as possible, the properties of all subnuclear particles coming out from the QGCW. One of these properties is the Time-Of-Flight (TOF).

THE MULTIGAP RESISTIVE PLATE CHAMBER – MRPC – the base of TOF

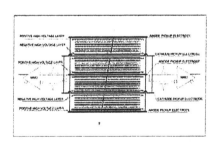 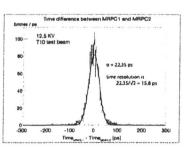

67/22/10   Erice 2010

The R&D initiatives mentioned above are only first steps towards more extended R&D works, proposed within the LAA long range R&D strategy in the framework of the Quark-Gluon-Colored-World project, initiated and guided by Prof. A. Zichichi.

References:

(1) Antonino Zichichi, Lecture at Erice School 2006
(2) N.Cabibbo and G. Parisi, Phys Lett B59 (1975) 67
(3) Summer Student Lectures 2010 CERN:
    From High-Energy Heavy-Ion Collisions to Quark Matter by Carlos Lourenco
(4) Paul Sorensen, Highlights from RHIC , Lecture at Erice School 2008
(5) J.M. Jowett, CERN: Quark Matter 2008, Japan, 7. Feb. 2008
(6) Lucio Rossi / CERN, private communication
(7) 6th International Zermatt Symposium Creativity in Science by Antonino Zichichi

Scientific Secretaries: V. Alba, S. Mironov

DISCUSSION

- *Burda:*

Could experiments with ions collisions at LHC energies tell us more about Gauge/Gravity correspondence than experiments at RHIC – energies?

- *Wenninger:*

I suppose we can get more information from the LHC experiment. We are going to use different time scales and plasma parameter on RHIC and LHC. This will help us.

- *Dennen:*

In the phase diagram for QGP,
- where does the diagram come from?
- could you summarize how much of the diagram has been probed experimentally?

- *Vidgor:*

I'm not sure which particular theoreticians made this. I mean that very little of that phase diagram is based on hard theory, there are very sophisticated lattice QCD calculations which carried out at the left hand extreme, so lattice QCD in its normal usage is quite applicable at zero density. There is very active field right now and one tries to push the lattice calculations towards the critical point but there is technical problem which is that normally in lattice calculations you use a Monte Carlo approach to evaluate the integral which is involved in action. And at non zero density the approximations really breaks down. So you can not use normal lattice approaches. So people are working on other approximations to work their way up towards the critical point. Right now there is area of lattice calculations which tries to estimate the non-zero value of density in which critical point will occur. They differs by factors 2 or 3 in their predictions for the chemical potential. At the right hand extreme where you are dealing with neutron there are no really hard calculations. Approximation to QCD which should be valid at high density but there are not rigorous calculation of the sort that lattice QCD is. In terms of what we have tested experimentally RHIC now has pretty good data near the left hand extreme. RHIC has just started in the past year an energy scan for low energy where we tried to make measurements of the baryon chemical potential. On the map there are data from XPS & AGS where we have primarily temperature of kinetic freeze out where hadron spectrum are frozen. Most of the diagram is unmapped experimentally and most of it contains speculations from the point of view of theory. There are good theoretical arguments why at the high density there should be first order phase transition. So somewhere there is a line of first order phase transition which is quite close to the critical point but nobody really knows where it is.

- *Wenninger:*

I put the transparency up because you can find most of it on the web page of Erice schools of past years.

- *Wenninger:*

They will change a lot of the hardware and in 2016 they will prepare to increase the luminosity. This is very good that LHC is starting now, because we have a great opportunity now, and the future is completely opened for young scientists. And we only need the people to stay on the field for the future progress. And this is a great idea of Zichichi to prepare the future today. That is the reason why he established this school.

- Mao:

Why the baryon chemical potential at LHC is approaching 0 (much smaller than RHIC) since we all know the medium generated at LHC is much hotter and denser compared to RHIC? How do we decide the LHC point on phase diagram?

- Vidgor:

It is simply because net baryon density goes to zero with increased energy. RHIC already extract thermo dynamical potential from the thermal model fits.

- Mao:

I want to know how we can measure experimentally?

- Vidgor:

The way it will be done in LHC is also by making thermal model fit. That fit gives you what is called chemical freeze out.

- Wenninger:

Chemical potential is decided by thermal model based on the measurements of particle production ratio. According to thermal model, higher temperature will give a smaller chemical potential, which is model dependent somehow.

- Miyazaki:

What is the definition of t^0_{QGP} in the table? It has an opposite behavior of what one would expect.

- Wenninger:

It is defined as start of thermalization. But all these times are quite approximately.

- Korthals-Altes:

Is the FAIR facility not meant to measure (hopefully!) this critical end point?

- Wenninger:

Yes, it will try. As its energy is lower than that at RHIC and LHC it will scan an area that will include the critical end point.

The LHC and Beyond – Status, Results and Perspectives

Rolf-Dieter Heuer
Director-General
CERN, CH-1211 Geneva 23, Switzerland

Abstract

This paper presents the status, physics results and perspectives for the Large Hadron Collider and outlines options for high-energy colliders at the energy frontier for the years to come. First exciting physics results from the LHC experiments were presented at the International Conference on High Energy Physics in July 2010. The immediate plans for CERN include the exploitation of the LHC at its nominal design luminosity and energy as well as upgrades to the LHC and its injectors. This may be followed by a linear electron-positron collider, based on the technology being developed for the Compact Linear Collider and for the International Linear Collider, or by a high-energy electron-proton machine, the LHeC. This paper also provides options for future directions, all of which have a unique value to add to experimental particle physics, and concludes by outlining key messages for the way forward.

The Large Hadron Collider

The Large Hadron Collider (LHC) [1] is primarily a proton-proton collider (see Figure 1) with a design centre-of-mass energy of 14 TeV and nominal luminosity of 10^{34} cm^{-2} s^{-1}, and will also be operated in heavy-ion mode. The high collision rates and the tens of interactions per crossing result in an enormous challenge for the detectors and for the collection, storage and analysis of the data.

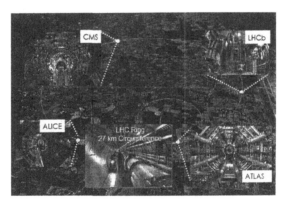

Figure 1: The LHC accelerator and experiments. There are also three smaller experiments – LHCf, MoEDAL and TOTEM.

By colliding unparalleled high-energy and high-intensity beams, the LHC will open up previously unexplored territory at the TeV scale in great detail, allowing the experiments to probe deeper inside matter and providing further understanding of processes that occurred very early in the history of the Universe.

LHC Operations

The start-up of the LHC on 10 September 2008 was a great success for both the accelerator and the experiments. Circulating beams were established rapidly and the beams were captured by the radiofrequency system with optimum injection phasing and with the correct reference. The incident of 19 September 2008, caused by a faulty inter-magnet bus-bar splice, resulted in significant damage in Sector 3-4 of the accelerator. Actions were taken immediately to repair the damage and to introduce measures to avoid any re-occurrence. The damaged thirty-nine main dipole magnets and fourteen quadrupole magnets were removed and replaced. Fast pressure release valves (DN200) were added on the main magnets, an improved anchoring on the vacuum barriers was introduced around the ring, and an enhanced quench protection system was implemented. This has resulted in a major amount of work and any remaining risks to the LHC, due to the shortcomings of copper-stabilizer joints of the main LHC magnets, are minimized by limiting the top beam energy in the first years of LHC operation.

Excellent progress was made in the above-mentioned repair, consolidation and improvement work, and first collisions at the LHC were recorded by the experiments on 23 November 2009 at a centre-of-mass energy of 900 GeV. During this first physics run at the end of 2009, the LHC accelerator performed exceptionally and the readiness of the experiments and the computing was excellent, resulting in impressive preliminary results provided already at an open seminar held at CERN on 18 December 2009 and the prompt publication of the first physics results by year's end.

First LHC beams for 2010 were available on 27 February for commissioning the accelerator with beam. This was followed by first physics collisions at 7 TeV centre-of-mass energy on 20 March (see Figure 2) and by the first physics runs with a stronger focusing at the interaction points. During the 2009 and 2010 LHC physics runs, data has been collected at 900 GeV, 2.36 TeV and 7 TeV centre-of-mass energies. The LHC instantaneous luminosity has increased by several orders of magnitude since the beginning of the 2010 run.

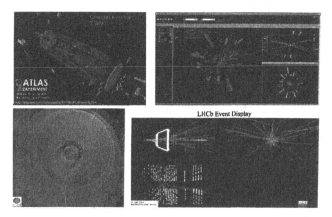

Figure 2: First collisions at 7 TeV centre-of-mass energy.

CERN has taken the following decisions that will allow the LHC to provide substantial physics in 2010-2011 and be technically capable of operating at the design energy and high intensities as of 2013:

- The LHC will be operated at 3.5 TeV/beam during 2010 and 2011, with a target integrated luminosity of 1 fb^{-1} and with a heavy-ion run at the end of both years.

- This extended operations period will be followed by a long shutdown (of the order of about 15 months) in 2012-2013 to repair and consolidate the inter-magnet copper-stabilizers to allow for safe operation at 7 TeV/beam for the lifetime of the LHC.

- In the shadow of the inter-magnet copper stabilizer work, the installation of the fast pressure release valves will be completed and between two and five magnets which are known to have problems for high energy will be repaired or replaced. In addition, SPS upgrade work will be carried out.

Present Physics Results

Several presentations were made by the LHC collaborations at the International Conference on High Energy Physics (ICHEP-2010), which was held in July 2010 in Paris. The experiments have successfully been recording data with increasing instantaneous luminosities of the LHC with excellent efficiency and the computing infrastructure has been facilitating in a very effective way the swift analysis of the data. Salient physics results presented included:

- The ATLAS and CMS experiments aim to elucidate the nature of electroweak symmetry breaking, for which the Higgs mechanism and the accompanying Higgs boson(s) are presumed to be responsible, and to also search for new particles, such as those predicted by Supersymmetry (SUSY). ATLAS and CMS are currently re-discovering the Standard Model and have observed the known particles of the Standard Model, including the W and Z bosons and the top quark. ATLAS and CMS are now in the position of discovering new physics at the LHC.

- The primary objective of the ALICE experiment is the study of the state of matter at high temperature and energy density in heavy-ion collisions – the so-called quark-gluon plasma. In addition, ALICE has collected data in LHC proton-proton collision mode to obtain reference data for their heavy-ion programme. At ICHEP-2010, ALICE presented results on charged particle production multiplicities and the ratio Λ-to-K_0^s, amongst other physics results.

- The LHCb experiment has the primary objective of searching for new physics through CP violation and rare decays of heavy-flavour particles, primarily B- and D- mesons. At ICHEP-2010, LHCb presented results on open charm, prompt J/ψ and bottom-antibottom production cross-sections at 7 TeV centre-of-mass energy, amongst other results. It is expected the LHCb can perform excellent measurements in the charm sector with an integrated luminosity of 50 pb^{-1} and with 1 fb^{-1} exciting prospects for the discovery of new physics may be possible in $B_s \rightarrow \mu\mu$ and $B_s \rightarrow J/\psi \, \phi$.

Many more physics analyses are ongoing in all four of these experiments.

Physics Prospects

In order to make significant inroads into the Standard Model Higgs Boson search, sizeable integrated luminosities of several fb^{-1} are needed. However, even with 1 fb^{-1} per experiment, discovery of the Standard Model Higgs Boson is still possible in certain mass regions beyond the lower limit of 114.4 GeV from direct searches at LEP2. Preliminary studies indicate that at a centre-of-mass energy of 7 TeV, combining the results from ATLAS and CMS would provide a 3σ sensitivity for 300 pb^{-1} per experiment for a Standard Model Higgs Boson mass of 160 GeV, and will exclude the Standard Model Higgs Boson between 145 GeV and 180 GeV for 1 fb^{-1} per experiment. Exclusion of the full mass range down to the LEP2 lower limit requires ~1.5 fb^{-1} per experiment at 14 TeV centre-of-mass energy, while the discovery of a the Standard Model Higgs Boson at the LEP2 lower limit requires 10 fb^{-1} per experiment at 14 TeV centre-of-mass energy.

The reach for additional new physics at the LHC is considerable already at LHC start-up. In SUSY theory, due to their high production cross-sections, squarks and gluinos can be produced in significant numbers even at modest luminosities. This would enable the LHC to start probing the nature of dark matter. The LHC discovery reach for SUSY particles is up to masses of about 400 GeV for 100 pb^{-1} and up to 800 GeV for 1 fb^{-1} per experiment at 7 TeV centre-of-mass energy. The mass discovery reach for the new heavy bosons Z' and W' is 1.5 TeV and 1.9 TeV, respectively, for 1 fb^{-1} per experiment at 7 TeV centre-of-mass energy.

The LHC will also provide information on the unification of forces, the number of space-time dimensions and on matter-antimatter asymmetry. With the heavy-ion collision mode, the LHC will probe the formation of the quark-plasma at the beginning of the Universe.

LHC Upgrade and Consolidation

The coming years will lay the foundation for the next decades of high-energy physics at CERN. The research programme until around 2030 is determined by the full exploitation of the LHC physics potential, consisting of the design luminosity and the high-luminosity upgrade (HL-LHC), together with focused R&D for a Linear Collider (machine and detectors) and for superconducting higher-field magnets for a higher-energy proton collider (HE-LHC), if necessitated by the physics. These initiatives will position CERN as the Laboratory at the energy frontier.

The strategy for the LHC for the coming years is the following:

- Exploitation of the physics potential of the LHC up to design conditions in the light of running experience and by optimizing the schedule for physics;

- Preparation of the LHC for a long operational lifetime through appropriate modifications and consolidation to the machine and detectors and through the build-up of an adequate spares inventory;

- Improvement to the reliability of the LHC through the construction of LINAC4 [2], which will reduce the risk to LHC operation by replacing the ageing LINAC2, that first came into operation in 1978;

- The R&D and subsequent implementation necessary for a significant luminosity increase of the LHC beyond the design luminosity, i.e. HL-LHC, if necessitated by the physics and/or running experience; in particular it includes the focusing elements in the interaction regions and the upgrades of the injector chain;

- LHC detector modifications to make optimal use of the design LHC luminosity;

- The detector R&D necessary for the luminosity upgrade HL-LHC and the corresponding modifications of the existing LHC experiments.

This strategy is also driven by the necessity to bring the LHC injector chain and the technical and general infrastructure up to the high standards required for a world laboratory in order to ensure reliable operation of the CERN complex.

The ambitious longer-term plans aim at a total integrated luminosity of the order of 3000 fb^{-1} (on tape) by the end of the life of the LHC in 2030. This implies an annual luminosity of about 250-300 fb^{-1} in the second decade of running the LHC. It also calls for a new strategy to optimize the integrated luminosity useful for physics. Therefore, the LHC operation schedule will henceforth be over a two-year cycle, with a short technical stop around Christmas at the end of the first year and a longer shutdown following the end of the second year. Such a schedule is more efficient for the operation of a superconducting accelerator.

In light of the above developments, the following strategy has been introduced:

- The Chamonix LHC Performance Workshop in January 2010 identified the need for a complete refurbishment of all copper-stabilizer joints of the main LHC magnets for safe running at 7 TeV/beam. The copper-stabilizer repair is scheduled throughout 2012 (long shutdown).

- To ensure reliable operation of the LHC in the coming years, there is a need to consolidate intensively the existing LHC injector chain. This is due to the fact that even if approved soon, the low-power superconducting proton linac LP-SPL and PS2 would realistically be available in 2020 at the earliest.

- In order to optimize the strategy towards the HL-LHC, with the goal of maximizing the integrated luminosity useful for physics, CERN has set up a task force. A preliminary recommendation from this task force is to delay the first inner triplet replacement to a single HL-LHC upgrade around 2020. The complete HL-LHC upgrade needs a much clearer definition of implementation objectives based on the requirements of the experiments, such as the use of crab cavities, in order for the LHC to operate reliably at luminosities of about 5×10^{34} cm^{-2} s^{-1}. This may include the option of luminosity leveling to ensure a high luminosity lifetime.

- Furthermore, the bottlenecks of the injector chain need to be tackled and hence upgrades are being studied with a view to increasing the extraction energy of the PS Booster as well as upgrades to the SPS, with the latter currently being a significant bottleneck for increasing the LHC intensity beyond design.

The Way Forward and the European Strategy for Particle Physics

The LHC will provide a first indication of any new physics at energies of the Terascale. Many of the open questions left by the LHC and its upgrades may be addressed best by an electron-positron collider, based on technology developed by the Compact Linear Collider (CLIC) [3] and the International Linear Collider (ILC) [4] collaborations. Moreover, the option of a high-energy electron-proton collider (LHeC) [5] is being considered for the high-precision study of QCD and of high-density matter.

Great opportunities are in store at the TeV scale and a fuller understanding of Nature will come about through a clearer insight at this energy level. As in the past, there is a synergy between collider types – proton-proton, electron-positron and electron-proton – that can be used to advantage. The discovery of the Standard Model over the past few decades has advanced through the synergy of hadron-hadron (e.g. SPS and the Tevatron), lepton-hadron (HERA) and lepton-lepton colliders (e.g. LEP and SLC). Such synergies should be continued in the future and thus a strategy has been developed along these lines. An upgrade to the LHC will not only provide an increase in luminosity delivered to the experiments, but will also provide the occasion to renew the CERN accelerator complex. The ILC could be constructed now whereas further R&D is needed for CLIC. There is a drive to converge towards a single electron-positron linear collider project. The above effort on accelerators should advance in parallel with the necessary detector R&D. First results from the LHC will be decisive in indicating the direction that particle physics will take in the future.

European particle physics is founded on strong national institutes, universities and laboratories, working in conjunction with CERN. The increased globalization, concentration and scale of particle physics require a well-coordinated European strategy. This process started with the establishment of the CERN Council Strategy Group, which organized an open symposium in Orsay in 2006, a final workshop in Zeuthen in May 2006 and with the strategy document being signed unanimously by Council in July 2006 in Lisbon [6]. CERN considers that experiments at the high-energy frontier to be the premier physics priority for the coming years. This direction for future colliders at CERN follows the priorities set in

2006 by the CERN Council Strategy Group. The European Strategy for Particle Physics includes several key areas of research, all in line with CERN's plans for the future directions.

The years 2010 and 2011 are seeing the start of the LHC physics exploitation leading to important input for the update of the European strategy for particle physics planned for 2012.

Key Messages

Particle physics will need to adapt to the evolving situation. Facilities for high-energy physics (as for other branches of science) are becoming larger and more expensive. Funding for the field is not increasing and the timescale for projects is becoming longer, both factors resulting in fewer facilities being realized. Moreover, many laboratories are changing their missions.

All this leads to the need for more co-ordination and more collaboration on a global scale. Expertise in particle physics needs to be maintained in all regions, ensuring the long-term stability and support through-out. It would be necessary to engage all countries with particle physics communities and to integrate the communities in the 'particle-physics-developing' countries. The funding agencies should in their turn provide a global view and synergies between various domains of research, such as particle physics and astroparticle physics, should be encouraged.

Particle physics is now entering a new and exciting era. The start-up of the LHC allows particle physics experiments at the highest collision energies. The expectations from the LHC are great, as it could provide revolutionary advances in the understanding of the microcosm and a fundamental change to our view of the early Universe. Due to the location of the LHC, CERN is in a unique position to contribute to further understanding in particle physics in the long term.

Results from the LHC will guide the way in particle physics for many years. It is expected that the period of decision-making concerning the energy frontier will be in the next few years. Particle physics is now in an exciting period of accelerator planning, design, construction and exploitation and will need intensified efforts in R&D and technical design work to enable the decisions for the future course and global collaboration coupled with stability of support over long time scales.

The particle physics community needs to define now the most appropriate organizational form and needs to be open and inventive in doing so, and it should be done in a dialogue between the scientists, funding agencies and politicians. It is mandatory to have accelerator laboratories in all regions as partners in accelerator development, construction, commissioning and exploitation. Furthermore, planning and execution of high-energy physics projects today require world-wide partnerships for global, regional and national projects, namely for the whole particle physics programme. The exciting times ahead should be used to advantage to establish such partnerships.

Fascinating Science

With the largest and most complex scientific equipment, the LHC accelerator and experiments are today attracting immense attention and the LHC is possibly the most-watched scientific endeavour. The LHC is in the spotlight of not only the scientific community but also of the general public and the international media. It has become so due to its fascinating and forefront science, which addresses long-standing questions of human-kind with vanguard technologies. Moreover, the LHC stimulates general interest, increases knowledge, educates and trains the scientists and engineers of tomorrow and drives innovation and technology. This current interest should be used to promote the field of particle physics.

Conclusions

In this paper we have reported on the status, physics results and future prospectives for the LHC and provided a description of the driving factors for the LHC physics programme and for future proton and lepton colliders. In the coming years, the ordered priorities are the full exploitation of the LHC, together with preparation for a possible luminosity upgrade and the consolidation and optimization of the CERN infrastructure and the LHC injectors. It will be necessary to keep under review the physics drivers for future proton accelerator options and it will be necessary to compare the physics opportunities offered by proton colliders with those available at a linear electron-positron collider and electron-proton collider. The R&D associated with future colliders should continue in parallel.

Acknowledgements

I would like to thank the organizers for the invitation to make this contribution and for the excellent organization of the very interesting conference, which included first results from the LHC. I would also like to thank Emmanuel Tsesmelis for his assistance in preparing this contribution.

Bibliography

[1] LHC Design Report, Volumes I, II and III, http://lhc.web.cern.ch/lhc/LHC-DesignReport.html

[2] http://linac4.web.cern.ch/linac4

[3] http://clic-study.web.cern.ch/CLIC-Study

[4] http://www.linearcollider.org/cms

[5] http://lhec.web.cern.ch/lhec

[6] The European Strategy for Particle Physics, http://council-strategygroup.web.cern.ch/council-strategygroup/Strategy_Statement.pdf

P.J. ODDONE

"Highlights from FERMILAB"

The Slides of the Lecture can be found at:

http://www.ccsem.infn.it/issp2010/

Scientific Secretaries: P. Frolov, I. Dubovyk

DISCUSSION I

- Inguglia:
You have shown a plot where we have access to leptonic CPV at Minos. Please can you explain how to access to CPV observables?

- Oddone:
Indeed, we don't have direct access to measurements showing CPV, but we have access to a measure of the oscillation parameters, the mixing angle θ_{23} and the mass difference, for both neutrinos and anti-neutrinos. We have indications that these values differ for neutrinos and antineutrinos by about 2.5 sigma. [Note added in proof: the difference disappeared with additional data taking in the subsequent year].

- Inguglia:
You have shown a plot for anomalous CPV at D0 at 3.2 σ. Do you have any confirmation of that signal from CDF?

- Oddone:
The 2 experiment indeed have different capabilities and, of course, a measurement of the same signal with CDF would be a very interesting result. For the moment, we don't have confirmation, but hopefully in few months we can confirm it.

- Haidt:
$H \rightarrow ll + \alpha$ apply kinematic fitting?

- Oddone:
I will talk a little bit more about it tomorrow. In general, kinematic fitting for the decays that are most prevalent in the Tevatron cannot be done because we have missing neutrinos in the b or anti-b decays of the Higgs and in general missing energy and momentum in the overall event. For example, the Higgs events in the Tevatron are studied in the associated production with a Z or W boson. We do not even observe the total energy of these events so kinematic fitting is impossible. Fitting can be done for part of the events such as for instance a real Z boson when it decays into charged leptons.

- Schmidt-Sommerfeld:
Did you mention the neutrino anti-neutrino difference any speculations for explanations?

- Oddone:
If they are really different, that would imply a violation of CPT, but that is very unlikely, as CPT is a cornerstone of QFT. What is more likely, is that neutrino and antineutrinos interact differently with matter that surrounds us, which is only made of matter and no antimatter. The most conservable interpretation is clearly a statistical fluctuation.

- *Cifarelli:*

How realistic is the zero background assumption in the COUPP search for WIMPs?

- *Oddone:*

Of course we don't know. Getting to zero background is the whole game. It very much depends on what you postulate the backgrounds to be and what ability you have to eliminate and estimate remaining background. For instance, you have to start with a very low neutron background. You are going to ask the questions: if I see 2, 3, 4 or 5 from multiple neutron scatterings, what is the probability with that distribution that I get 1 bubble event that would mimic a WIMP signal? That would be the background, and you want neutron fluxes to be low enough to make that number essentially zero. So, that's the game.

- *Mulhearn:*

Will you run the Tevatron for three more years, please?

- *Oddone:*

We will see. Some people in this lecture hall said to me, that if we don't run the Tevatron, this will break their hearts. [Note added in proof: the extension of the Tevatron was not funded]

- *Mulhearn:*

To what extent do you see this as a zero sum game?

- *Oddone:*

It cannot be a zero sum game. We gave priority to the new projects and the Tevatron extension should not come at the expense of these new programs. At the present time, we must meet the requirements of our main financial sponsor (Department of Energy). We understand our responsibility and hope that the Government will complete the financing of our project.

- *Inguglia:*

Please can you explain better the neutral current background?

- *Oddone:*

It is a background which is small when we detect muons due to the charge current interaction, but which may become important in *e*-appearance, and it is related with θ_{13}. A very large θ_{13} would lead to a higher number of *e*-appearance events and hence a lower signal-to-background ratio

- *Grelli:*

Why Fermilab is sending the neutrino beam in a region with some nuclear reactor close by and not in a free region?

- *Oddone:*

There is no need to worry about nuclear reactors since the energy of the neutrinos from reactors is very low. The large energy neutrinos that we work with are easily distinguished. They can be confused with neutrinos induced by cosmic rays, but that background gets eliminated by timing cuts.

- *Mao:*

Can you explain how can you reconstruct neutrino energy if we don't know the reference of true energy from the neutrino $\nu^{\pm}$ beam?

- Oddone:

There are two energies to talk about: the energy of the produced neutrinos that we know from the characteristics of the beam and the detected energy if the detector is a calorimeter sensitive to the total energy. In producing neutrinos we take a high energy proton beam (120Gev) and hit a graphite target. This target produces all sorts of particles: pions, kaons, etc. After the target there are some electromagnetic devises, called horns. These devises create magnetic fields that focus the particles of the right charge. Doing so, we can select particles that are going to decay into neutrino or antineutrino. Following the horns we have a long decay pipe, where pions and kaons decay to muons and neutrinos. After this we absorb all charged particles in the rock. So we can have clean beams of neutrinos or antineutrinos. This gives us beams with a broad spectrum of energies centered at a few GeV. So we do not know the energy of the neutrinos precisely. We can narrow the distribution of energies if we go off-axis. This energy is reconstructed in the detector, but only with the particles that we can fully absorb in the detector. We don't measure the energy precisely because generally there are missing neutrinos.

Scientific Secretaries: D. Gerbaudo, A. Miyazaki

DISCUSSION II

- *Inguglia:*
    Could you please say something more about the large effect that supersymmetry would induce on *Kaon* decays. You showed this in your slides on page 51-55.

- *Oddone:*
    Within the standard model one can predict the decay rate with a 3% accuracy. Additional diagrams, where the supersymmetric particles would appear in the loop, can induce a rather large perturbation. Furthermore, these additional contributions would appear in different regions of the plot showing the correlation between the decays $K^0_L \to pi0$ v *v-bar* and $K^+ \to pi+$ v *v-bar*. This would even allow to discriminate between different values of parameters for a particular supersymmetric theory. Of course one would need a significant number of events to get enough discriminating power.

- *Inguglia:*
    I still do not understand how one can separate events due to the standard model WW process from the ones due to supersymmetric particles. In both cases one can only detect the two neutrinos.

- *Oddone:*
    One cannot determine what happens for one single event. What one can do is to precisely calculate the rate from the diagrams. In this particular case, when you consider the supersymmetric hypothesis, you have additional contributions from diagrams involving supersymmetric particles. You can make assumptions about the couplings for these particles, and determine the rate, which will depend on your assumptions. However, if you have several rare decays, you can then put a constraint and select the subset of the possible supersymmetric models which agree with your observations.

- *Inguglia:*
    Isn't this measurement limited by the fact that one can only detect the two neutrinos by measuring a single quantity, the missing energy?

- *Oddone:*
    This is true, but one can already extract very useful information from the spectrum of the missing energy. For example the typical energy for this kaon experiment has typically a few hundred MeV.

- *Inguglia:*
    Which one is the least expensive between: μ collider, ILC, or CLIC?

- *Oddone:*
    We don't know yet. There have been extensive cost studies for the ILC, but we do not know yet exactly the price for the muon collider nor for CLIC. From this point of view, one advantage of the muon collider is that muons don't radiate, so this kind of collider can be fairly compact.

However, if one already knows the energy of the new interesting physics, and if this energy is low enough, then the ILC represents the easiest solution. On the other hand, for a machine like CLIC or a muon collider, we would need quite a few years of R&D. These will be necessary if we need to explore energies of few TeV.

- Grelli:
Few years ago there was a proposal for an experiment at BNL to study $K^0_L \rightarrow$ pi0 ν *ν-bar*. The issue with that proposal was that it was extremely expensive. Do you think that such an experiment would be possible now?

- Oddone:
That is correct. Such an experiment would be expensive. A single rare decay experiment is not sufficient to interpret the underlying physics. You would need several of them. In order to be able to discriminate between different theories, one would need a measurement with about a 3% precision, which is the level of precision for calculations of rate in the Standard Model.

- Preghenella:
The two experiments NOvA and T2K are high power and off-axis neutrino experiments. What is the main difference in the physics reach of these two experiments?

- Oddone:
The parameters of the two experiments are different. T2K has a baseline of about 200km, and it uses a relatively low energy beam. This setup is good for determining the angle sin(theta13). In the case of NOvA, the energy is higher, typically 2 or 3 GeV, and the baseline is 780km. This makes NOvA sensitive to the mass hierarchy of neutrinos. The other thing that NOvA will do very well is comparing neutrinos with anti neutrinos. In this context, effects like the mild two-sigma one observed by MINOS, would become more evident with NOvA.

- Gerbaudo:
The strongest argument for an extended Tevatron run is the search for the Higgs. However, at the LHC, there are a number of measurements that one either cannot do, or that are quite difficult to perform. Could you please say something about these legacy measurements?

- Oddone:
There are a set of measurements, for example the top mass, the W mass, and limits on the CP violation for charm, that will probably remain for a long time as the most accurate. For most of the other things, the LHC will produce them abundantly, so I think that, eventually, they will be measured very well also at the LHC. The most difficult issue for the LHC will be whether the largest backgrounds will allow the LHC to do this well or not. It will probably take some time to understand.

- Schmidt-Sommerfeld:
Why would it be harder to find a Higgs with a mass of 100 GeV than one with a mass of 130 GeV?

- Oddone:
If you look at the cross section plot you can see that, in particular for the associated production that matters at low mass, the cross section is higher at lower masses. So at lower masses you get more Higgs bosons produced.

- Mulhearn (comment):

It is correct, it is a cross section effect. However, if one goes further down in mass, then one runs into more Z background events.

- Mao:

You showed a plot where the limit is going down when increasing the integrated luminosity. Why does the expected limit decrease when the luminosity increases?

- Oddone:

Each curve shows how you increase the sensitivity by simply adding more data at a given mass of the Higgs, without any improvement to the analysis. This increase is a statistical increase that goes like one over the square root of the integrated luminosity. As you understand better your data, and you can broaden up your acceptance, you learn how to deal with your backgrounds, and you become more sensitive. In fact we are at the standard model limit thanks to the improvements that were introduced.

- Mao:

I don't understand the plot on the kaon decay rates; how can you understand whether a kaon decay is due to the standard model, or from some process beyond the standard model?

- Oddone:

You can calculate, within the standard model, the rate of one of these processes very precisely. If the rate that you observe is different from the predicted one, then you can make hypotheses for the reason why this is different from what you expect, and this is usually due to non-SM contribution to the decay diagrams. Then you can make predictions, for example, about the existence of a mass resonance that you can then verify at the LHC.

- Mao:

What is the difference between the ILC and CLIC? do they have the same physics goal?

- Oddone:

The ILC is quite expensive because of the superconducting-cavity technology, and is also limited in the energy that it can reach within a limited length, given a limited gradient. CLIC is trying to build a more compact machine by using a different technology with a quite large gradient. It is also a pulsed machine, and the structures are smaller so you do not need the huge cryogenic infrastructure needed for the ILC. But there are many technical problems associated with CLIC that we do not know yet how we will be able to solve. For example alignment, big amount of power in a small space, etc. The physics goals are the same, except for the fact that they are designed to look at different energies.

- Dunin-Barkowski:

What is the best signal to background ratio which can be achieved at the Tevatron for a Higgs boson?

- Oddone:

I don't know this number right now, but we can estimate it from the sensitivity plot.

- Mulhearn+Gerbaudo (comment):

If one considers just the number of events at final selection, then one typically has few thousands of background events and tens of signal events. This gives a S/B on the order of 1%. However, the significant quantity in this context is the bin-by-bin signal-to-background ratio for the distribution of the final discriminant. This quantity depends on the binning of the final discriminant, but it can be greater than one for the bins with the greatest signal content.

Anti- and Hypermatter Research at the Facility for Antiproton and Ion Research FAIR

J. Steinheimer*[1], Z. Xu[2], P. Rau[3,4], C. Sturm[5,6], and H. Stöcker[3,4,5]

[1]Lawrence Berkeley National Laboratory, Berkeley, CA 94720, USA
[2]Physics Department, Brookhaven National Laboratory, Upton, NY 11973, USA
[3]Frankfurt Institute for Advanced Studies, Ruth-Moufang-Str. 1, 60438 Frankfurt am Main, Germany
[4]Institut für Theoretische Physik, Goethe-Universität, Max-von-Laue-Str. 1, 60438 Frankfurt am Main, Germany
[5]GSI Helmholtzzentrum für Schwerionenforschung GmbH, Planckstr. 1, 64291 Darmstadt, Germany
[6]Institut für Kernphysik, Goethe Universität, Max-von-Laue Str. 1, 60438 Frankfurt am Main, Germany

Within the next six years, the Facility for Antiproton and Ion Research (FAIR) is built adjacent to the existing accelerator complex of the GSI Helmholtz Center for Heavy Ion Research at Darmstadt, Germany. Thus, the current research goals and the technical possibilities are substantially expanded. With its worldwide unique accelerator and experimental facilities, FAIR will provide a wide range of unprecedented fore-front research in the fields of hadron, nuclear, atomic, plasma physics and applied sciences which are summarized in this article. As an example this article presents research efforts on strangeness at FAIR using heavy ion collisions, exotic nuclei from fragmentation and antiprotons to tackle various topics in this area. In particular, the creation of hypernuclei, metastable exotic multi-hypernuclear objects (MEMOs) and antimatter is investigated.

1   The FAIR Project

The Facility for Antiproton and Ion Research, FAIR [1–3], will provide an extensive range of particle beams from protons and their antimatter partners, antiprotons, to ion beams of all chemical elements up to the heaviest one, uranium, with in many respects world record intensities. As a joint effort of 16 countries the new facility builds, and substantially expands, on the present accelerator system at GSI, both in its research goals and its technical possibilities. Compared to the present GSI facility, an increase of a factor of 100 in primary beam intensities, and up to a factor of 10000 in secondary radioactive beam intensities, will be a technical property of the new facility.

---

*jsfroschauer@lbl.gov

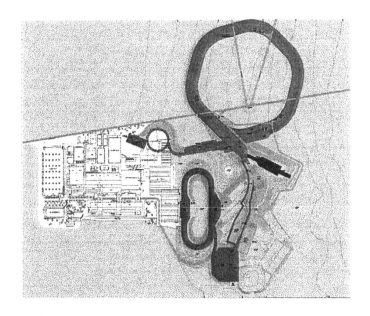

Figure 1: The existing GSI facility is shown on the left. Displayed in color is the so called Modularized Start Version of FAIR including module 0, 1, 2 and 3. Coloring: the 100 Tm super conducting synchrotron SIS100 (module 0) - green; the experimental area for CBM/HADES (module 1) - red; the NuSTAR facility including the Super-FRS (module 2) - yellow; the Antiproton facility including the PANDA experiment (module 3) - orange. The additional experimental area above ground for the APPA community (module 1) is not shown.

After the official launch of the project on November 7th, 2007, on October 4th, 2010, nine countries[1] signed the international agreement on the construction of FAIR. Civil work for the first buildings of FAIR will start during this year and first beams will be delivered in 2018. The start version of FAIR, the so-called *Modularized Start Version* [4, 5], includes a basic accelerator SIS100 (module 0) as well as three experimental modules (module 1-3) as it is illustrated in Fig. 1. The superconducting synchrotron SIS100 with a circumference of 1100 meters and a magnetic rigidity of 100 Tm is at the heart of the FAIR accelerator facility. Following an upgrade for high intensities, the existing GSI accelerators UNILAC and SIS18 will serve as an injector. Adjacent to the SIS100 synchrotron are two storage-cooler rings and experiment stations, including a superconducting nuclear fragment separator (Super-FRS) and an antiproton production target. The Modularized Start Version secures a swift start of FAIR with outstanding science potential for all scientific pillars of FAIR within the current funding commitments. Moreover, after the start phase and as additional funds become available the facility will be upgraded by experi-

---

[1] In alphabetical order: Finland, France, Germany, India, Poland, Romania, Russia, Slovenia and Sweden

mental storage rings enhancing capabilities of secondary beams and upgraded by SIS300 providing particle energies 20-fold higher compared to those achieved so far at GSI.

2 The Experimental Program of FAIR

The main thrust of FAIR research focuses on the structure and evolution of matter on both a microscopic and on a cosmic scale. The approved FAIR research program embraces 14 experiments, which form the four scientific pillars of FAIR and offers a large variety of unprecedented forefront research in hadron, nuclear, atomic and plasma physics as well as applied sciences. Already today, over two 2500 scientists and engineers are involved in the design and preparation of the FAIR experiments. They are organized in the experimental collaborations APPA, CBM, NuSTAR, and PANDA.

2.1 APPA – Atomic Physics, Plasma Physics and Applications

Atomic physics with highly charged ions [6] will concentrate on two central research themes: a) the correlated electron dynamics in strong, ultra-short electromagnetic fields including the production of electron-positron pairs and b) fundamental interactions between electrons and heavy nuclei - in particular the interactions described by Quantum Electrodynamics, QED. Here bound-state QED in critical and supercritical fields is the focus of the research program. In addition, atomic physics techniques will be used to determine properties of stable and unstable nuclei and to perform tests of predictions of fundamental theories besides QED.

For Plasma physics the availability of high-energy, high-intensity ion-beams enables the investigation of High Energy Density Matter in regimes of temperature, density and pressure not accessible so far [7]. It will allow probing new areas in the phase diagram and long-standing open questions of basic equation of state (EoS) research can be addressed. The biological effectiveness of high energy and high intensity beams was never studied in the past. It will afford to investigate the radiation damage induced by cosmic rays and protection issues for the Moon and Mars missions. Furthermore, the intense ion-matter interactions with projectiles of energies above 1 GeV/u will endorse systematic studies of material modifications.

2.2 CBM/HADES – Compressed Baryonic Matter

Violent collisions between heavy nuclei promise insight into an unusual state in nature, that of highly compressed nuclear matter. In addition to its relevance for understanding fundamental aspects of the strong interaction, this form of matter may exist in various so far unexplored phases in the interior of neutron stars and in the core of supernovae. The mission of high-energy nucleus-nucleus

collision experiments worldwide is to investigate the properties of strongly interacting matter under these extreme conditions. At very high collision energies,
as available at RHIC and LHC, the measurements concentrate on the study of
the properties of deconfined QCD matter at very high temperatures and almost
zero net baryon densities. Results from lattice QCD indicate that the transition from confined to deconfined matter at vanishing net baryon density is
a smooth crossover, whereas in the region of high baryon densities, accessible
with heavy-ion reactions at lower beam energies, a first-order phase transition
is expected [8]. Its experimental confirmation would be a substantial progress
in the understanding of the properties of strongly interacting matter.

Complementarily to high-energy nucleus-nucleus collision experiments at
RHIC and LHC, the CBM experiment [9, 10] as well as HADES [11, 12] at
SIS100/300 will explore the QCD phase diagram in the region of very high
baryon densities and moderate temperatures by investigating heavy-ion collision in the beam energy range 2–35 AGeV. This approach includes the study
of the nuclear matter equation-of-state, the search for new forms of matter, the
search for the predicted first order phase transition to the deconfinement phase
at high baryon densities, the QCD critical endpoint, and the chiral phase transition, which is related to the origin of hadron masses. In the case of the predicted
first order phase transition, basically one has to search for non-monotonic behavior of observables as function of collision energy and system size. The CBM
experiment at FAIR is being designed to perform this search with a large range
of observables, including very rare probes like charmed hadrons. Produced near
threshold, their measurement is well suited to discriminate hadronic from par-
tonic production scenarios. The former requires pairwise creation of charmed
hadrons, the latter the recombination of c-quarks created in first chance collisions of the nucleus-nucleus reaction. Ratios of hadrons containing charm
quarks as a function of the available energy may provide direct evidence for a
deconfinement phase.

The properties of hadrons are expected to be modified in a dense hadronic
environment which is eventually linked to the onset of chiral symmetry restora-
tion at high baryon densities and/or high temperatures. The experimental verification of this theoretical prediction is one of the most challenging questions in
modern strongly interacting matter physics. The dileptonic decays of the light
vector mesons (ρ,ω,ϕ) provide the tool to study such modifications since the
lepton daughters do not undergo strong interactions and can therefore leave the
dense hadronic medium essentially undistorted by final-state interaction. For
these investigations the ρ meson plays an important role since it has a short
lifetime and through this a large probability to decay inside the reaction zone
when created in a nucleus-nucleus collision. As a detector system dedicated
to high-precision di-electron spectroscopy at beam energies of 1–2 AGeV, the
modified HADES detector at SIS100 will measure e^+e^- decay channels as well
as hadrons [13, 14] up to 10 AGeV beam energy. Complementarily, the CBM
experiment will cover the complete FAIR energy range by measuring both the
e^+e^- and the $\mu^+\mu^-$ decay channels.

Most of the rare probes like lepton pairs, multi-strange hyperons and charm will be measured for the first time in the FAIR energy range. The goal of the CBM experiment as well as HADES is to study rare and bulk particles including their phase-space distributions, correlations and fluctuations with unprecedented precision and statistics. These measurements will be performed in nucleus–nucleus, proton–nucleus, and proton–proton collisions at various beam energies. The unprecedented beam intensities will allow studying extremely rare probes with high precision which have not been accessible by previous heavy-ion experiments at the AGS and the SPS.

2.3 NuSTAR – Nuclear Structure, Astrophysics and Reactions

The main scientific thrusts in the study of nuclei far from stability are aimed at three areas of research: (i) the structure of nuclei, the quantal many-body systems built by protons and neutrons and governed by the strong force, toward the limits of stability, where nuclei become unbound, (ii) nuclear astrophysics delineating the detailed paths of element formation in stars and explosive nucleosynthesis that involve short-lived nuclei, (iii) and the study of fundamental interactions and symmetries exploiting the properties of specific radioactive nuclei.

The central part of the NuSTAR program at FAIR [15, 16] is the high acceptance Super-FRS with its multi-stage separation that will provide high intensity mono-isotopic radioactive ion beams of bare and highly-ionized exotic nuclei at and close to the driplines. This separator, in conjunction with high intensity primary beams with energies up to $1.5A$ GeV, is the keystone for a competitive NuSTAR physics program. This opens the unique opportunity to study the evolution of nuclear structure into the yet unexplored territory of the nuclear chart and to determine the properties of many short-lived nuclei which are produced in explosive astrophysical events and crucially influence their dynamics and associated nucleosynthesis processes.

2.4 PANDA – AntiProton ANnihilation in Darmstadt

The big challenge in hadron physics is to achieve a quantitative understanding of strongly interacting complex systems at the level of quarks and gluons. In $p\bar{p}$-annihilation, particles with gluonic degrees of freedom as well as particle-antiparticle pairs are copiously produced, allowing spectroscopic studies with unprecedented statistics and precision. The PANDA experiment at FAIR [17–19] will bring new fundamental knowledge in hadron physics by pushing the precision barrier toward new limits. The charmonium ($c\bar{c}$) spectroscopy will take advantage by precision measurements of mass, width, decay branches of all charmonium states. Particular emphasis is placed on mesons with open and hidden charm, which extends ongoing studies in the light quark sector to heavy quarks, and adds information on contributions of the gluon dynamics to hadron masses. The search for exotic hadronic matter such as hybrid mesons or heavy glueballs gains enormously by precise scanning of resonance curves of narrow states as well. Recently, this field has attracted much attention with the surprise observation at electron-positron colliders of the new X, Y and Z states with masses around 4 GeV. These heavy particles show very unusual properties, whose theoretical interpretation is entirely open. Additionally the precision gamma-ray spectroscopy of single and double hypernuclei will allow extracting information on their structure and on the hyperon-nucleon and hyperon-hyperon interaction.

3   The creation of antimatter

The history of antimatter is a brief and fascinating history of scientific discoveries. In 1928, Dirac predicted the existence of negative energy states of electrons based on the application of symmetry principles to quantum mechanics. The states were recognized as antimatter partner of electrons (positrons) discovered by Anderson in the cosmic rays in 1932. The constructions of accelerators have provided the necessary energy and luminosity for the discoveries of heavier antimatters. The extension of Diracs theory implied the existence of antimatter protons and neutrons, and both particles were discovered at Bevatron in 1955. The scientific investigation of antimatter has three major focuses since then:

a) Antiparticles are produced as by-products of high-energy particle collisions. Many particle and antiparticle pairs are created in such collisions through strong or electromagnetic processes. Antiparticles are merely part of the energy and chemical (baryon, isospin or lepton) conservation laws.

b) Precise measurements of particle and antiparticle properties, which can provide insights into the fundamental CPT conservation and baryon asymmetry in the Universe.

c) Constructing more complex system of antimatter.

It is clear that all these three topics are related to each other with a different emphasis for each subject. Although it may sound trivial to define what antimatter is, its definition is not without controversy. There are particles and antiparticles (such as μ^+ and μ^-), which annihilate when put together. However, neither of them annihilates the ordinary matter. Antonino Zichichi (2008) argues that there is a basic difference between antiparticle and antimatter, and even anti-hydrogen is not antimatter. In this proceeding, we mainly focus on constructing more complex systems of antimatter: antinuclei and antihypernuclei.

After the discoveries of antiprotons and antineutrons, one of the important questions was whether the building blocks in the antimatter world have the same force to glue together the antinucleons into nuclei and eventually anti-atoms by adding positrons. Figure 2 depicts the history of the discoveries of antimatter. We note that the antimatter project span eight decades with four decades per step in our discoveries. There are effectively three periods in these 80 years. The first discovery was made in the cosmic ray in 1932. The second period of discovery was between 1955 to 1975 when the fixed target accelerators provided increasing intensity and energy for producing heavier and heavier antimatter.

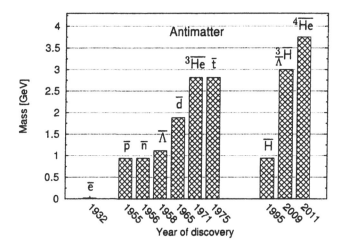

Figure 2: Discovery year of the antimatter and its associated mass.

The third period was made possible with high energy relativistic heavy-ion collider at RHIC and at the LHC. At the same time, the technology advance also enables us to decelerate antiproton beams and trap antimatter hydrogen. The necessity of the long term commitment was expressed by Walter Greiner (2001) in 'Fundamental Issues in the Physics of Elementary Matter': "The extension of the periodic system into the sectors of hypermatter (strangeness) and antimatter is of general and astrophysical importance. [...] The ideas proposed here, the verification of which will need the *commitment for 2-4 decades of research, could be such a vision* with considerable attraction for the best young physicists." [20]

Understanding the asymmetry of antimatter and matter is one of the frontiers of modern physics. Nuclei are abundant in the universe, but antinuclei with $|A| \geq 2$ have not been found in nature. Relativistic heavy ion collisions, simulating the condition at the early universe, provide an environment with abundant antinucleons and antihyperons and produce antinuclei and antihypernuclei by coalescing them together [21]. This offers the first opportunity for discovery of antihypernuclei [22] and heavier antinuclei [23] having atomic mass numbers (or baryon numbers) $|A| > 2$. The production of antimatter nuclei can be explained by coalescence of antiprotons and antineutrons close in position and momentum. Figure 3 compiles all the antideuteron production in $e^{+}e^{-}$, γp, pp, pA and AA collisions [24]. The results are shown for $\bar{d}/\bar{p}$ ratio as a function of beam energy. One can see that this ratio increases from 10^{-5} at low energy to 10^{-3} at high energy. Each additional antinucleon into the heavier antimatter decreases its production rate by that same penalty factor. At a center of mass energy of 100 GeV and above, this factor is relatively flat at slightly below 10^{-3}. It is interesting to note that this effective measure of antibaryon density shows no difference among pp, pA and AA collisions. In heavy ion collisions, more antiprotons are produced in each collision than in pp collisions. However, if more pp collisions are collected to match the amount of antiproton

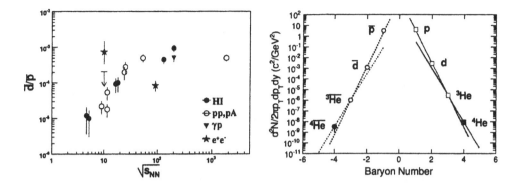

Figure 3: (Color online) Left: Antideuteron and antiproton yield ratio as a function of $\sqrt{S_{NN}}$ of the colliding target and projectile. Right: Matter and antimatter invariant differential yields as a function of the baryon number. The solid and dashed lines are a fit to experimental data. The dotted and dashed dotted lines are extrapolations from yields of $|A| = 3$ and $|A| = 4$.

yields in heavy-ion collisions, one can essentially produce the same amount of heavy antimatter in pp and heavy-ion collisions. Now we understand that there are two deciding facts that RHIC discovered the last two heavy antimatters: sufficient energy to provide the highest antibaryon density for antinuclear production, and high luminosity heavy-ion collisions for effective data collection and particle identification.

Figure 3 shows the matter and antimatter yields as a function of baryon numbers as measured by the STAR Collaboration at RHIC [23]. The fit lines yield the production reduction rate by a factor of 1.6×10^{-3} (1.1×10^{-3}) for matter (antimatter) for each additional nucleon (antinucleon). The sensitivity of current and planned space based charged particle detectors is below what would be needed to observe antihelium produced by nuclear interactions in the cosmos. This implies that any observation of antihelium or even heavier antinuclei in space would indicate the existence of a large amount of antimatter elsewhere in the universe. In particular, finding antimatter ^{4}He in the cosmos is one of the major motivations for space detectors such as the Alpha Magnetic Spectrometer [25]. We have shown that antimatter ^{4}He exists and provided a measure of the background rate in nuclear collisions for possible future observations in cosmic radiation.

The next stable antimatter nucleus would be $A = 6$ (^{6}He; ^{6}Li). However, the penalty factor on the production rate for an additional antinucleon is about 1500 as shown in Fig. 3. This means that the $A = 6$ antinuclei are produced at a rate 2×106 lower than that of an $A = 4$ antialpha particle. Unless production mechanisms or collider technology change dramatically, it is unlikely that $A = 6$ antinuclei can be produced in collider or fixed-target experiments (STAR 2011). On the other hand, the ratio of the ^{4}He/^{3}He $= 3.1 \times 10^{-3}$ and $^4\overline{\text{He}}/^3\overline{\text{He}} = 2.4 \times 10^{-3}$. There is a factor of 2 higher yield of $|A| = 4$ over $|A| = 3$

than the extrapolation from the fit. The excess is visible even in a log-scale plot of 13 orders of magnitude. This ratio is also much higher than that shown in Fig. 3 for the $|A| = 2$ over $|A| = 1$. It has been argued that a more economic way of producing heavier antimatter and/or nuclear matter containing large amount of strange quark contents is through excitation of complex nuclear structure from the vacuum or through strangeness distillation from a QGP. Is this enhanced yield an indicative of a new production mechanism or a minor deviation due to trivial configuration of nuclear binding? Where do we go from here into the future in search and construction of heavier and more exotic antimatter? The indicative enhancement of higher antialpha yields suggests that even higher enhanced yields of heavier antimatter. Besides the possible high yields of $|A| = 6$ antimatter, the heaviest antimatter that can be produced and detected with a tracking detector in high-energy accelerators are likely to be $A = 4$ or 5 unstable antinuclei: $^4\mathrm{He}^* \to \mathrm{t} + \mathrm{p}$, $^4\mathrm{Li} \to ^3\mathrm{He} + \mathrm{p}$, and $^5\mathrm{Li} \to ^4\mathrm{He} + \mathrm{p}$. New trigger scheme and high data acquisition rate have been proposed to improve the effective data taking rate by two orders of magnitude in STAR during the heavy ion collisions [26]. This should confirm if the enhancement indeed exists and provide a possible path for discovering even heavier antimatter. In addition, as mentioned in the previous section, the antimatter yield reduction factor is similar in $p + p$ and AA collisions. One expects that the penalty factor to persist for antimatter heavier than antideuteron in $p+p$ collisions. A comparison between the antimatter yields as shown in Fig. 3 in $p+p$ and $A+A$ collisions will provide a reference for whether the enhancement seen in antialpha production in AA collisions is due to new production mechanism. Both RHIC and LHC have sufficient luminosity in $p + p$ collisions to produce antialpha. The only experimental issue is how to trigger and identify those particles. STAR has proposed a new trigger and TPC readout schemes for heavy antimatter search by using the Electromagnetic Calorimeter (EMC) for charged hadrons and only readout small sector of TPC associated with that struck EMC.

4 Hypermatter

Relativistic heavy ion collisions are an abundant source of strangeness. As strange quarks have to be newly produced during the hot and dense stage of the collision, they are thought of carrying information on the properties of the matter that was created [27]. Exotic forms of deeply bound objects with strangeness have been proposed [28] as states of matter, either consisting of baryons or quarks. The H di-baryon was predicted by Jaffe [29] and later, many more bound di-baryon states with strangeness were proposed using quark potentials [30, 31] or the Skyrme model [32]. However, the non-observation of multi-quark bags, e.g. strangelets is still one of the open problems of intermediate and high energy physics. Lattice calculations suggest that the H-dibaryon is a weakly unbound system [33], while recent lattice studies report that there could be strange di-baryon systems including Ξ's that can be bound [34]. Because of the size of these clusters lattice studies are usually very demanding

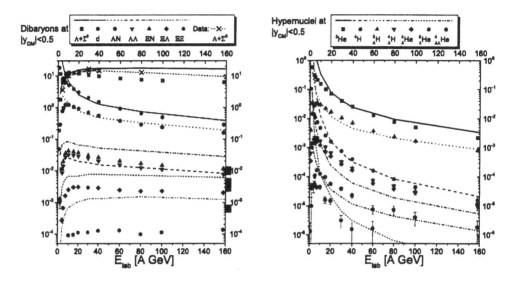

Figure 4: (Color online) Left: Yields of different di-baryons in the mid rapidity region ($|y| < 0.5$) of most central collisions of Pb+Pb/Au+Au. Shown are the results from the thermal production in the UrQMD hybrid model (lines) as compared to coalescence results with the DCM model (symbols). The small bars on the right hand axis denote results on di-baryon yields from a previous RQMD calculation at $\sqrt{s_{NN}} = 200$ GeV [54]. In addition, the black lines and symbols depict results for the production rate of Λ's from both models, compared to data (gray crosses) from [47–49]. Right: Yields of different (hyper-)nuclei in the mid rapidity region ($|y| < 0.5$) of most central collisions of Pb+Pb/Au+Au. Shown are the results from the thermal production in the UrQMD hybrid model (lines) as compared to coalescence results with the DCM model (symbols).

on computational resources and have large lattice artifacts, therefore an experimental confirmation of such a state would be an enormous advance in the understanding of the hyperon interaction.

Hypernuclei are known to exist and be produced in heavy ion collisions already for a long time [35–38]. The interest in the field of hypernuclear physics was fueled by the recent discoveries of the first anti-hypertriton [39] and anti-α [40] (the largest antiparticle cluster ever reported). Metastable exotic multi-hypernuclear objects (MEMOs) as well as purely hyperonic systems of Λ's and Ξ's were introduced in [41, 42] as the hadronic counterparts to multi-strange quark bags [43, 44].

In the work presented in this section we will focus on the production of hypernuclei in high energy collisions of Au+Au ions. In such systems strangeness is produced abundantly and is likely to form clusters of different sizes. Hypernuclear clusters can emerge from the hot and dense fireball region of the reaction. In this scenario the cluster is formed at, or shortly after, the (chemical-)freeze out of the system. A general assumption is, that these clusters are then formed

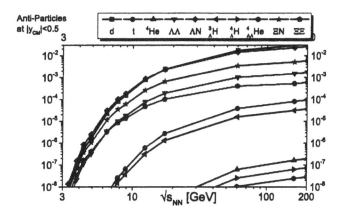

Figure 5: (Color online) Yields of antiparticle clusters with $|y| < 0.5$ of most central collisions of Pb+Pb/Au+Au as a function of $\sqrt{s_{NN}}$. Shown are only the results from the thermal production in the UrQMD hybrid model (lines with symbols).

through coalescence of different newly produced hadrons. To estimate the production yield we can employ thermal production of clusters from a fluid dynamical description to heavy ion collisions. Though thermal production differs significantly in its assumptions from a coalescence approach one would expect to obtain different results, depending on the method used. However it can be shown that both approaches can lead to very similar results [45]. More detailed information on the calculations performed for the results in this section can be found in [46].

Figure 4 shows our results for the mid rapidity yields ($|y| < 0.5$) of di-baryons and hypernuclei as a function of the beam energy E_{lab}. In our calculations we considered most central ($b < 3.4$ fm) Pb+Pb/Au+Au collisions at $E_{\text{lab}} = 1$ - $160A$ GeV. In addition, Fig. 4 shows the Λ yield (black lines and squares) for the model compared to data [47–49]. In these figures, the UrQMD hybrid model calculations are shown as lines. At lower energies the cluster production should be suppressed additionally due to the non-equilibrium of strangeness. In the thermal calculations restrictions of energy and momentum conservation, resulting in a phase space reduction for produced strange particles, strongly decreases strange particle yields [50–52]. This behavior was also observed in a core-corona implementation in the hybrid model [53].

Di-baryon production rates have been calculated in a coalescence approach using the RQMD model for $\sqrt{s_{NN}} = 200$ GeV collisions of Au nuclei [54]. To relate our calculations to these results, they are indicated as the colored bars on the right axis of Fig. 4. The RQMD model used was in particular tuned to reproduce multi strange particle yields (such as the Ξ) and the results are therefore close to the ones obtained with our thermal/hydrodynamic approach. When the beam energy of the collisions is increased, the system created becomes almost net-baryon free. This means that the probability to create an antipar-

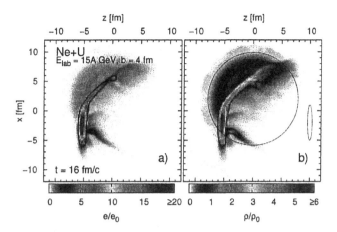

Figure 6: (Color online) Density distributions of the energy a) and baryon number b) in units of the nuclear ground state values e_0, ρ_0 in the reaction plane of the asymmetric collision Ne+U with an impact parameter $b = 4$ fm and $E_{\mathrm{lab}} = 15A$ GeV. The blue lines in b) show the initial setup of the colliding Ne and U nuclei for the calculation with a hydrodynamic model.

ticle cluster approaches that of the particle cluster. Figure 5 shows the results for antiparticle cluster production at mid-rapidity ($|y| < 0.5$) in collisions of Pb+Pb/Au+Au at center of mass energies of $\sqrt{s_{NN}} = 3$ - 200 GeV. The yields of the antiparticle clusters show a monotonous increase with beam energy. They show that, at the highest RHIC energy (and at the LHC) the reconstruction of $^4_\Lambda$He might be a feasible task.

As another promising experimental tool for the production of antimatter clusters and hypernuclei we propose collisions of asymmetric sized nuclei, e.g. Ne+U, Ca+U. In Fig. 6 we show the energy (a) and baryon number (b) density distributions in the reaction plane at $t = 16$ fm of a Ne+U collision at $E_{\mathrm{lab}} = 15A$ GeV with an impact parameter $b = 4$ fm as calculated with a hydrodynamic model. At the collision zone a highly compressed Mach shock wave is created and propagates through the target nucleus causing a directed cone like emission of particles [55]. Depending on the underlying EoS [56] high densities above the phase transition are reached already at rather small beam energies. In this very dense and hot shock zone, antibaryons are abundantly produced. Due to the directed, correlated emission through the surface, many of those $\bar{p}$ and $\bar{n}$ are able to escape annihilation and coalesce to antimatter clusters. Similarly, multi hyperon clusters form and can be detected downstream.

5 Summary

After about ten years of negotiations, R&D and writing reports, on 4th of October nine countries finally signed the international agreement on the construction of the Facility for Antiproton and Ion Research (FAIR). Construction of the first FAIR buildings will start in 2012 and first beams will be delivered in 2018. The initial version of FAIR, the so-called *M*odularized Start Version, includes the superconducting synchrotron SIS100 and three experimental modules to perform experiments for all research pillars. It will allow to carry out an outstanding and world-leading research program in hadron, nuclear, atomic and plasma physics as well as applied sciences. Due to the high luminosity, exceeding current facilities by orders of magnitude, it will be possible to conduct experiments that could not be done anywhere yet. FAIR will expand the knowledge in various scientific fields beyond current frontiers. Moreover, the exploitation of exiting strong cross-topical synergies promise novel insights.

Earlier status reports for FAIR including previous versions of this article can be found e.g. in Refs. [2, 57–67] and studies on strangeness at FAIR energies are found in Refs. [46, 68–71]

6 Acknowledgments

This work was supported by HGS-HIRe and the Hessian LOEWE initiative through the Helmholtz International center for FAIR (HIC for FAIR). Computational resources were provided by Frankfurt Center for Scientific Computing (CSC). J. S. acknowledges a Feodor Lynen fellowship of the Alexander von Humboldt foundation.

References

[1] H.H. Gutbrod *et al.* (Eds.), *FAIR Baseline Technical Report*, (2006).
[2] W. F. Henning, Nucl. Phys. A **805**, 502 (2008).
[3] H. Stöcker, Proc. of 11th Europ. Part. Acc. Conf., Genoa, Italy (2008).
[4] C. Sturm, B. Sharkov and H. Stöcker, Nucl. Phys. A **834**, 682c (2010).
[5] Green Paper of FAIR: *The Modularized Start Version*, www.gsi.de/documents/DOC-2009-Nov-124-1.pdf (2009)
[6] Th. Stöhlker *et al.*, Nucl. Instrum. Meth. B **261**, 234 (2007).
[7] I.V. Lomonosov and N.A. Tahir, Nucl. Phys. News 16 **1**, 29 (2006).
[8] B. Friman, C. Höhne, J. Knoll, S. Leupold, J. Randrup, R. Rapp, P. Senger (Editors), Lect. Notes Phys. **814**, 1 (2011)
[9] P. Senger *et al.* [CBM Collab.], Phys. Part. Nucl. **39**, 1055 (2008).
[10] P. Senger *et al.* [CBM Collab.], Prog. Part. Nucl. Phys. **62**, 375 (2009).
[11] G. Agakishiev *et al.* [HADES Collab.], Eur. Phys. J. A **41**, 243 (2009).
[12] J. Stroth *et al.* [HADES Collab.], Prog. Part. Nucl. Phys. **62**, 481 (2009).
[13] I. Fröhlich [HADES Collab.], arXiv:0906.0091 [nucl-ex].
[14] P. Tlusty *et al.* [HADES Collab.], AIP Conf. Proc. **1322**, 116 (2010).
[15] R. Krücken *et al.* [NuSTAR Collab.], J. Phys. G **31**, S1807 (2005).
[16] B. Rubio and T. Nilsson [NuSTAR Collab.], Nucl. Phys. News **16**, 9 (2006).
[17] Physics Performance Report for PANDA, arXiv:0903.3905v1 [hep-ex].

[18] K. Fohl *et al.* [PANDA Collab.], Eur. Phys. J. ST **162**, 213 (2008).

[19] J. S. Lange [PANDA Collab.], Int. J. Mod. Phys. A **24**, 369 (2009).

[20] W. Greiner, Int. J. Mod. Phys. E **5**, 1 (1996).

[21] J. Adams *et al.* [STAR Collab.], Nucl. Phys. A **757**, 102 (2005).

[22] B. I. Abelev [STAR Collab.], Science **328** 58 (2010).

[23] H. Agakishiev *et al.* [STAR Collab.], Nature **473**, 353 (2011).

[24] H. Liu and Z. Xu, arXiv:nucl-ex/0610035.

[25] S. P. Ahlen, V. M. Balebanov, *et al.*, Nucl. Instrum. Meth. A **350**, 351 (1994).

[26] STAR decadal plan 2011, `www.bnl.gov/npp/docs/STAR_decadal_Plan_Final.pdf`.

[27] P. Koch, B. Müller and J. Rafelski, Phys. Rept. **142**, 167 (1986).

[28] A. R. Bodmer, Phys. Rev. D **4**, 1601 (1971).

[29] R. L. Jaffe, Phys. Rev. Lett. **38**, 195 (1977).

[30] J. T. Goldman *et al.*, Phys. Rev. Lett. **59**, 627 (1987).

[31] J. T. Goldman *et al.*, Mod. Phys. Lett. A **13**, 59 (1998).

[32] B. Schwesinger, F. G. Scholtz and H. B. Geyer, Phys. Rev. D **51**, 1228 (1995).

[33] I. Wetzorke and F. Karsch, Nucl. Phys. Proc. Suppl. **119**, 278 (2003).

[34] S. R. Beane *et al.* [NPLQCD Collab.], arXiv:1109.2889 [hep-lat].

[35] P. Braun-Munzinger and J. Stachel, J. Phys. G **21**, L17 (1995).

[36] J. K. Ahn *et al.*, Phys. Rev. Lett. **87**, 132504 (2001).

[37] H. Takahashi *et al.*, Phys. Rev. Lett. **87**, 212502 (2001).

[38] A. Andronic, P. Braun-Munzinger, J. Stachel and H. Stöcker, Phys. Lett. B **697**, 203 (2011).

[39] STAR Collab., Science **328**, (2010) 58.

[40] STAR Collab., Nature **473** (2011) 353.

[41] J. Schaffner, H. Stöcker and C. Greiner, Phys. Rev. C **46**, 322 (1992).

[42] J. Schaffner *et al.*, Phys. Rev. Lett. **71**, 1328 (1993).

[43] E. P. Gilson and R. L. Jaffe, Phys. Rev. Lett. **71**, 332 (1993).

[44] J. Schaffner-Bielich, C. Greiner, A. Diener and H. Stöcker, Phys. Rev. C **55**, 3038 (1997).

[45] J. Steinheimer *et al.* to be published.

[46] J. Steinheimer *et al.*, Phys. Lett. B **676**, 126 (2009).

[47] S. Ahmad *et al.*, Phys. Lett. B **382**, 35 (1996).

[48] A. Mischke *et al.* [NA49 Collab.], J. Phys. G **28**, 1761 (2002).

[49] C. Alt *et al.* [NA49 Collab.], Phys. Rev. C **78**, 034918 (2008).

[50] F. Becattini and U. W. Heinz, Z. Phys. C **76**, 269 (1997).

[51] J. Cleymans, K. Redlich and E. Suhonen, Z. Phys. C **51**, 137 (1991).

[52] A. Andronic, P. Braun-Munzinger and J. Stachel, Nucl. Phys. A **772**, 167 (2006).

[53] J. Steinheimer and M. Bleicher, Phys. Rev. C **84**, 024905 (2011).

[54] J. Schaffner-Bielich, R. Mattiello and H. Sorge, Phys. Rev. Lett. **84**, 4305 (2000).

[55] P. Rau *et al.*, arXiv:1003.1232 [nucl-th].

[56] P. Rau, J. Steinheimer, S. Schramm and H. Stöcker, Phys. Rev. C **85**,

025204 (2012)

[57] W. F. Henning, AIP Conf. Proc. **773**, 3 (2005).

[58] H. H. Gutbrod, Nucl. Phys. A **752**, 457 (2005).

[59] G. Rosner, AIP Conf. Proc. **870**, 575 (2006).

[60] G. Rosner, Nucl. Phys. Proc. Suppl. **167**, 29 (2007).

[61] W. F. Henning, J. Phys. G **34**, S551 (2007).

[62] I. Augustin, CERN Cour. **47N4**, 23 (2007).

[63] I. Augustin, H. H. Gutbrod, D. Kramer, K. Langanke and H. Stöcker, arXiv:0804.0177 [hep-ph].

[64] H. Stöcker and C. Sturm, Nucl. Phys. A **862-863**, 92 (2011).

[65] C. Sturm and H. Stöcker, Phys. Part. Nucl. Lett. **8**, 865 (2011).

[66] H. Stöcker and C. Sturm, Nucl. Phys. A **855**, 506 (2011).

[67] J. Steinheimer, Z. Zhu, P. Rau, C. Sturm and H. Stöcker Proc. Int. Sym. on Exciting Phys. Makutsi, South Africa (2011).

[68] J. Steinheimer, H. Stöcker, I. Augustin, A. Andronic, T. Saito and P. Senger, Prog. Part. Nucl. Phys. **62**, 313 (2009).

[69] H. Stöcker, I. Augustin, J. Steinheimer, A. Andronic, T. Saito and P. Senger, Nucl. Phys. A **827**, 624C (2009).

[70] J. Steinheimer, I. Augustin, A. Andronic, T. Saito, P. Senger and H. Stöcker, J. Phys. G **36**, 064036 (2009).

[71] J. Steinheimer, C. Sturm, S. Schramm and H. Stöcker, J. Phys. G **37**, 094026 (2010).

Scientific Secretaries: A. Ortiz Velasquez, O. Hrycyna

DISCUSSION

- *Inguglia:*

You presented some ideas about the future research at FAIR. In which processes the CP-violation in the charm sector will be studied?

- *Stoecker:*

In the $D\overline{D}$ system: search for flavor ($D\overline{D}$) mixing in rare D decays produced at the $D\overline{D}$ threshold. Formed near threshold, no asymmetries are expected in the production process and the observation of one D meson tells about the quantum numbers of the other one at production in a charge symmetric environment (flavor tagging).

- *Gerbaudo:*

Could you say something about what properties of the isotopes produced at GSI you will be able to study?

- *Stoecker:*

Production of rare isotopes is of course what we also try to do at FAIR. At the production energy of GSI of 1-2 AGeV a fragmentation cone of unstable nuclear beams moving almost with the speed of light is formed via kinematical focusing. The picture here shows the present setup at GSI, which consists of a synchrotron of 210m circumference and the 70m long fragment separator. What you can do is to bring this in to the magnets and study for instance excited states within exotic nuclei, which will behave differently when there is very large neutron access. Neutrons can then swing into different modes than the protons and you can study giant resonant excitations and single particle properties similar to nuclei but at a very different environment with strong neutron access. It also allows you to get interesting information on the properties of neutron rich nuclei, e.g. separation energies and other things that are important for astrophysics. Resonance neutrons swinging against nucleons and you can study the properties of neutron skins when you have very high neutron densities. In the core of the nucleus the neutron number is more or less equal to the proton number but outside pure neutron-proton matter. In the core you have a different charge to mass ratio than in the surface. And the skin and also the halos for which you can have very light nuclei like for example a lithium nucleus three protons in the center and then its halo of five additional neutrons which move in a rather big distance in a kind of a skin around the protons. You have other changes of the shell structure, I mentioned it very briefly, we know that in the nucleus there are shells like in atomic shells and there are certain shells which are relevant also for production of the very heavy elements. In normal burning processes inside stars you can get only up to iron but you can not go up to uranium. You must have this additional passage far out in the very neuron-rich region of the nuclear chart during the explosion process of supernovae to get there. We know that nuclei exhibit a certain shell structure with shell closures at 2, 8, 20, 28, 50, etc.; however these shells may not exist if there are many more neutrons there. This is the question we have seen some simulations done with characteristic mean field calculations which show that this shell may actually exist for very low charge states also. It could be, nobody knows, that here we could make nuclei which have two to three times more neutrons than protons.

- Pagnutti:

What is the difference between the QGP produced at FAIR and the ones produced at RHIC and LHC?

- Stoecker:

With FAIR it becomes possible to reach higher baryon densities than those produced by LHC and RHIC. Basically the net baryon density or baryon chemical potential is zero at LHC. With FAIR it is possible to access areas which are in density higher than SPS at CERN had achieved. A first order phase transition into the QGP is expected in this area of the phase diagram.

- Preghenella:

CBM will run without a trigger. How demanding is that way of operation for the data-acquisition system (DAQ)?

- Stoecker:

This is development work, the data rates is of the order of 30 times higher than at LHC. The problem is not only the high rate but also the kinematical focusing into the forward direction.

- Mao:

In your project, you are trying to explore a different regime on phase diagram compared to LHC. How can you constrain the T but keep the chemical potential increasing quickly to reach your goal?

- Stoecker:

When we accelerate the beams we always increase T and the chemical potential, but cannot keep T constant.

- Schmidt-Sommerfeld:

Why do you not strip off more electrons off U^{27+} to get beams with higher energies?

- Stoecker:

Because then you cannot have high intensity beams due to coulomb repulsion, which makes the beam unstable.

- Ponomarev:

Some people believe that there are islands of stability of nuclei with extra large atomic numbers. What do you think about that ?

- Stoecker:

This is of course a hot topic at GSI. You mean very large charge like Z=110 and 120 - super heavy elements. We have actually carried out experiments to produce those very heavy elements at GSI for many years. Actually it was the main motivation to found GSI more than 40 years ago. Not in very hard but in very soft collisions these superheavy elements are produced by fusing a led nucleus which is a very stable double shell-stabilized nucleus with e.g. a nickel nucleus. Nickel is 28, led is 82 protons, added together you get nuclei with 110 protons. These nuclei have been produced very rarely and unfortunately they decay in fractions of a second by alpha decay. When you produce those nuclei you pull them trough the spectrometer which is similar to this monster here but with much lower energy. Produced one every two weeks roughly, observed in a silicon detector, you wait let's say a microsecond or a second and suddenly this nucleus spits out six alpha particle one after another and decays. By observing the characteristic energies of the alpha particle, which of course decrease if the nucleus has less protons, you can identify very clearly the mother nucleus. This was successfully done for element 107 (Bohrium after Niels Bohr), 108 (Hassium after the state of Hesse), 109 (Meitnerium after Lise Meitner), 110 (Darmstadtium after Darmstadt), 111 (Roentgenium after Wilhelm Conrad Rontgen) and 112 (Copernicium after Nicolaus Copernicus). These events cannot be observed in very high energy reactions because they are destroying nuclei. If you want to fuse the nuclei normally you have to go to energies just at the Coulomb barrier. We do have a program, which will not be a part of the FAIR project, but at GSI to upgrade the linear accelerator. We are hunting now element 120 in the collaboration between Oak Ridge National Lab, Vanderbilt University, Joint Institute for Nuclear Research in Dubna (Russia) and GSI.

- Cerchiai:

What is the reason behind creating the FAIR project as a company (GmbH) ?

- Stoecker:

This is a political question which should better be answered by Rolf-Dieter Heuer. To establish the FAIR GmbH was a political decision made in 2004-2005. The reasoning behind that is that Germany and also our international partners are interested in legal a framework set by the German law and, accordingly, in an organizational structure like a GmbH, in particular also to limit the risks for the shareholders. The limited liability company is nothing special for German research institutions, GSI is also a GmbH.

- Grelli:

In the new detectors planned at GSI will be used RPCs. Are they done with new materials or are just bakelite like in CMS ?

- Stoecker:

There is a new program development in particular in Poland and China dealing with high track densities. I am much more worried about the area around the vertex detector. Yes, there are new developments. We are still 8 years away from the running of these detectors so we have still time to work on this issue.

S. VIGDOR

"Highlights from BNL-RHIC"

The Slides of the Lecture can be found at:

http://www.ccsem.infn.it/issp2010/

Scientific Secretaries: N. Krupina, D. Ponomarev

DISCUSSION I

- *Schmidt-Sommerfeld:*
N=4 SYM is dual to a classical gravity theory. How can a quantum theory be related to a classical one?

- *Vigdor:*
I am not a theorist, so I am not the most qualified to answer this question about AdS/CFT duality. One way of thinking is, in my understanding at least, that the strongly coupled Yang-Mills theory to which gravity is dual, is one in which the number of colors tends toward infinity. So, in that limit you are dealing with a sort of classical limit. It is not QCD. It is the strongly interacting gauge theory. It is equivalent to having infinite number of colors.

- *Duff:*
I agree that when 1/N corrections are taken into account, both sides of the duality are quantum. It is only in the large N limit that one side looks classical.

- *Haidt:*
Is the temperature in a relativistic hydrodynamics model to be treated as 4th component of a vector or it is a scalar?

- *Vigdor:*
Again, I could use some help from theorists in the audience to answer this question.

- *Somebody else:*
It is a well-known discussion. Temperature is well-defined for the system in equilibrium, that is it requires the coordinate frame. I think one should avoid this question first of all.

- *Burda:*
Could we investigate the so-called chiral magnetic effect on RHIC? If yes, do the results of experiments agree with lattice calculations and theoretical predictions?

- *Vigdor:*
I will discuss this issue in depth in tomorrow's lecture, so here is a brief answer. The chiral magnetic effect basically is a mechanism for violating parity locally in the strong interaction, associated with sphaleron transitions -- the transitions among alternative vacuum states which differ from one another by a leftward or rightward twist in the gluon field. In principle, if you have enough statistics in a single event you could actually look for parity violating signal in a single event. But, certainly, at RHIC the expected magnitude of the effect is too small to see in a single event. So you have to look for this effect averaged over many events. When you look for this effect averaged over many events the characteristic of the sphaleron transitions in the chiral magnetic effect is that you can get an electric dipole moment effectively in a given event, for a given bubble. But the sign and the direction of electric dipole moment with respect to the magnetic field in the collision can change from bubble to bubble or from event to event. So if you average over many events, you can`t do something as simple as looking for a correlation between a charge preference and the direction of the magnetic field, since that correlation could be opposite in different events.

But what has been looked for so far in measurements at RHIC are charged particle correlations. That is, you look for a correlation that is suggestive of an electric dipole moment, in the sense that you see a preference for like-sign charges to come out on the same side of the reaction plane and for opposite-sign charges to come out on opposite sides. That correlation was predicted by the chiral magnetic effect and the measurement for those correlations see a clearly non-zero effect very consistent qualitatively with the prediction. Because you are looking at a two-particle correlation, it is intrinsically a parity-even signal, even though it may have a parity-violating origin.

Because it is a parity-even signal, there are conceivable mundane effects that could mock it up. So right now we have an observable which is consistent with the prediction of the chiral magnetic effect. I will talk tomorrow about many layers of subsequent follow-up experiments that we are going to do to address contaminating sources, or that we are very actively trying to pursue until somebody can come up with a clever idea -- nobody has yet -- of a truly parity-odd experimental signature. That would be the best experiment to do.

One of the questions I asked myself very early is whether LHC will be in a better situation simply because you produce more particles, therefore you are likely to produce more bubbles in any given collision event. I believe that there is effectively a "no win" theorem because the effect is small already in RHIC. What happens if you get more bubbles in a given event is that you get a random walk in the direction of the chiral magnetic effect. So the net effect in a single event can grow up as the square root of the number of bubbles, rather than proportional to the number of bubbles. The relative size of the effect in the event therefore goes as $1/\sqrt{N}$. But the statistical precision improves by the same factor, so you neither gain nor lose. The single-event effect remains similarly just beyond the statistical grasp of the experiment.

- Mulhearn:

How do you reconstruct observables like T, energy density, shear viscosity?

- Vigdor:

Most can not be directly measured, but are inferred from reconstructed observables using tunable models adjusted to fit the data.

- Mulhearn:

How predictive are these models?

- Vigdor:

As more constraints are added, models are becoming more predictive.

- Mao:

For flow measurements the observables are particle correlations, I am wondering how could we separate the contributions from flow and d-jets events correlations measurements?

- Vigdor:

In principle, we can (and do) apply different approaches to study flow by reconstructing the reaction plane using different detector subsystems, or by using 4-particle rather than 2-particle correlations to change the sensitivity to non-flow effects. But still it is difficult to completely separate them, so it is an open question in RHIC, which is reflected in the systematic errors.

- Pagnutti:

Can RHIC control the coupling of QGP during the energy scan experiment, and could this be used to verify the AdS/CFT value of $1/4\pi$?

- Vigdor:

RHIC can't control coupling, but it will change if the energy changes. Lattice QCD theorists suggest that it may be an accident that RHIC is currently probing a regime where AdS/CFT appears to be applicable – the lattice calculations suggest a very different temperature-dependence of observables, consistent with a fundamentally weak coupling that accidentally crosses the entropy density characteristic of the AdS/CFT approach at the current RHIC energies.

- Bettini:

You showed a plot of the energy density against temperature of the medium. It shows a curve, which saturates, and a number of points with error bars, which I guess to be theoretical calculations. Are there any "experimental" points available for comparison?

- Vigdor:

The only experimental information above the critical temperature, as I answered to the previous question, is from the higher RHIC energies run so far. We extract the initial temperature and the initial energy density by doing model comparisons to the data. So, at this point we have an idea of what the initial temperature is, but only within a factor of two because it depends critically on what thermalization time is assumed. And we can from hydrodynamic models also extract the energy density. So the experimental points have quite significant systematic error bars right now at the RHIC energy. Clearly at the higher temperatures only the early Universe has done the experiments to date. Until we measure at LHC, we do not have other experimental points to put on that plot at higher temperature. The points there are the results of lattice QCD-inspired calculations.

- Mulhearn:

What was the challenge overcome to produce these beautiful plots of reconstructed tracks in extremely crowded events?

- Vigdor:

It was surprisingly easy with the tracking detectors designed for the RHIC environment. There is some loss of tracking efficiency very near the event vertex, due to the very high density of ionization there, for example, in the STAR TPC gas. But the basic tracking worked very well, very early on. The many pileup events actually make pp collisions more challenging for proper vertex reconstruction. The reason for that is that luminosity is much higher in pp collisions, so one sees partial track segments not emerging from a common vertex, arising from collisions at different beam crossings within the TPC drift time period.

DISCUSSION II

- Preghenella:

You said the monojet structure observed in *d-Au* collisions is due to initial state effects. Previous measurements about jet-quenching in *Au-Au* collisions use *d-Au* collisions to claim that jet-quenching is not due to initial state effects. Could you please comment on that?

- Vigdor:

You are absolutely correct that in the past, there has been a measurement, which looked somewhat analogous to the present one, which we used to provide an example to what we call jet-quenching. However, those were measurements at mid-rapidity. In heavy-ion collisions, when one collides gold nuclei, and looks at mid-rapidity, in the middle region of the detector, around 90 degrees, when you see a high-p_T hadron on one side, the coincident away-side hadron tends to disappear. When you look in the *d-Au,* which should be sensitive to initial state effects, it is there, and you see the full di-jet correlation. This is true in the middle angle range -- mid-rapidity -- where you are sensitive to not-very-small-*x* partons. The difference in the present measurement is that here we measure the true far forward hadrons, it is looking at very asymmetric parton collisions. This way, we are probing very-low-*x* partons inside the gold nuclei, so here we do see what we interpret as an initial-state effect, which is not really jet-quenching, and which is interpreted now – and we shall see if that interpretation holds – as scattering from a coherent gluon field rather than a single gluon. So, instead of seeing a well-defined away-side jet, the energy is spread over many small jets. This has some similarity in the observation to what one sees in heavy-ion collisions for jet quenching, but the interpretation is quite different.

- Noferini:

I remember that in the past, STAR has seen a bigger correlation in jets for opposite sign pairs (in the 2 to 4 GeV region). In your lecture, you have shown an opposite behaviour due to the charge distribution in nucleus-nucleus collisions. Can you explain how these two apparently opposite forms of behaviour can be explained in a common way?

- Vigdor:

If I understand the first half of the question correctly, what you are referring to is that in a normal jet structure, for high-p_T hadrons there is a correlation between the leading and the next-to-leading hadrons, that they tend to be of opposite charge. The reason for this is simply that in this case strings are being broken, and thus pairs of $q\bar{q}$ s are being produced, and then the q becomes part of the leading particle, and the $\bar{q}$ that of the next-to-leading one, which gives us a preferred correlation of unlike charges in a jet. What I have shown in the lecture was a measurement not for jets, but rather for the bulk of particles, one that integrates over all particles in the event, which are dominated by soft particles at a 99% level. These particles are not jet fragments, they are rather part of the background. So what is seen clearly here, in this measurement, is a correlation between like-sign soft particles. These prefer to emerge in the same direction, either above or below the reaction plane. This is simply a different observation, because it is not concentrating on the ~1% of the particles that are jet fragments.

- Schmidt-Sommerfeld:

The temperature at RHIC is at an energy scale where the strong coupling constant is not really perturbatively small. Why were you surprised to find a strongly coupled system?

- Vigdor:

I can only give my interpretation of the subject. The temperature at RHIC is in the range where $\alpha_s \approx 0.3$ approximately (i.e. in the middle range). The basic interpretation is based on lattice calculations, that above the transition temperature, the system achieves an entropy density within 80% of the Stefan-Boltzmann limit for an ideal (weakly coupled) gas. In the early days, before the exploitation of the AdS/CFT duality, most theorists interpreted that so that 80% is close to 100%, so the coupling must be weak. I am not sure if there was much more to it. In the wake of the AdS/CFT duality, where the entropy density of an infinitely strongly coupled system turns out to be 75% of the Stefan-Boltzmann limit, suddenly 80% seems to be very close to 75%, so in retrospect one should not be surprised.

- Comment from the audience.:

At twice the critical temperature, $g^2(2T_C) = 2.17$.

- Vigdor:

Ok, so it is even further than middle region. Most theorists were surprised. Nobody predicted that what we would see is so strongly coupled, but in the aftermath, by exploitation of the Maldacena duality, it is suddenly obvious, and people are not surprised anymore. In this sense, this is a historical question.

- Mao:

If the gluon field becomes saturated in a nucleon, then more low-x gluons are in the saturated region, so we should see an enhancement at the low-p_T range. Why did RHIC measure quite the opposite (i.e. stronger suppression at low-p_T)?

- Vigdor:

If the gluon density saturates, then it is smaller than what one would expect, by extrapolating from the nucleon-nucleon case to that of nuclei, so a suppression of hadrons can be observed in the gluon saturation region. When a given hadron is examined, it is assumed that it comes from a jet. In this way a given momentum hadron can originate from a range of jets, that can vary in jet fraction, so a given p_T of a hadron can correspond to a range of p_T values for the jet. But on the average, lower-p_T hadrons come from lower-p_T jets, which means that they are probing lower-x gluons, which may be in the saturated region. The lower p_T in the experiment is clearly observed to be more suppressed, and in the calculations this is also true basically, just because in these measurements one is sampling the low-x regime where the gluon density is assumed to saturate. This is consistent with the expectation of gluon saturation. Both observables are consistent with what one would expect assuming gluon saturation. Neither of them, or not even a combination of the two, can be taken as proof though, because there are potentially other explanations.

- Mao:

I still do not understand everything, because I think that gluon saturation happens for low-x gluons, that will fragment to low-p_T hadrons, right?

- Vigdor:

Yes.

- Mao:

If that's so, if you can compare the high-p_T range to the low-p_T one, you should see the enhancement.

- Vigdor:

Ok. The high-p_T range here is not very relevant. The same measurement has been made with *d-Au*, and what one compares, is the same p_T-range for that and *pp*. The suppression behaviour is the same whether you are comparing the peripheral *d-Au* or *pp* events. The saturation means that the gluon density in a nucleus is lower than what one would have guessed by extrapolating from what it is inside the proton, and therefore what is seen is a suppression when the gluon density becomes saturated. You are correct that the gluon densities are higher at lower x, but here one is not looking at the absolute gluon densities, but instead compares the results to those of a *pp*-collision experiment, and because of the coherent effects at low x in a nucleus, the saturation shows up at a Bjorken x value, which is much higher in a nucleus than in a nucleon. Equivalently, the gluon densities in a nucleus in the saturated region are lower than what one would have guessed by doing a linear extrapolation from their values in a proton, thus, saturation gives rise to a suppression.

- Bettini:

It seems to me, that the test for local P-violation might be done by looking for the correlations between different particles of the same/opposite sign (e.g. pions and kaons). In this way, one could avoid some of the "mundane effects". Can you do that?

- Vigdor:

Sure, absolutely. This first measurement was done without particle identification simply because that gives the best statistics, by the inclusion of all particles. But we do have particle identification, so we can look for the same correlation with different particle types, so that is an obvious next step. The use of particle identification certainly rules out some other explanations, like Bose-Einstein correlations (which cannot explain the effect anyhow, because these are correlations at very low relative momentum, and we know that the correlation here is not peaked at low relative momentum). Particle identification would still rule out some other explanations anyhow, so it is a good suggestion, and it will be done.

Highlights from Gran Sasso

Lucia Votano

Laboratori Nazionali del Gran Sasso dell'Istituto Nazionale di Fisica Nucleare, Italy
E-mail: lucia.votano@lngs.infn.it

1. Introduction

When we look back at the late 70s, at the moment when the proposal to create the laboratories under the Gran Sasso massif was submitted by the President of INFN, Antonino Zichichi, inevitably we notice that astroparticle physics was already taking the first steps towards the extraordinary developments that we have seen in the following decades.

The last thirty years have seen enormous progress in terms of knowledge and understanding of the phenomena studied in this sector of physics which is at the crossroads among elementary particle physics, astrophysics and cosmology. An ever increasing number of physicists have started working in this sector attracted and fascinated by the prospect of trying to find answers to fundamental questions in particle physics and astrophysics without having to use accelerators and being able to study interactions at very high energies which are still inaccessible even to the most powerful particle accelerators. We furthermore notice the growth in quality of the technologies used in the experiments in the underground laboratories.

Technologies developed in accelerator apparatus were imported, but successively the search for very rare events, the need of increasing sensitivity and efficiency, the complexity of the analysis have called for the development of ever more cutting edge technologies. At present the experimental apparatus dedicated to astroparticle physics have mass and dimensions and technological complexity comparable to that used in the LHC. Taking an overall view of the scientific production in the last 30 years, the increased presence of astroparticle physics in the most quoted scientific articles is immediately visible.

Astroparticle physics would not have been able to make such a massive and rapid progress without the great infrastructures necessary for this kind of study: the facilities of the underground laboratories. Having planned and built such a large and well equipped laboratory as early as the late 1970s allowed Italy to take a leading role in this process.

At present the scientific motivations for such big and complex underground infrastructures are even more valid and new halls are being excavated in existing laboratories and new projects are under considerations.

Underground laboratories are the main infrastructures for astroparticle and neutrino physics, to explore the highest energy scales not accessible with accelerators, by searching for extremely rare phenomena. The searches for rare events like $2\beta 0\nu$ or proton decay, the study of weak interactions from cosmic or neutrinos from accelerators, the detection of dark matter candidates and the nuclear astrophysics studies require low-background environments and can be realized only in a full shielded underground environment. Thanks to the rock coverage and the corresponding reduction in the cosmic ray flux and c.r.-spallation

Figure 1: Sketch of the underground Laboratory and the Gran Sasso massif.

induced neutrons, underground laboratories provide the necessary low-background environment to investigate these processes. The main characteristics of the Gran Sasso Laboratory will be described together with an overview of its broad scientific activities; the highlights from neutrino physics in particular will be emphasized.

2. Gran Sasso National Laboratory

The INFN Gran Sasso National Laboratory (LNGS) is the largest underground laboratory in the world for astroparticle physics. It is one of the four INFN national laboratories and it is a worldwide facility for scientists working in one of the twenty experiments at Gran Sasso. Located between L'Aquila and Teramo, about 120 km far from Rome, the underground structures are on a side of the highway tunnel (10 km long) which crosses the Gran Sasso massif, towards Rome, and consist of three huge experimental halls (each one 100 m long, 20 m large and 18 m high) and service tunnels, for a total volume of $\sim$180.000 m^3 and a surface of $\sim$18.000 m^2.

The easy access to the Laboratory from the highway allows the transport of huge and heavy apparatus directly into the halls, the constant provision of supplies necessary for the functioning both of the experiments and of the Laboratory itself, and finally it ensures an easy turnover of the personnel working underground. The 1400 m of rock overhead are such a cover able to reduce the cosmic rays flux by a million times; moreover, the flux of neutrons in the underground halls is about a thousand times less than on the surface, due to the very small amount of Uranium and Thorium of the Dolomite calcareous rock of the mountain.

The halls are equipped with complex technical and safety plants necessary to carry out the experimental activities as well as to ensure proper working conditions to the personnel involved in them. For instance the ventilation plants allow an air exchange in less than 3 hours in normal conditions and keep the radon content in the air in the 20-80 Bq/m^3 range. On the outside, next to the tollgate of Assergi on the highway A24, an area of more than 23 acres belonging to a National Park of exceptional environmental and naturalistic value on the slopes of Gran Sasso hosts the laboratories of chemistry, electronic, mechanical design and workshops, the Computing Centre, the Directorate and the various Offices. Presently the LNGS staff consists of 90 people, and more than 950 scientists from 29 different countries (see figure 2) visit the Laboratory every year to perform their research activities.

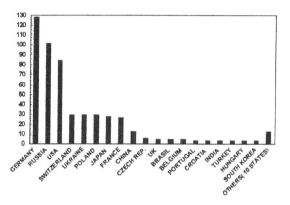

Figure 2: Distribution by country of the researchers involved in the experiments at Gran Sasso.

3. Physics at LNGS

Many of the experiments currently under way or soon to start at LNGS are looking at fundamental and fascinating questions which are still unanswered. What is the Universe made up of and in particular Dark Matter? In this area are we any nearer to open a door to physics that goes beyond the Standard Model of particle physics? What is the intimate nature and what are the latest characteristics of the neutrino and what role have they played in the evolution of the Universe? What do we know about the interior of the Sun and the stars, of the mechanisms that produce energy and how they evolve and die? What do we know about the interior of our own Earth, can we find out more through the study of neutrinos?

3.1. Study of dark matter

Several experimental pieces of evidence (the first and most famous is the observation of the spiral galaxies rotation speed) indicate the existence in the Universe of an amount of mass larger than the one observable by means of telescopes. It is called Dark Matter because it neither emits nor absorbs radiation and thus it is invisible to our eyes and instruments. It is supposed to be five times more abundant than ordinary matter, which constitutes only 5% of our Universe.

One of the best known hypotheses is the assumption of the existence of massive particles that constitute a widespread halo permeating our galaxy as well as others. Such particles, whether they exist or not, interact very weakly with the ordinary matter, therefore their detection could occur exclusively in apparatus shielded against cosmic rays and natural radioactivity, as in the underground halls of LNGS, with experiments very sensitive to the weak interactions among these particles and the ordinary matter atoms. There are different techniques that could be able to discover Dark Matter.

At LNGS we have 4 experiments devoted to hunt dark matter candidates through their direct detection, each one by means of a different technology. These experiments put Gran Sasso Laboratory at the forefront in this kind of study.

DAMA/LIBRA is an experiment that has been operational since 2003, consisting of $\sim$250 kg of NaI(Tl) extremely radio-pure crystals. Results so far published have confirmed the annual modulation of very low energy signals induced in the detector, already observed in the previous experiment DAMA of lower mass. Such modulation is identical to the one that the relative motion of the Earth through the huge amount of dark matter halo of our galaxy is supposed to cause. It is worth mentioning that the annual modulation is a model independent signature of dark matter detection. The interesting result has produced a lively debate inside the scientific community as well as the production of theoretical models able to reconcile such results with the absence of positive signals by other experiments.

The experiment Warp 100 is a cryogenic detector that uses two-phases argon (liquid and gas) as active volume and is based on a double detection technique, the scintillation in liquid argon that produces the primary signal and the ionization in the gaseous state. The ratio between the two different signals produced and the analysis of the primary signal shapes are means to discriminate between the signal produced by a dark matter particle and other background signals. The experiment is in commissioning phase.

The experiment Xenon100 is also a two-phases detector (liquid and gas) that measures concurrently the scintillation and ionization produced by the interaction of the particles with the xenon contained in the apparatus. The technique is analogous to the one of WARP, while the target is different. The apparatus contains 170 kg of xenon, 65 kg of which constitute the active part while the remaining ones act as a shield.

With only 11.2 live days of unblinded data, the experiment demonstrated a background-free operation for a net exposure exceeding that of the former experiment Xenon10. The background inside a 30 kg fiducial volume is two orders of magnitude below the previous experiment. This new result puts a preliminary upper limit of $2.7 \times 10^{-44} cm^2$ (90 % C.L.) at a WIMP mass of 46 GeV/c^2 for spin independent interactions, and $3.7 \times 10^{-44} cm^2$ at 100 GeV/c^2 for spin dependent interactions. The experiment Xenon100 will run for at least one year and will search for the annual modulation signal observed by the DAMA/LIBRA experiment. In order to reach a lower sensitivity of about $5 \times 10^{-47} cm^2$ and/or to conform a possible detection in Xenon100, the collaboration proposes to build a detector with a total mass of 2.4 tons of LXe at LNGS.

The experiment CRESST is based on the bolometer technique with $CaWO_4$ crystals at a temperature of 10 mK as well as on the simultaneous detection of scintillation light and the heat resulting by the interaction of a particle with the crystals. The experiment started a new run in 2009 with 10 complete detector modules and is running stably. The analysis of the first 330 kg per day of data has started and preliminary results look promising. Furthermore research and development activity goes on in an ad hoc cryostat, meant to finalize an experiment proposal of larger mass.

3.2. Neutrino Physics

The study of the intrinsic properties of neutrinos is of prime interest in elementary particle physics. Neutrino physics has provided by now the only experimental evidence of phenomena not included in the Standard Model of elementary particles by measuring neutrino oscillations and showing that they do have a mass, although very small. In order to properly comprehend the mechanism of the oscillation, which allows a neutrino to change its flavour into another, it is necessary to measure the elements of the mixing matrix. Various neutrino sources, both natural (the Sun or other stars) and artificial (particle accelerators) are used.

Moreover, the study of a phenomenon as the neutrino-less double-beta decay could allow us to find out if the neutrino overlaps with its antiparticle, thus providing a very significant answer towards the comprehension of the evolution of the Universe. Massive Majorana neutrinos may have played a relevant role to explain the slight asymmetry between matter and antimatter, which has generated the Universe of matter, the way we know it. Neutrino physics is therefore a window open to a new theory of elementary particles and to the comprehension of the evolution of the Universe.

Finally, neutrinos from the cosmos are very important messengers which transport information fundamental to understand the functioning of the stars as energy sources, their evolution and what happens when they "turn off". LNGS activities range among various aspects of the neutrino physics study.

4. Borexino

The analysis of neutrinos coming from the Sun contributes to the study of the internal functioning (i.e. nuclear fusion reactions) of the nearest star. Starting with the experiment Gallex-GNO, ended in 2002, LNGS has continued on this path with the experiment Borexino, currently operational.

Borexino is a large unsegmented liquid scintillator detector built for the observation and real time measurement of low-energy solar neutrinos. The inner part of the detector is a Stainless Steel Sphere (SSS) of diameter 13.7 m that represents both the container of the scintillator and the mechanical support of the 2200 PMTs. Within the sphere, two nylon vessels separate the scintillator volume in three shells of increasing radiopurity. The inner one, of diameter 8.5 m, contains the liquid scintillator solution, namely a mixture of pseudocumene (PC) as a solvent and fluor PPO (2,5-Diphenyloxazole) as a solute at a concentration of 1.5 g/l; the second and the third shell contain PC with a small amount of DMP (Dimethylphthalate) added as a light quencher.

The SSS is enclosed within a large tank filled with ultrapure water that represents a powerful shielding against external background and is used as a Cherenkov muon counter and muon tracker. Key features of the PC/PPO solution are high scintillation yield (10^4 *photons/MeV*), high light transparency (mean free path typically 8m) and fast decay time (3 ns). The experiment can then rely on an optimal energy and spatial resolution together with a very low threshold (40 KeV).

The main goal of the experiment is the detection of the monochromatic ^{7}Be neutrinos; however the extremely high radiopurity of the detector has resulted in a broadening of the scientific goals. Borexino now aims at the spectral study of other solar neutrino components, such as the CNO, pep and, possibly, pp.

Low energy neutrinos of all flavours are detected by their elastic scattering of electrons while electron antineutrinos by means of their inverse beta decay on protons or carbon nuclei. As the shape of the energy spectrum of recoil electrons is the only signature available for neutrinos, γ or β background events due to natural radioactivity cannot be distinguished from the signal on an event by event analysis, but only through their spectral shape. This fact, together with the low event rate expected, demands an extreme radiopurity of the detector and actually the inner core of Borexino is 9-10 orders of magnitude less radioactive than anything else on Earth.

The ^{7}Be neutrino flux is obtained by fitting the energy spectrum of the events reconstructed inside a fiducial volume defined via software and specifically by fitting the α-subtracted spectrum in the region 100-800 p.e.. After the first result published in 2007 with a 30% precision, the best value for the interaction rate of ^{7}Be neutrino published in 2008 is $49 \pm 3(stat) \pm 4(syst)$ *counts/(day* $\times$ *100t)* [1].

The expected signal for non oscillated solar neutrinos in the high metallicity is $74\pm$ *counts/day* $\times$ *100t*, while in the MSW-LMA scenario of solar neutrino oscillation it is 48 ± 4 *counts/(day* $\times$ *100t)*, in very good agreement with the Borexino measurement. In this scenario, neutrino oscillations are dominated by matter effects above 3 MeV and by vacuum effects below 0.5 MeV. The ^{7}Be neutrinos lie in the lower edge of this transition region.

The contributions to the uncertainty of the measurement are related to statistics, as well as to systematic effects due to the accuracy of the definition of the fiducial volume, to the precise knowledge of the energy scale and of the overall response function of the detector. After the completion of the calibration campaign performed with radioactive sources deployed in approximately one hundred positions inside the scintillator active volume and using 3 more years of data collection, Borexino has recently published a new result at the level of 5%.

Borexino has also been able, for the first time in a liquid scintillator detector, to detect ^{8}B neutrinos above 3 MeV [2]. The first simultaneous measurement of the neutrino survival probability in the vacuum and in the matter-enhanced oscillations regions is of great importance and could be improved in the near future.

Thanks to the extreme high radiopurity achieved, the high photon yield, and the large number of free target protons, Borexino is also a very sensitive detector for antineutrinos in the MeV energy range where geo neutrinos lie.

Geo neutrinos ($\bar{\nu}_e$) are electron antineutrinos produced in β decays of ^{40}K and of several nuclides in the chains of long-lived ^{238}U and ^{232}Th present in the Earth. They are direct messengers of the abundance and distribution of radioactive elements within our planet. By measuring their flux and spectrum it is possible to reveal the distribution of the natural radioactive isotopes and to assess the radiogenic contribution to the total heat balance of the Earth. Nowadays the existence of large mass scintillation detectors made their detection

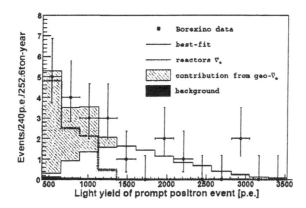

Figure 3: Light yield spectrum for the positron prompt events of the 21 $\bar{\nu}_e$ candidates (thick solid line). Thin solid line: reactor-$\bar{\nu}_e$ signal from the fit. Dotted line (red): geo-$\bar{\nu}_e$ signal resulting from the fit. The darker area isolates the contribution of the geo-$\bar{\nu}_e$ in the total signal.

feasible. KamLAND has opened the way to geo neutrino detection in 2005 while in 2010 the geo neutrino observation has definitely been confirmed by Borexino [3]. The best estimate of the geo $\bar{\nu}_e$ rate has been determined with an unbinned maximum likelihood analysis of the twenty-one $\bar{\nu}_e$ candidates that passed all selections cuts, including the reactor antineutrinos that represent an irreducible source of background. The KamLAND 2008 and Borexino 2010 combined analysis gives an evidence at more than 5σ for the detection of geo neutrinos.

In the coming years present and future experiments will be able to determine with enough statistics the percentage of terrestrial heat power produced by the radioactive decay thus improving significantly our knowledge of the interior of the Earth.

5. The CNGS project and Opera

The project CNGS started in 2006. It consists of an artificial neutrino beam, produced by the protons accelerator SPS of CERN and directed towards Gran Sasso. The neutrino beam crosses the earth crust for 732 km. So far $\nu_\mu \leftrightarrow \nu_\tau$ oscillations have been claimed only indirectly in disappearance experiments. The goal of OPERA is the detection of neutrino oscillations in direct appearance mode through the the detection of the short lived τ lepton produced in the charge current interaction of ν_τ. The huge apparatus mainly consists of 150.000 'bricks' made up of lead layers interleaved with nuclear emulsions, historically called Emulsion Cloud Chamber (ECC). This is, in fact, the only kind of detector able to record the micrometric tracks left by the interaction of a τ neutrino with a nucleus of lead acting as target. The apparatus is then able to fulfill two contradictory requests, big mass to get neutrino interacting and micrometric spatial precision to detect the decay of τ particles.

OPERA is a hybrid detector made of a veto plane followed by two identical super modules (SM) each consisting of a target section of about 625 tons made of 75000 bricks, and of a scintillator Target Tracker Detector (TT) to trigger the read-out and localize neutrino interactions within the target, followed by a muon spectrometer.

A target brick consists of 56 lead plates of 1 mm thickness interleaved with 57 emulsion films and weight 8.3 Kg. In order to reduce the emulsion scanning load, a doublet of Changeable Sheets (CS) film interfaces has been used, glued to the downstream face of each brick. Charged particles from a neutrino interaction in a brick cross the CS and produce signals in the TT that allow the corresponding brick to be identified and extracted by and automated system. The emulsions extracted from the brick are sent to the automatic scanning microscopes at LNGS and at the various laboratories in Europe and in Japan.

The ν_μ beam from CERN is optimized for the observation of ν_τ CC interactions, the average energy is 17 GeV. With a total CNGS beam intensity of 22.5×10^{19} protons on target, about 24300 neutrino events would be collected. The experiment should observe 10 ν_τ CC events for the present Δm_{23}^2 allowed region with a background of less than one event. In 2006 and 2007 first interactions were recorded during the commissioning phase of the CNGS. Significant amounts of data were collected starting in 2008 and at the end of the 2010 run the total amount of collected neutrino interaction events in the target track were 9635.

The observation of a first ν_τ candidate event in the experiment has been reported in June 2010 based on the analysis of part of the data taken during 2008 and 2009 runs. The appearance of the τ lepton is identified in OPERA by the detection of its characteristic decay topologies, either in one prong (electron, muon or hadron) or in three prongs. A software algorithm selects CC and NC events inside the target. When a trigger in the electronic detectors is compatible with an interaction inside a brick, a software reconstruction program processes the electronic detector data to select the brick with the highest probability to contain the neutrino interaction vertex. The brick is removed from the target by the BMS and the validation of the interaction is achieved using the information of the CS films. After a brick has been validated, it is brought to the surface to be exposed to high energy cosmic rays for a precise film-to-film alignment. The emulsion films are developed and dispatched to the various scanning laboratories.

All tracks measured in the CS are sought in the most downstream films of the bricks up to the stopping point, considered as the signature of a primary or secondary vertex. A further phase of analysis is applied to located vertices to detect possible decay or interaction topologies on tracks attached to the primary vertex. When secondary vertices are found in the event, a kinematical analysis is performed, using particles angles and momenta measured in the emulsion films. The detection of decay topologies is triggered by the observation of a track with a large impact parameter with respect to the primary vertex.

Charmed particles have similar lifetimes as τ leptons and, if charged, share the same decay topologies. In the sample of ν_μ CC interactions that have been searched for, the number of charm decays observed is in agreement with the number expected from the

MC simulation. In the event samples analysed so far one candidate has been identified with measured characteristics fulfilling the selection criteria a priori defined for the ν_τ interaction search [4].

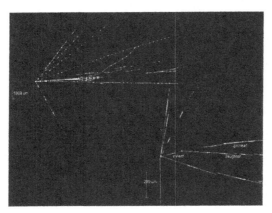

Figure 4: First candidate event of $\nu_\mu \leftrightarrow \nu_\tau$ oscillation in appearance mode.

The primary neutrino interaction consists of seven tracks, of which one shows a visible kink. Two electromagnetic showers caused by γ rays, associated with the event, have been located. The event passes all cuts, with the presence of at least one gamma pointing to the secondary vertex, and it is therefore a candidate to the $\tau \to 1$ prong hadron decay mode. The invariant mass of the two detected gammas is consistent with the π^0 mass value. The invariant mass of the π^0-$\gamma\gamma$ system has a value compatible with that of the ρ (770). The ρ appears in about 25% of the τ decays $\tau \to \rho(\pi^-\pi_0)\nu_\tau$.

The two main sources of background to this channel are:

- the decay of charmed particles produced in ν_μ interactions, estimated for the analyzed sample to $0.007 \pm 0.004(syst)$;

- the one prong inelastic interactions of primary hadrons produced in ν_μ interactions.

This has been evaluated with a FLUKA based Monte Carlo code and experimentally cross-checked following hadron tracks, produced in CC and NC interactions, from the primary vertex over a total length of 8.6 m. Finally, a brick has been exposed to a 4 GeV/c π^- beam and analysed. The contribution to background from re-interactions is estimated to 0.011. The probability that the expected $0.018 \pm 0.007(syst)$ background events into the 1-prong hadron channel fluctuate to 1 event is 1.8% (2.36σ). If one considers all τ decays modes which were included in the search, the probability to observe one event due to the background fluctuation is 4.5% and this corresponds to a significance of 2.01 σ.

6. Icarus

Another experiment able to detect CNGS beam is ICARUS [5], an innovative apparatus consisting of a big mass (about 600 tons) of liquid Argon, at a temperature of -186 ^oC. In particular conditions and by means of proper devices this liquefied gas is able to act as an excellent and unique particles detector, allowing a 3D reproduction of any interactions of charged particles inside its volume. ICARUS began to be operative in Spring 2010 and in May 2010 has registered the first events from CNGS neutrino interactions.

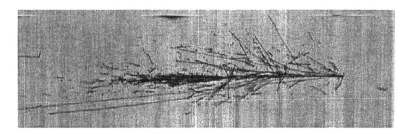

Figure 5: CNGS neutrino interaction in the Icarus experiment [6].

7. Lvd

The experiment LVD [7] is a permanent observatory for the low energy neutrinos coming from gravitational stellar collapses in our galaxy. An unexpected bunch of neutrinos accompany the last instants of life of massive stars before they completely turn off. The experiment consists of about 1000 tons of liquid scintillator in which neutrinos may interact, while the iron and stainless steal support structure is about 900 t. It has a modular structure, made of 840 identical scintillation counters. The counters are grouped in three big modules, called "towers", with independent data acquisition.

Besides interactions with protons and carbon nuclei in the liquid scintillator, LVD is also sensitive to interactions with the iron nuclei of the support structure. The experiment has been taking data, under different configurations, since 1992, reaching in 2001 its present and final configuration. Its modularity and rock overburden, together with the trigger strategy, make this detector particularly suited to disentangle on-line a cluster of neutrino signals from the background.

LVD is one of the largest liquid scintillator apparatus for the detection of stellar collapses and, besides SNO, SuperKamiokande and IceCube, it is a member of the SNEWS network, the early warning network of experiments looking for neutrino burst and quite important for astronomical observation of supernova through different messengers. No gravitational core-collapse has been detected by LVD during 14 years of data acquisition; this allows to put an upper limit of 0.18 events y^{-1} in our galaxy at the 90% C.L.

The experiment is also performing monitor measurements of the CNGS beam. Thanks to its wide area LVD can detect about 120 muons per day originated by ν_μ CC interactions in the rock. The LVD total mass is 2 kt, this allows to get 30 more CNGS events per day as

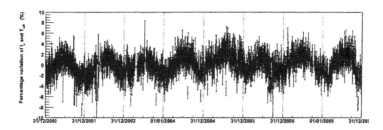

Figure 6: Analysis of the seasonal modulation of the cosmic muon flux in the LVD detector.

internal (NC + CC) ν_μ interactions, for a total of 150 events per day.

8. The research of neutrinoless double beta decay

The existence of neutrino oscillations has been clearly demonstrated in various experiments; however neutrino oscillations experiments are not enough to determine the nature of that mass (Dirac or Majorana) and to determine the absolute mass scale because they can measure only the absolute value of the difference of the square of the neutrino masses while two different hierarchical mass arrangements of neutrino masses (Direct and Inverted), besides the obvious quasi-degenerate option are possible.

Many theories beyond the standard Model suggest a mass generation mechanism that implies a Majorana character of neutrinos (that is particle and antiparticle coinciding). The neutrinoless beta decay process can occur only if neutrinos are massive Majorana particles and then it is a unique tool to test this hypothesis.

Some isotopes could undergo a rare process called "double beta decay" involving a change of the nuclear charge Z by two units. Naturally-occurring even-even nuclei exist, for which this is the only allowed decay mode and this rare process (for $^{76}Ge\ \tau = 1.5 \times 10^{21} y$) has been already observed. This is not the case for neutrinoless double beta decay, object nowadays of a renewed interest, thanks to the discovery of neutrino oscillations.

Neutrinoless double-beta decay is a process by which two neutrons in a nucleus undergo beta decay by exchanging a virtual Majorana neutrino, and each emitting an electron. This would violate lepton number conservation ($\Delta L = 2$) and would require neutrinos to be Majorana particles. There are several possible mechanisms leading to the $\beta\beta0\nu$ process: exchange of a light neutrino, right-handed weak currents, exchange of super-symmetric particles, and other non standard interactions. Independent of the leading term, the observation of $\beta\beta0\nu$ decay would unambiguously establish the Majorana nature of neutrinos. From an experimental point of view, the only available information is carried by the daughter nucleus and the two emitted electrons and then only the sum of the electron energies, single electron energy and angular distributions, identification and/or counting of the daughter nuclei can be measured.

In most cases the experimental signature rely on the electron sum energies and on the discrimination of the different distribution expected for: a continuous bell distribution for $\beta\beta2\nu$ and a sharp line at the transition energy for $\beta\beta0\nu$. Energy resolution and level of background measured are the critical issues. In 2001 evidence for a $\beta\beta0\nu$ signal has been claimed by a small subset (KHDK) of the HDM ("Heidelberg-Moscow") collaboration at Gran Sasso National Laboratory. The result is based on a re-analysis of the HDM data. The claim has raised criticism. However the existing experiments are not yet sensitive enough to confirm or reject the claim. Next generation experiments will have the required sensitivity and will be described in the following sections.

9. Gerda

The experiment GERDA [8] will use the same enriched Germanium crystals as Heidelberg - Moscow, but directly immersed in 60 m^3 of liquid Argon, as shielding against gamma radiation, the dominant background in earlier experiments. GERDA is designed to search for $\beta\beta0\nu$ decay of ^{76}Ge using high purity germanium detectors (HPGe), enriched ($\sim 85\%$) in ^{76}Ge, directly immersed in LAr which acts both as a shield and as a cooling medium. The cryostat is located in a stainless steel water tank providing an additional shield against external background.

GERDA experiment is foreseen to proceed in two phases. For Phase I, eight reprocessed enriched HPGe detectors from the past HdM experiments ($\sim$ 18 kg) and six reprocessed natural HPGe detectors ($\sim$ 15 kg) from the Genius Test-Facility will be deployed in strings. At the expected background rate of 10^{-2} $cts/kg \times keV \times y$ at the Q-value of the

Figure 7: View of the water tank, acting as muon VETO, of the GERDA experiment.

^{76}Ge decay (2039 keV) the resulting sensitivity for the half-life of the neutrinoless double beta decay is $2 \times 10^{25} y$ after a year of exposure. This is sufficient to confirm or refute the existing claim from KHDK. In Phase II new diodes, able to discriminate between single- and multi-site events, will be added to increase the active mass up to 37.5 kg.

A simultaneous background reduction down to 10^{-3} $cts/kg \times keV \times y$ will allow to increase the sensitivity to the half-life of the process by one order of magnitude for the total exposure of 100 kg y. In this case it would be also possible to probe the effective neutrino masses (m_{ee}) at the level of 150 meV.

The installation of the experiment in the Gran Sasso Laboratory has been completed and now the experiment is in the commissioning phase. The first non-enriched Ge detectors have been deployed for a technical run aimed to perform background evaluation.

10. Cuore

The experiment Cuore (Cryogenic Underground Detector for Rare Events) [9] is the most recent and ambitious development of the 'TeO$_2$ bolometers' technique, in which INFN has more than 20 years experience. In the bolometer detectors, the energy from particle interactions is converted into heat and measured via the resulting rise in temperature. Operated at a temperature of about 10 mK, these detectors have an energy resolution of a few keV over their energy range, extending from a few keV up to several MeV. The measured resolution in the region of interest (2527 keV) is ~keV FWHM; A background level of the order of 0.01 $cts/kg \times keV \times y$ is expected by extrapolating the CUORICINO background results and the dedicated CUORE R&D measurements. Considering the high mass of the experiment, the CUORE predicted limit, after 5 years of running sensitivity, for the half-life is 2.1 $\times 10^{26}y$ (90% C.L.) corresponding to m$_v \lesssim$ (24÷83) meV. CUORE will therefore allow a close look at the IH (Inverted Hierarchy) region of neutrino masses.

CUORE will consist of a rather compact cylindrical structure of 988 cubic natural TeO$_2$ crystals of 5 cm side (750 g), arranged into 19 separated towers (13 planes of 4 crystals each) and operated at a temperature of 10 mK. As well as being protected by the Gran Sasso rock, the apparatus is protected by a series of shields, including ancient Roman lead for the utmost reduction of residual radiation in the laboratory and to allow the detection of the very rare decay. Very recently the Laboratory has received additional 120 lead bricks (4 tons) from an ancient Roman ship that sunk off of the coast of Sardinia 2,000 years ago. It is great and unique that the most advanced and innovative technologies must rely on archaeology and on the technology of the ancient Romans.

CUORE will be installed in the Gran Sasso National Laboratory and will start the data taking in 2014. The first CUORE tower, CUORE-0, composed by 52 bolometers, is under preparation and will start the data taking in 2011.

Thanks to the bolometer versatility, alternative options with respect to TeO$_2$ are also possible. In particular, promising results have been recently obtained with scintillating bolometers which could allow to study in the future new $\beta\beta0\nu$ active isotopes with improved sensitivity. In this context it is worth mentioning the program LUCIFER, lately approved and funded by the European Community, which will be developed in the LNGS. The program aims at realizing a ZnSe crystals detector able to match the bolometer technique used in the experiment CUORE to the detection of the scintillating light proper of the dark matter detectors.

Figure 8: A prototype of a detector tower made of TeO_2 for the CUORE experiment.

11. Cobra

The experiment COBRA completes the outline of research activities on $\beta\beta0\nu$. The basic idea of the experiment is to use as detectors a kind of semiconductors of CdZnTe (CZT) with a low radioactive background level, good energy resolution and supposed to be used at room temperature. Its research and development program is expected to be completed in the next years thus arriving to a definitive experiment proposal.

12. Conclusions

Neutrino and astroparticle physics would not have been able to undergo the great progresses seen in the last thirty years without the necessary relevant infrastructures like underground laboratories. The present scientific program of LNGS is at the top of its life thanks to many competitive activities; experiments which have already been approved are in different phases of development and thus they have very different time scales. They will reach the expected research goals in time intervals ranging from a few years up to the end of the decade.

At present Gran Sasso owns a leadership in massive experiments with record performances from the point of view of a low level of background. It is a very competitive benefit compared with other laboratories and it has to be preserved and exploited in order to maximize the discovery potential in the field of double beta decay researches and dark matter. These line of research, together with the study of nuclear reactions of astrophysics interest, will surely be the main evolution of the scientific activity of the Laboratory in the next decade.

13. References

1) G. Bellini et al., Phys. Rev. D **82**(2010),033006
2) G. Bellini et al., Phys. Rev. D **82**(2010), 033006
3) The Borexino Collaboration Physics Letters B **687**(2010), 299-304
4) N. Agafonova et al., Physics Letters B **691**(2010), 138-145
5) C. Rubbia CERN-EP (1977), 77-08
6) The Icarus collaboration, private communication
7) F.Arneodo et al., Journal of Physics **136**(2005), 042082
8) S. Schönert et al., Nucl. Phys. B **145**(2005), 242
9) C. Arnaboldi et al., Phys. Rev. Lett **95**(2005), 14501

Scientific Secretaries: M. D. Azmi, R. Rescigno

DISCUSSION

- *Cifarelli:*

Since this is a School, could you please come back to the results on dark matter observed by DAMA/LIBRA experiments and illustrate in more details the somehow disputed modulation effect

- *Votano:*

It has been many years since the experiment is at Gran Sasso and it has obtained a very important result. The annual modulation is a clear and model independent signature of dark matter. It depends on the fact that the movement of the Earth and of the sun combined with respect to galaxy changes the velocity of the dark matter particle respect the detector. This is an important result but a confirmation is needed, also because the collaboration has already evaluated many possible sources of background. The problem is the limit fixed by other experiments that excluded the region obtained by this experiment in the hypothesis of WIMP candidates. Many people at LIBRA collaboration suggest to put the same experiment in another place because in this way it is possible to check against any other source of background showing an annual modulation. At LNGS there is also another experiment, XENON100, that has showed the first results. The background is at the moment the lowest respect to any other experiment. There are many dark matter candidates and also different hypotheses of interaction with ordinary matter. It cannot be excluded that a signal of the dark matter candidate can be seen in one molecule and not in xenon. So the best way is to confirm the DAMA/ in a different place.

- *Pagnutti:*

Can you determine supernova direction and what would the angular resolution of an event like Betelgeuse be? What information about supernovae can you determine? What about using spherical detector for finding direction?

- *Votano:*

The Scintillator experiments (LVD and Borexino) cannot determine direction for the inverse beta decay. They can detect excess of events and energy spectrum.

- *Alberte:*

How are the neutrino masses measured and what improvement in the understanding would the knowledge of the absolute mass scale give?

- *Votano*:

The masses of neutrinos are very essential numbers still missing in the lepton-neutrino sector beyond Standard Model. At LNGS this measurement can be performed by neutrino-less double beta decay experiments.

- *Dunin-Barkowski*

Could you, please, elaborate more on $0\nu\beta\beta$ experiments at LNGS? What are the statistics at the moment? What can be expected?

- *Votano*:

The best limit obtained up to now comes from the Heidelberg-Moscow experiment. A positive claim came from a subset of the Collaboration (Klapdor). A similar limit has been obtained by CUORICINO. There are various new generation experiments like CUORE, and GERDA at LNGS searching for neutrinoless double-beta decay. GERDA in one year run should be able to confirm or dismiss the Klapdor claim, while CUORE will start data taking in 2013.

- Gerbaudo:

You mentioned this morning the possibility of not having beam from CNGS during 2012. What is the impact of one year data taking on the timescale of this experiment.

- Votano:

The decision has not been made yet. Initial plan for CNGS and OPERA was 5 years. However one should keep in mind two points: the experiment up to 2009 got a number of pot equivalent to one year at nominal intensity, and at the end of 2010 the OPERA experiment will almost double the statistics. If the planned 5-years run will be extended too much, the experiment could have problems.

- Wenninger:

Will neutrino beams be available in 2012? Will the emulsion of OPERA suffer by delays of neutrino beam operations?

- Votano:

The first question has to be discussed at the CERN Council and DG. The nuclear emulsions could start to show aging effects and the efficiency will afterward be reduced.

-'t Hooft:

It should be easy to establish whether the muon oscillation that you mentioned is sidereal or solar. In either case, what is the explanation? The primaries must be very energetic, does the Sun emit such particles?

- Votano:

No, it's an atmospheric annual effect.

- Grelli:

Do you have an estimation (from MC) about how many years you need to obtain a reasonable significance on $0\nu\beta\beta$ decay. What about background from normal $2\nu\beta\beta$ decay?

- Votano:

We need some years to improve the present limits and actually $2\nu\beta\beta$ decay is the main source of background and the experiments need very good energy resolutions in order to discriminate it.

- Mulhearn:

Any future prospects for magnetic monopole searches?

- Votano:

No.

- Preghenella:

Can the OPERA experiment measure the oscillations parameter?

- Votano:

The present limits from the other experiments are better than what OPERA can do. I mean that this is not the main goal of the experiment.

S.C.C. TING

"Highlights from ISS-AMS"

The Slides of the Lecture can be found at:

http://www.ccsem.infn.it/issp2010/

CHAIRMAN: S.C.C.TING

Scientific Secretaries: A. Krislock, Y. Mao

DISCUSSION

- Preghenella:
What is the potential of discovering magnetic monopoles with AMS?

- Ting:
Magnetic monopoles is easily detected by AMS. Because it bends dramatically with magnetic field.

- Gerbaudo:
Can you say something about data flow of the AMS experiment? And how is data going to be analyzed?

- Ting:
Data are recorded at CERN and then distributed/sent to each participated institute, to make sure all the collaborators have chance to analyze the data.

- Gerbaudo:
Can you point out how much freedom of the ISS detector towards to the interesting objects in AMS?

- Ting:
You have no freedom. But for charged particles this is not very important. For gammas we have two precise star trackers and GPS system that help us precisely determine the origin.

- Alberte:
Could you please comment on the possibility of primordial black holes?

- Ting:
I have no comment.

- Mulhearn:
How do you optimize the design of a machine built to discover the unexpected?

- Ting:
AMS has a unique opportunity to measure charge particles at high energy in space. What we have done is to design maximum acceptance, the best resolution and rejection regardless of cost.

- Preghenella:
What do higher order corrections in the Standard Model predict for the neutrino magnetic moment?

- Lee:
The Standard Model predicts 10^{-19} from quantum fluctuations. For a non-Standard Model with new physics operators at a new scale around 1-10 TeV we expect 10^{-16} for Dirac neutrinos, and 10^{-14} for Majorana neutrinos. In a few years, we may be able to reach precisions of 10^{-13} or 10^{-14}.

- Inguglia:

How much time will it take to perform the experiment and get the first results?

- Lee:

It will take about 10 years from planning until the first results. This is not a precision experiment. You have to control and understand your background.

- Inguglia:

Any ideas on how to control the background?

- Lee:

We are going to low-threshold for very small recoil of electron energy. First at a few keV, then down to a few hundred eV. Nobody has gone to this region before. So, there are many things which look like signal that we need to understand and eliminate one by one. Also, we need to use pulse shape. Different electron or nucleon recoils have different pulse shapes. So, we need to be fast and record every pulse, not just the total charge.

- Mao:

From which measured results will you conclude that the neutrino has substructures inside?

- Lee:

As I said before, we expect the neutrino to be elementary. In the Standard Model it should have a magnetic moment of 10^{-19}, so we should not see it. If there is a new scale around a few TeV, we see 10^{-16} if it is a Dirac neutrino and 10^{-14} if it is Majorana. So, if we instead see it at 10^{-13}, that means we see that the neutrino may not be elementary.

- Mao:

If you see what?

- Lee:

If we see a signal instead of a bound for the neutrino magnetic moment.

- Preghenella:

Suppose the neutrino is not a point-like particle. What could it be made of?

- Lee:

There are some models. There are 2 ways we know until now that a new scale can be generated. You can break some gauge symmetry spontaneously, or have asymptotic freedom like the strong interaction. So, if we see a new scale, we have to ask: Is it broken or unbroken? If broken, what is the new symmetry? If it is unbroken, maybe it is some new type of strong interaction. For example, Technicolor was one such theory. So, many things can happen. We don't know which will occur. We just look to the experiments for the unexpected. Of course, once you see something, more experiments must be done.

- Inguglia:

What do you think of the possibility of substructure in other elementary particles? Also, other elementary particles may be easier to deal with. Should other experiments probe substructures in them as well?

- Lee:

I just told you of 2 kinds of new physics scales. For QCD, we have quarks and gluons which are elementary, but they are elementary to all scales. If you look inside a quark, it is made of quarks, anti-quarks and gluons. The same is true for gluons. So for asymptotically free theories, there is no need to ask what is more elementary. Of course, you can have something different, but you don't have to look for it in an asymptotically free theory. However, for the weak interaction, it is different.

- Inguglia:

For example, muons?

- Lee:

Yes. You can look for the anomalous magnetic moment for electrons and muons, and people do that, currently to precisions of 10^{-10} or 10^{-11}. In the muon (g-2), you see something. But it is difficult, since it already has a magnetic moment and you have to look for a small difference. If you look at a neutrino, it is supposed to be zero. And we are reaching the same precision now for neutrinos as for electrons.

- Inguglia:

What do you think about searching for new structures in muon colliders?

- Lee:

I hope it can be built. There are many difficulties to overcome. Talking to scientists, even in the US, they don't believe it can be done in a short time. It depends on you guys.

Four-qubit entanglement: lessons of a black hole

L. Borsten[*]

Blackett Laboratory, Imperial College London, Prince Consort Road, London SW7 2BZ, UK

Proceedings of the International School of Subnuclear Physics, Erice

March 3, 2011

Abstract

We invoke the black-hole/qubit correspondence to derive the classification of four-qubit entanglement. The U-duality orbits resulting from timelike reduction of string theory from $D = 4$ to $D = 3$ yield 31 entanglement families, which reduce to nine up to permutation of the four qubits. Based on work done in collaboration with D. Dahanayake, M. J. Duff, A. Marrani and W. Rubens.

1 Introduction

These are interesting times. We stand at two frontiers, each heralding a new era of understanding at its most fundamental level. As the sun rises over the age of *quantum information,* new trails through the jungle that is particle physics are navigated at the latest generation of high energy colliders.

Quantum information theory (QIT) is the study of information processing systems which rely on the fundamental properties of quantum mechanics [1]. There is an expectation that quantum mechanics may be exploited to perform computational tasks beyond the capability of any, even idealistic, purely classical device. This idea enjoys a certain poetic turn; just as the conventional microchip meets its fundamental limit, fixed by the onset of quantum noise at the atomic scale, the very same quantum phenomena open the door to new, superior, modes of computation. Perhaps the single most important quantum information theoretic resource is the phenomenon of quantum entanglement. The quantum states of two or more entangled objects must be described with reference to each other, even though the individual objects may be spatially separated. This leads to classically unexplainable, but experimentally observable, quantum correlations between the spatially separated systems - "spooky" action at a distance as Einstein called it. Quantum entanglement is vital to the emerging technologies of quantum computing, communication and cryptography. One of the longest standing open problems in QIT is the proper characterisation of multipartite entanglement. It is of utmost importance from both a foundational and a technological perspective.

In quite separate developments black holes have occupied an equally privileged position in our various attempts to unify the fundamental interactions including quantum gravity. While general relativity refuses to succumb to quantum rule, black holes raise quandaries that strike at the very heart of quantum theory. Without a proper theory of quantum gravity, such paradoxes will continue to haunt us. M-theory, which grew out of pioneering work on supergravity

---

[*]leron.borsten@imperial.ac.uk

and superstring theory, is a promising approach to quantum gravity. Living in eleven space-time dimensions, it encompasses and connects the five consistent 10-dimensional superstring theories, as well as 11-dimensional supergravity and, as such, has the potential to unify the fundamental forces into a single consistent framework. However, M-theory is fundamentally non-perturbative and consequently remains largely mysterious, offering up only disparate corners of its full structure. Black holes and, more generally, black p-branes offer unique insights into its deep structure. Whatever form M-theory finally takes, black holes will be an essential ingredient.

For the most part these important endeavors in quantum information and gravity have led separate lives. However, the present work reports on a curious and unexpected interplay between these seemingly disparate themes. It constitutes one corner of the *black-hole/qubit correspondence*: a relationship between the entanglement of qubits, the basic units of quantum information, and the entropy of black holes in M-theory. This story began in 2006 [2] when it was observed that the entanglement shared by three qubits and the entropy of the *STU* black hole [3,4,5], that appears in the compactification of M-theory to four dimensions, are both given by the same mathematical object, *Cayley's hyperdeterminant* [6]. It was soon realised, there is in fact a one-to-one correspondence between the classification of 3-qubit entanglement [7] and the classification of extremal *STU* black holes [8]. Further work [9, 10, 11, 12, 13, 14, 15, 16, 17, 18, 19, 20, 21] has led to a more complete dictionary translating a variety of phenomena in one language to those in the other. It seems that we are, as yet, only glimpsing the tip of an iceberg.

This contribution reports on a recent application [22] of the black-hole/qubit correspondence to the much more difficult problem of classifying 4-qubit entanglement, currently an active area of research in QIT as experimentalists now control entanglement with four qubits [23]. The key technical ingredient is the Kostant-Sekiguchi theorem [24, 25], which provides the link between the black holes and qubits. Our main result, summarized in Table 2, is that there are 31 entanglement families which reduce to nine up to permutations of the four qubits. From Table 1 we see that the nine agrees with [26, 27] while the 31 is new.

As befits a proceedings on high energy physics we begin, in section 2, with an elementary introduction to entanglement in QIT, with particular emphasis on *Stocastic Local Operations and Classical Communication* and the status of 4-qubit entanglement classification. In section 3 we briefly review black holes in supergravity and, in particular, the role of time-like dimensional reduction and nilpotent orbits. In section 4 we invoke the Konstant-Sekiguchi theorem, which maps the black hole solutions to the 4-qubit entanglement classes. In doing so we derive a classification of 4-qubit entanglement by studying the black holes of the *STU* model.

2 Entanglement

The advent of quantum theory precipitated a reassessment of "reality" itself. Of course, what is meant by reality is, and has always been, a philosophically perilous line of inquiry. In the context of classical physics, however, we may make a fairly uncontentious first approximation. The fundamental substance of objects in classical physics is constituted by their observable properties; "things" in themselves are defined as the bearers of said properties [28,29]. Essential to such a prescription of reality is the crucial assumption that all physical attributes do indeed possess definite, if not definitely known, values at all times [28, 29, 30]. These ideas form the crux of what may be called a realist philosophy of nature. This realist stance is naturally embodied by the mathematical (set-theoretic) formalism underpinning classical theory.

A further requirement that one might demand of a sensible description of reality is *locality* (or perhaps more precisely in this context *separability*). Indeed, this concept was something Einstein regarded as worth fighting for, as he describes to Max Born [31],

"I just want to explain what I mean when I say that we should try to hold on to physical reality. We all of us have some idea of what the basic axioms of physics will turn out to be... whatever we regard as existing (real) should somehow be localised in time and space. That is, the real in part of space A should (in theory) somehow 'exist' independently of what is thought of as real in space B. When a system in physics extends over the parts of space A *and* B, then that which exists in B should somehow exist independently of that which exists in A. That which really exists in B should therefore not depend on what kind of measurement is carried out in part of space A; it should also be independent of whether or not any measurement at all is carried out in space A."

The union of these ideas is typically referred to as *local realism*. The pinnacle of classical physics, Einstein's general theory of relativity, is a complete local realist theory. On the other hand, the orthodox Copenhagen interpretation of quantum physics rejects local realism in all its parts. As for realism, it is not possible to speak of a physical system as actually "possessing" values for all their physical observables at a given time, and that these values are intrinsic and independent of the measurement setup used to reveal them [29, 30]. This is not merely a philosophical whim, it is a mathematical consequence of the Kochen-Specker theorem [32]. Furthermore, we lose our grip on locality thanks to our curious friend, entanglement.

Given the startling nature of these two statements it comes as no surprise that over the years many a physicist and philosopher has taken exception, not least of all Einstein. This led to the now famous Bohr-Einstein dialog, with Einstein fighting the corner of local realism and Bohr that of the Copenhagen interpretation. Its culmination was the seminal 1935 work by Einstein, Podolsky, and Rosen (EPR) [33]. They correctly concluded that assuming local realism the quantum mechanical wave function cannot be a complete description of reality. They speculated on the existence of a more fundamental underlying (classical) theory that towed the line of local realism. However, such questions remained a matter of philosophical preference, seemingly inaccessible to experiment. All this changed in 1964 when Bell introduced his now famous inequality [34]. In one fell swoop, entanglement had been elevated from a conceptual puzzle to an experimental observable confronting the very assumptions of local realism. This was Bell's great insight - to derive from EPR's criteria something which could be used to experimentally check the phenomenological viability of local realism. Moreover, Bell had laid the foundations of entanglement as a quantum information theoretic resource. For example, in 1991 Ekert [35] used the Bell inequality as the basis for a new secure means of quantum communication.

It was subsequently realised that there are physically distinct forms of entanglement. For example, multipartite states provide so-called Bell inequalities without the inequality [36]. Indeed, as quantum information theory developed, entanglement became an essential computing *resource*, that may be created, destroyed and manipulated. All this motivated a need to properly quantify and classify entanglement. Conventionally, the state of a composite system is said to be entangled if it cannot be written as a tensor product of states of the constituent subsystems. However, this particular measure is perhaps insufficient to really capture the various subtleties of entanglement. For example, there are two totally non-separable 3-qubit states that have physically distinct entanglement properties [7]. Is there a more illuminating notion of entanglement? Let us take our cues from experiment. We do not actually observe the tensor product structure, even though it underpins our theoretical understanding. What we do observe are correlations between spatially separated systems that admit no classical explanation. This motivates the more general and quantum information theoretic notion of entanglement as correlations between constituent pieces of a composite system that are of a quantum origin [37]. The question now is, how does one differentiate between classical correlations and those correlations which may be attributed to genuine quantum phenomena? A quantum information theoretic perspective provides a precise solution: classical correlations are *defined* as those which may be generated by *Local Operations and Classical Communication* (LOCC) [38, 39, 37]. Any classical correlation may be experimentally established using LOCC. Conversely, all correlations unobtainable via LOCC are regarded as *bona fide* quantum entanglement.

The LOCC paradigm is quite intuitive. Heuristically, given a composite quantum system with its components spread among different laboratories around the world we allow each experimenter to perform any quantum operation or measurement on their component locally in their laboratory. These local operations cannot establish any correlations, classical or quantum. However, the experimenters may communicate any information they see fit via a classical channel (carrier pigeon, smoke signals, e-mail). Any number of LO and CC rounds may be performed. In this manner one may set-up arbitrary classical correlations. However, since all information exchanged between the separated parties at any point was intrinsically classical, LOCC cannot create genuine quantum correlations.

Two quantum states of a composite system are then said to be *stochastically* LOCC (SLOCC) equivalent if and only if they may be probabilistically interrelated using LOCC. Since LOCC cannot create entanglement two SLOCC-equivalent states must possess the same "amount" of entanglement. For more details see [37, 40] and the references therein.

Let us make this a little more precise by focusing on the specific case of multi-qubit systems. What is a qubit? Quantum information can live in a quantum mechanical superposition. Hence, the qubit is a quantum superposition of the classical binary digits "0" and "1". The particular physical realisation (there are many: photon polarisations, quantum dots, trapped ions, mode splitters, to name but a few) of the qubit is not important, any two state quantum system will do. Hence, qubits are simply denoted abstractly as elements of the 2-dimensional Hilbert space $\mathbb{C}^2$, equipped with the conventional norm, where the two basis states are labelled $|0\rangle$ and $|1\rangle$. An n-qubit bit string $|\Psi\rangle$ lives in the n-fold tensor product of $\mathbb{C}^2$:

$$\begin{aligned}
|\Psi\rangle &= a_{A_1 \ldots A_n} |A_1\rangle \otimes |A_2\rangle \otimes \ldots |A_n\rangle \\
&= a_{A_1 \ldots A_n} |A_1 A_2 \ldots A_n\rangle,
\end{aligned} \tag{1}$$

where $a_{A_1 \ldots A_n} \in \mathbb{C}$ and we sum over $A_1, \ldots, A_n = 0, 1$. In [7] it was argued that two states of an n-qubit system are SLOCC-equivalent if and only if they are related by $[SL(2, \mathbb{C})]^{\otimes n}$, under which $a_{A_1 \ldots A_n}$ transforms as the fundamental $(\mathbf{2}, \mathbf{2}, \ldots, \mathbf{2})$ representation. It may be usefully thought of as the "gauge" group of n-qubit entanglement. Hence, the space of physically distinct n-qubit entanglement classes (or orbits) is given by,

$$\frac{\mathbb{C}^2 \otimes \mathbb{C}^2 \otimes \ldots \mathbb{C}^2}{SL_1(2, \mathbb{C}) \times SL_2(2, \mathbb{C}) \times \ldots SL_n(2, \mathbb{C})}. \tag{2}$$

When classifying entanglement it is this space we wish to understand.

This very quickly becomes a difficult task. Although two and three qubit entanglement is well-understood, the literature on four qubits can be confusing and seemingly contradictory, as illustrated in Table 1. This is due in part to genuine calculational disagreements, but in part to the use of distinct (but in principle consistent and complementary) perspectives on the criteria for classification. On the one hand there is the "covariant" approach which distinguishes the SLOCC orbits by the vanishing or not of $[SL(2, \mathbb{C})]^{\otimes n}$ covariants/invariants. This philosophy is adopted for the 3-qubit case in [7, 46], for example, where it was shown that three qubits can be tripartite entangled in two inequivalent ways, denoted W and GHZ (Greenberger-Horne-Zeilinger). The analogous 4-qubit case was treated, with partial results, in [47]. On the other hand, there is the "normal form" approach which considers "families" of orbits. An arbitrary state may be transformed into one of a finite number of normal forms. If the normal form depends on some of the algebraically independent SLOCC invariants it constitutes a family of

Table 1: Various results on four-qubit entanglement.

Paradigm	Author	Year	Ref	result mod perms	result incl. perms
	Wallach	2004	[41]	?	90
	Lamata et al	2006	[42]	8 genuine,5 degenerate	16 genuine,18 degenerate
classes	Cao et al	2007	[43]	8 genuine,4 degenerate	8 genuine,15 degenerate
	Li et al	2007	[44]	?	≥ 31 genuine,18 degenerate
	Akhtarshenas et al	2010	[45]	?	11 genuine, 6 degenerate
	Verstraete et al	2002	[26]	9	?
families	Chterental et al	2007	[27]	9	?
	String theory	2010		9	31

orbits parametrized by these invariants. On the other hand a parameter-independent family contains a single orbit.

This philosophy is adopted for the 4-qubit case $|\Psi\rangle = a_{ABCD}|ABCD\rangle$ in [26, 27]. Up to permutation of the four qubits, these authors found 6 parameter-dependent families called G_{abcd}, L_{abc_2}, $L_{a_2b_2}$, $L_{a_20_{3\oplus\bar{1}}}$, L_{ab_3}, L_{a_4} and 3 parameter-independent families called $L_{0_{3\oplus\bar{1}}0_{3\oplus\bar{1}}}$, $L_{0_{5\oplus\bar{3}}}$, $L_{0_{7\oplus\bar{1}}}$. For example, a family of orbits parametrized by all four of the algebraically independent SLOCC invariants is given by the normal form G_{abcd}:

$$\frac{a+d}{2}(|0000\rangle + |1111\rangle) + \frac{a-d}{2}(|0011\rangle + |1100\rangle)$$
$$+\frac{b+c}{2}(|0101\rangle + |1010\rangle) + \frac{b-c}{2}(|1001\rangle + |0110\rangle). \tag{3}$$

To illustrate the difference between these two approaches, consider the separable EPR-EPR state $(|00\rangle + |11\rangle) \otimes (|00\rangle + |11\rangle)$. Since this is obtained by setting $b = c = d = 0$ in (3) it belongs to the G_{abcd} family, whereas in the covariant approach it forms its own class. Similarly, a totally separable A-B-C-D state, such as $|0000\rangle$, for which all covariants/invariants vanish, belongs to the family L_{abc_2}, which also contains genuine four-way entangled states. These interpretational differences were also noted in [42]. As we shall see, our black hole perspective lends itself naturally to the "normal form" framework.

3 Nilpotent orbits and black holes

We consider $D = 4$ supergravity theories in which the moduli parameterize a symmetric space of the form $M_4 = G_4/H_4$, where G_4 is the global U-duality group and H_4 is its maximal compact subgroup. After a further time-like reduction to $D = 3$ the moduli space becomes a pseudo-Riemannian symmetric space $M_3^* = G_3/H_3^*$, where G_3 is the $D = 3$ duality group and H_3^* is a non-compact form of the maximal compact subgroup H_3. One finds that geodesic motion on M_3^* corresponds to stationary solutions of the $D = 4$ theory [48,49,50,51,52]. These geodesics are parameterized by G_3 Lie algebra valued matrices of Noether charges Q. The problem of classifying the spherically symmetric extremal (non-extremal) black hole solutions consists of classifying the nilpotent (semisimple) orbits of Q (nilpotent means $Q^k = 0$ for some $k \in \mathbb{N}$).

Specifically, we consider the STU model [3,4,5]: $\mathcal{N} = 2$ supergravity coupled to three vector multiplets. It has three complex scalars denoted S, T and U, which parameterize the symmetric coset space,

$$[\mathrm{SL}(2, \mathbb{R}) \times \mathrm{SL}(2, \mathbb{R}) \times \mathrm{SL}(2, \mathbb{R})]/[\mathrm{SO}(2) \times \mathrm{SO}(2) \times \mathrm{SO}(2)]. \tag{4}$$

The static black hole solutions of the STU model are characterized by their mass and a maximum of 8 charges (four electric and four magnetic), $1 + 3$ from the gravity and vector multiplets respectively, plus their magnetic duals. Hence, the Hawking temperature and

Bekenstein-Hawking entropy are functions of the mass and the 8 charges. Through scalar-dressing, these charges can be grouped into the $\mathcal{N} = 2$ central charge z and three "matter charges" z_a ($a = 1, 2, 3$), which exhibit a triality (corresponding to permutation of three of the qubits). The black holes are divided into extremal or non-extremal according as the temperature is zero or not. The orbits are nilpotent or semisimple, respectively. Depending on the values of the charges, the extremal black holes are further divided into small or large according as the entropy is zero or not. The small ones are termed lightlike, critical or doubly critical according as the minimal number (under U-duality) of representative electric or magnetic charges is 3, 2 or 1. The lightlike case is split into one 1/2-BPS solution, where the charges satisfy $z_1 = 0, |z|^2 = 4|z_2|^2 = 4|z_3|^2$ and three non-BPS solutions, where the central charges satisfy $z = 0, |z_1|^2 = 4|z_2|^2 = 4|z_3|^2$ or $z_2 = 0, |z_3|^2 = 4|z_1|^2 = 4|z|^2$ or $z_3 = 0, |z_2|^2 = 4|z_1|^2 = 4|z|^2$. The critical case splits into three 1/2-BPS solutions with $z = z_a \neq 0, z_b = z_c = 0$ and three non-BPS cases with $z = z_a = 0, z_b = z_c \neq 0$, where $a \neq b \neq c$. The doubly critical case is always 1/2-BPS with $|z|^2 = |z_1|^2 = |z_2|^2 = |z_3|^2$ and vanishing sum of the z_a phases. The large black holes may also be 1/2-BPS or non-BPS. One subtlety is that some extremal cases, termed "extremal", cannot be obtained as limits of non-extremal black holes.

Performing a time-like reduction one obtains a pseudo-Riemannian symmetric moduli space,

$$SO(4,4)/[SO(2,2) \times SO(2,2)]. \tag{5}$$

Hence, the extremal solutions are classified by the nilpotent orbits of $SO(4,4)$ acting on its adjoint representation. Here we consider the finer classification, obtained from the nilpotent orbits of $SO_0(4,4)$, where the 0 subscript denotes the identity component. These orbits may be labeled by "signed" Young tableaux, often referred to as ab-diagrams in the mathematics literature. See [53] and the references therein. Each signed Young tableau, as listed in Table 2, actually corresponds to a single nilpotent $O(4,4)$ orbit of which the $SO_0(4,4)$ nilpotent orbits are the connected components. Since $O(4,4)$ has four components, for each nilpotent $O(4,4)$ orbit there may be either 1, 2 or 4 nilpotent $SO_0(4,4)$ orbits. This number is also determined by the corresponding signed Young tableau. If the middle sign of every odd length row is "$-$" ("$+$") there are 2 orbits and we label the diagram to its left (right) with a I or a II. If it only has even length rows there are 4 orbits and we label the diagram to both its left and right with a I or a II. If it is none of these it is said to be stable and there is only one orbit. The signed Young tableaux together with their labellings, as listed in Table 2, give a total of 31 nilpotent $SO_0(4,4)$ orbits. The matching of the extremal classes to the nilpotent orbits is given in Table 2. We also supply the complete list of the associated cosets in Table 2, some of which may be found in [51].

4 The entanglement classification of four qubits

To relate the extremal black hole solutions to the entanglement classes of four qubits we invoke the Kostant-Sekiguchi theorem [24, 25]. Let $\mathfrak{g}_{\mathbf{R}}$ be the real Lie algebra of a semi-simple group $G_{\mathbf{R}}$, with a Cartan decomposition $\mathfrak{g}_{\mathbf{R}} = \mathfrak{k}_{\mathbf{R}} + \mathfrak{p}_{\mathbf{R}}$, where $\mathfrak{k}_{\mathbf{R}}$ is the Lie algebra of a subgroup $K_{\mathbf{R}} \subset G_{\mathbf{R}}$. Let $\mathfrak{k}_{\mathbf{C}}$, $\mathfrak{p}_{\mathbf{C}}$ and $K_{\mathbf{C}}$ be the respective complexifications of $\mathfrak{k}_{\mathbf{R}}$, $\mathfrak{p}_{\mathbf{R}}$ and $K_{\mathbf{R}}$. Then the nilpotent orbits of $G_{\mathbf{R}}$ in $\mathfrak{g}_{\mathbf{R}}$ are in one-to-one correspondence with the nilpotent orbits of $K_{\mathbf{C}}$ in $\mathfrak{p}_{\mathbf{C}}$. In particular, there is a natural one-to-one correspondence between the nilpotent orbits of $SO_0(n,m)$ in $\mathfrak{so}(n,m)$ and $SO(n,\mathbf{C}) \times SO(m,\mathbf{C})$ in $(\mathbf{v}_n, \mathbf{v}_m)$, where $\mathbf{v}$ denotes the vector representation.

Noting the convenient isomorphism $SO(2,2) \cong SL(2,\mathbb{R}) \times SL(2,\mathbb{R})$, the moduli space G_3/H_3^* of the time-like reduced STU model may be rewritten as $SO(4,4)/[SL(2,\mathbb{R})]^4$, which yields the

Table 2: Each black hole nilpotent $SO_0(4,4)$ orbit corresponds to a 4-qubit nilpotent $[SL(2,\mathbb{C})]^4$ orbit.

description	Young tableaux (STU black holes)	$SO_0(4,4)$ coset	$[SL(2,\mathbb{C})]^4$ coset	$\dim_{\mathbb{R}}$	nilpotent rep (Four qubits)		family
trivial	trivial	$\dfrac{SO_0(4,4)}{SO_0(4,4)}$	$\dfrac{[SL(2,\mathbb{C})]^4}{[SL(2,\mathbb{C})]^4_\tau}$	1	0	$\in$	G_{abcd}
doubly-critical $\tfrac{1}{2}$ BPS		$\dfrac{SO_0(4,4)}{[SL(2,\mathbb{R})\times SO(2,2;\mathbb{R})]\times[(2,4)^{[1]}\oplus 1^{[2]}]}$	$\dfrac{[SL(2,\mathbb{C})]^4}{[SO(2,\mathbb{C})]^4\times\mathbb{C}^4}$	10	$|0110\rangle$	$\in$	L_{abc_2}
critical, $\tfrac{1}{2}$ BPS and non-BPS	$\left(\;{}_{I,II}\right)$	$\dfrac{SO_0(4,4)}{SO(3,2;\mathbb{R})\times[(5\oplus1)^{[2]}]}$ $\dfrac{SO_0(4,4)}{SO(2,3;\mathbb{R})\times[(5\oplus1)^{[2]}]}$ $\dfrac{SO_0(4,4)}{Sp(4,\mathbb{R})\times[(5\oplus1)^{[2]}]}$	$\dfrac{[SL(2,\mathbb{C})]^4}{SO(3,\mathbb{C})\times\mathbb{C}\times[SO(2,\mathbb{C})\times\mathbb{C}}$	12	$|0110\rangle + |0011\rangle$	$\in$	$L_{a_2b_2}$
lightlike $\tfrac{1}{2}$ BPS and non-BPS	$\left(\;{}_{I,II}\right)$ $\left(\;{}_{I,II}\right)$	$\dfrac{SO_0(4,4)}{SL(2,\mathbb{R})\times[(2\times2)^{[1]}\oplus(3\times1)^{[2]}\oplus2^{[3]}]}$	$\dfrac{[SL(2,\mathbb{C})]^4}{[SO(2,\mathbb{C})\times\mathbb{C}]\times\mathbb{C}^2}$	16	$|0110\rangle + |0101\rangle + |0011\rangle$	$\in$	$L_{a_20_3\oplus\bar1}$
large non-BPS $z_H\neq0$		$\dfrac{SO_0(4,4)}{SO(1,1;\mathbb{R})\times SO(1,1;\mathbb{R})\times[((2,2)\oplus(3,1))^{[2]}\oplus1^{[4]}]}$	$\dfrac{[SL(2,\mathbb{C})]^4}{\mathbb{C}^3}$	18	$\tfrac{1}{\sqrt2}(|0001\rangle + |0010\rangle - |0111\rangle - |1011\rangle)$	$\in$	L_{ab_3}
"extremal"	$\left(\;{}_{I,II}\right)$ $\left(\;{}_{I,II}\right)$	$\dfrac{SO_0(4,4)}{SO(1,1;\mathbb{R})\times[1^{[2]}]\oplus s^{[4]}\oplus1^{[6]}}$ $\dfrac{SO_0(4,4)}{SO(1,2;\mathbb{R})\times[1^{[2]}]\oplus s^{[4]}\oplus1^{[6]}}$ $\dfrac{SO_0(4,4)}{Sp(2,\mathbb{R})\times[1^{[2]}]\oplus s^{[4]}\oplus1^{[6]}}$	$\dfrac{[SL(2,\mathbb{C})]^4}{SO(2,\mathbb{C})\times\mathbb{C}}$	20	$i|0001\rangle + |0110\rangle - i|1011\rangle$	$\in$	L_{a_4}
large $\tfrac{1}{2}$ BPS and non-BPS $z_H=0$	$\left(\;{}_{I,II}\right)$ $\left(\;{}_{I,II}\right)$	$\dfrac{SO_0(4,4)}{SO(2,\mathbb{R})\times SO(2,\mathbb{R})\times[(2,3)\oplus(3,1))^{[2]})\oplus1^{[4]}]}$	$\dfrac{[SL(2,\mathbb{C})]^4}{SO(2,\mathbb{C})\times\mathbb{C}}$	18	$|0000\rangle + |0111\rangle$	$\in$	$L_{0_3\oplus\bar1\,0_3\oplus\bar1}$
"extremal"	$\left(\;{}_{I,II}\right)$	$\dfrac{SO_0(4,4)}{\mathbb{R}^3(2)\oplus\mathbb{R}^{[4]}\oplus\mathbb{R}^2(6)}$	$\dfrac{[SL(2,\mathbb{C})]^4}{\mathbb{C}}$	22	$|0000\rangle + |0101\rangle + |1000\rangle + |1110\rangle$	$\in$	$L_{0_5\oplus3}$
"extremal"	$\left(\;{}_{I,II}\right)$	$\dfrac{SO_0(4,4)}{\mathbb{R}^{(2)}\oplus\mathbb{R}^2(6)\oplus\mathbb{R}^{[10]}}$	$\dfrac{[SL(2,\mathbb{C})]^4}{Id}$	24	$|0000\rangle + |1011\rangle + |1101\rangle + |1110\rangle$	$\in$	$L_{0_7\oplus\bar1}$

Cartan decomposition

$$\mathfrak{so}(4,4) \cong [\mathfrak{sl}(2,\mathbb{R})]^4 \oplus (\mathbf{2},\mathbf{2},\mathbf{2},\mathbf{2}). \tag{6}$$

The relevance of (6) to four qubits was pointed out in [17] and recently spelled out more clearly by Levay [21] who relates four qubits to $D = 4$ STU black holes. The Kostant-Sekiguchi correspondence then implies that the nilpotent orbits of $SO_0(4,4)$ acting on its adjoint representation are in one-to-one correspondence with the nilpotent orbits of $[SL(2,\mathbb{C})]^4$ acting on its fundamental $(\mathbf{2},\mathbf{2},\mathbf{2},\mathbf{2})$ representation and, hence, with the classification of 4-qubit entanglement.

It follows that there are 31 nilpotent orbits for four qubits under SLOCC. For each nilpotent orbit there is precisely one family of SLOCC orbits since each family contains one nilpotent orbit on setting all invariants to zero. The nilpotent orbits and their associated families are summarized in Table 2, which is split into upper and lower sections according as the nilpotent orbits belong to parameter-dependent or parameter-independent families.

If one allows for the permutation of the four qubits the connected components of each $O(4,4)$ orbit are re-identified reducing the count to 17. Moreover, these 17 are further grouped under this permutation symmetry into just nine nilpotent orbits. It is not difficult to show that these nine cosets match the nine families of [26,27], as listed in the final column of Table 2 (provided we adopt the version of L_{ab_3} presented in [27] rather than in [26]). For example, the state representative $|0111\rangle + |0000\rangle$ of the family $L_{0_{3\oplus\bar{1}}0_{3\oplus\bar{1}}}$ is left invariant by the $[SO(2,\mathbb{C})]^2 \times \mathbb{C}$ subgroup, where $[SO(2,\mathbb{C})]^2$ is the stabilizer of the three-qubit GHZ state [46]. In contrast, the four-way entangled family $L_{0_{7\oplus\bar{1}}}$, which is the "principal" nilpotent orbit [25], is not left invariant by any subgroup. Note that the total of 31 does not follow trivially by permuting the qubits in these nine. Naive permutation produces far more than 31 candidates which then have to be reduced to SLOCC inequivalent families.

There is a satisfying consistency of this process with respect to the covariant approach. For example, the covariant classification has four biseparable classes A-GHZ, B-GHZ, C-GHZ and D-GHZ which are then identified as a single class under the permutation symmetry. These four classes are in fact the four nilpotent orbits corresponding to the families $L_{0_{3\oplus\bar{1}}0_{3\oplus\bar{1}}}$ in Table 2, which are also identified as a single nilpotent orbit under permutations. Similarly, each of the four A-W classes is a nilpotent orbit belonging to one of the four families labeled $L_{a_20_{3\oplus\bar{1}}}$ which are again identified under permutations. A less trivial example is given by the six A-B-EPR classes of the covariant classification. These all lie in the single family $L_{a_2b_2}$ of [26], which is defined up to permutation. Consulting Table 2 we see that, when not allowing permutations, this family splits into six pieces, each containing one of the six A-B-EPR classes. Finally, the single totally separable class A-B-C-D is the single nilpotent orbit inside the single family L_{abc_2} which maps into itself under permutations.

Acknowledgements

It is a pleasure to thank all the organisers of the school for a scientifically stimulating week set in beautiful surroundings. I would also like to extend my gratitude to S. Ferrara for illuminating correspondence. Finally, many thanks to my collaborators on this project, D. Dahanayake, M.J. Duff, A. Marrani and W. Rubens, it has been a pleasure as always.

References

[1] M. A. Nielsen and I. L. Chuang, *Quantum Computation and Quantum Information*. Cambridge University Press, New York, NY, USA, 2000.

[2] M. J. Duff, "String triality, black hole entropy and Cayley's hyperdeterminant," *Phys. Rev.* **D76** (2007) 025017, arXiv:hep-th/0601134.

[3] M. J. Duff, J. T. Liu, and J. Rahmfeld, "Four-dimensional string-string-string triality," *Nucl. Phys.* **B459** (1996) 125–159, arXiv:hep-th/9508094.

[4] K. Behrndt, R. Kallosh, J. Rahmfeld, M. Shmakova, and W. K. Wong, "*STU* black holes and string triality," *Phys. Rev.* **D54** no. 10, (1996) 6293–6301, arXiv:hep-th/9608059.

[5] S. Bellucci, S. Ferrara, A. Marrani, and A. Yeranyan, "*STU* black holes unveiled," *Entropy* **10** no. 4, (2008) 507–555, arXiv:0807.3503 [hep-th].

[6] A. Cayley, "On the theory of linear transformations." *Camb. Math. J.* **4** (1845) 193–209.

[7] W. Dür, G. Vidal, and J. I. Cirac, "Three qubits can be entangled in two inequivalent ways," *Phys. Rev.* **A62** no. 6, (2000) 062314, arXiv:quant-ph/0005115.

[8] R. Kallosh and A. Linde, "Strings, black holes, and quantum information," *Phys. Rev.* **D73** no. 10, (2006) 104033, arXiv:hep-th/0602061.

[9] P. Lévay, "Stringy black holes and the geometry of entanglement," *Phys. Rev.* **D74** no. 2, (2006) 024030, arXiv:hep-th/0603136.

[10] M. J. Duff and S. Ferrara, "E_7 and the tripartite entanglement of seven qubits," *Phys. Rev.* **D76** no. 2, (2007) 025018, arXiv:quant-ph/0609227.

[11] P. Lévay, "Strings, black holes, the tripartite entanglement of seven qubits and the Fano plane," *Phys. Rev.* **D75** no. 2, (2007) 024024, arXiv:hep-th/0610314.

[12] M. J. Duff and S. Ferrara, "E_6 and the bipartite entanglement of three qutrits," *Phys. Rev.* **D76** no. 12, (2007) 124023, arXiv:0704.0507 [hep-th].

[13] P. Lévay, "A three-qubit interpretation of BPS and non-BPS *STU* black holes," *Phys. Rev.* **D76** no. 10, (2007) 106011, arXiv:0708.2799 [hep-th].

[14] L. Borsten, D. Dahanayake, M. J. Duff, W. Rubens, and H. Ebrahim, "Wrapped branes as qubits," *Phys. Rev. Lett.* **100** no. 25, (2008) 251602, arXiv:0802.0840 [hep-th].

[15] L. Borsten, "$E_{7(7)}$ invariant measures of entanglement," *Fortschr. Phys.* **56** no. 7–9, (2008) 842–848.

[16] P. Lévay, M. Saniga, and P. Vrana, "Three-qubit operators, the split Cayley hexagon of order two and black holes," *Phys. Rev.* **D78** no. 12, (2008) 124022, arXiv:0808.3849 [quant-ph].

[17] L. Borsten, D. Dahanayake, M. J. Duff, H. Ebrahim, and W. Rubens, "Black Holes, Qubits and Octonions," *Phys. Rep.* **471** no. 3–4, (2009) 113–219, arXiv:0809.4685 [hep-th].

[18] P. Levay, M. Saniga, P. Vrana, and P. Pracna, "Black Hole Entropy and Finite Geometry," *Phys. Rev.* **D79** (2009) 084036, arXiv:0903.0541 [hep-th].

[19] L. Borsten, D. Dahanayake, M. J. Duff, and W. Rubens, "Superqubits," *Phys. Rev.* **D81** (2010) 105023, arXiv:0908.0706 [quant-ph].

[20] P. Levay and S. Szalay, "The attractor mechanism as a distillation procedure," arXiv:1004.2346 [hep-th].

[21] P. Levay, "STU Black Holes as Four Qubit Systems," arXiv:1004.3639 [hep-th].

[22] L. Borsten, D. Dahanayake, M. J. Duff, A. Marrani, and W. Rubens, "Four-qubit entanglement from string theory," *Phys. Rev. Lett.* **105** (2010) 100507, arXiv:1005.4915 [hep-th].

[23] E. Amselem and M. Bourennane, "Experimental four-qubit bound entanglement," *Nat. Phys.* **5** (2009) 748–752.

[24] J. Sekiguchi, "Remarks on real nilpotent orbits of a symmetric pair," *J. Math. Soc. Japan* **39** no. 1, (1987) 127–138. http://dx.doi.org/10.2969/jmsj/03910127.

[25] D. H. Collingwood and W. M. McGovern, *Nilpotent orbits in semisimple Lie algebras.* Van Nostrand Reinhold mathematics series. CRC Press, 1993.

[26] F. Verstraete, J. Dehaene, B. De Moor, and H. Verschelde, "Four qubits can be entangled in nine different ways," *Phys. Rev.* **A65** no. 5, (2002) 052112, arXiv:quant-ph/0109033.

[27] O. Chterental and D. Ž. Djoković, "Normal Forms and Tensor Ranks of Pure States of Four Qubits," in *Linear Algebra Research Advances*, G. D. Ling, ed., ch. 4, pp. 133–167. Nova Science Publishers Inc, 2007. arXiv:quant-ph/0612184.

[28] M. Redhead, *Incompleteness, nonlocality, and realism: a prolegomenon to the philosophy of quantum mechanics*. Oxford University Press, 1989.

[29] C. J. Isham, *LECTURES ON QUANTUM THEORY: Mathematical and Structural Foundations*. Imperial College Press, London, 1995.

[30] A. Doring and C. Isham, "'What is a Thing?': Topos Theory in the Foundations of Physics," arXiv:0803.0417 [quant-ph].

[31] M. Born and I. Born, *The Born-Einstein letters*. Walker and Co., London, 1971.

[32] S. Kochen and E. P. Specker, "The problem of hidden variables in quantum mechanics," *Journal of Mathematics and Mechanics* **17** no. 1, (1967) 59–87.

[33] A. Einstein, B. Podolsky, and N. Rosen, "Can quantum-mechanical description of physical reality be considered complete?," *Phys. Rev.* **47** no. 10, (1935) 777–780.

[34] J. S. Bell, "On the Einstein-Podolsky-Rosen paradox." *Physics* **1** (1964) no. 3, 195.

[35] A. K. Ekert, "Quantum cryptography based on Bell's theorem," *Phys. Rev. Lett.* **67** no. 6, (Aug, 1991) 661–663.

[36] D. M. Greenberger, M. A. Horne, A. Shimony, and A. Zeilinger, "Bell's theorem without inequalities," *Am. J. of Phys.* **58** (1990) 1131–1143.

[37] M. B. Plenio and S. Virmani, "An introduction to entanglement measures," *Quant. Inf. Comp.* **7** (2007) 1, arXiv:quant-ph/0504163.

[38] C. H. Bennett, D. P. DiVincenzo, J. A. Smolin, and W. K. Wootters, "Mixed-state entanglement and quantum error correction," *Phys. Rev. A* **54** no. 5, (Nov, 1996) 3824–3851.

[39] C. H. Bennett, S. Popescu, D. Rohrlich, J. A. Smolin, and A. V. Thapliyal, "Exact and asymptotic measures of multipartite pure-state entanglement," *Phys. Rev.* **A63** no. 1, (2000) 012307, arXiv:quant-ph/9908073.

[40] R. Horodecki, P. Horodecki, M. Horodecki, and K. Horodecki, "Quantum entanglement," *Rev. Mod. Phys.* **81** no. 2, (Jun, 2009) 865–942, arXiv:quant-ph/0702225.

[41] N. Wallach, "Quantum Computing and Entanglement for Mathematicians," in *Representation Theory and Complex Analysis*, vol. 1931 of *Lecture Notes in Mathematics*, pp. 345–376. Springer Berlin / Heidelberg, 2008. http://dx.doi.org/10.1007/978-3-540-76892-0_6.

[42] L. Lamata, J. León, D. Salgado, and E. Solano, "Inductive entanglement classification of four qubits under stochastic local operations and classical communication," *Phys. Rev.* **A75** no. 2, (2007) 022318, arXiv:quant-ph/0610233.

[43] Y. Cao and A. M. Wang, "Discussion of the entanglement classification of a 4-qubit pure state," *Eur. Phys. J.* **D44** (2007) 159–166.

[44] D. Li, X. Li, H. Huang, and X. Li, "SLOCC classification for nine families of four-qubits," *Quant. Info. Comp.* **9** (2007) 0778–0800, arXiv:0712.1876 [quant-ph].

[45] S. J. Akhtarshenas and M. G. Ghahi, "Entangled graphs: A classification of four-qubit entanglement," arXiv:1003.2762 [quant-ph].

[46] L. Borsten, D. Dahanayake, M. J. Duff, W. Rubens, and H. Ebrahim, "Freudenthal triple classification of three-qubit entanglement," *Phys. Rev.* **A80** (2009) 032326, arXiv:0812.3322 [quant-ph].

[47] E. Briand, J.-G. Luque, and J.-Y. Thibon, "A complete set of covariants of the four qubit system," *J. Phys.* **A36** (2003) 9915–9927, arXiv:quant-ph/0304026.

[48] P. Breitenlohner, D. Maison, and G. W. Gibbons, "Four-Dimensional Black Holes from Kaluza-Klein Theories," *Commun. Math. Phys.* **120** (1988) 295.

[49] M. Gunaydin, A. Neitzke, B. Pioline, and A. Waldron, "Quantum Attractor Flows," *JHEP* **09** (2007) 056, arXiv:0707.0267 [hep-th].

[50] E. Bergshoeff, W. Chemissany, A. Ploegh, M. Trigiante, and T. Van Riet, "Generating Geodesic Flows and Supergravity Solutions," *Nucl. Phys.* **B812** (2009) 343–401, arXiv:0806.2310 [hep-th].

[51] G. Bossard, Y. Michel, and B. Pioline, "Extremal black holes, nilpotent orbits and the true fake superpotential," *JHEP* **01** (2010) 038, arXiv:0908.1742 [hep-th].

[52] G. Bossard, H. Nicolai, and K. S. Stelle, "Universal BPS structure of stationary supergravity solutions," *JHEP* **07** (2009) 003, arXiv:0902.4438 [hep-th].

[53] Z. D. Djokovic, N. Lemire, and J. Sekiguchi, "The closure ordering of adjoint nilpotent orbits in $\mathfrak{so}(p,q)$," *Tohoku. Math J.* **53** (2000) 395–442.

P. BURDA

"A simple way to take into account back reaction on pair creation"

This contribution was not received

Search for a high-mass Higgs Boson at the Tevatron

Davide Gerbaudo

Princeton University, Princeton, NJ 08540 USA

On behalf of the D0 Collaboration

Abstract

We present recent developments on the search for the standard model (SM) Higgs boson at the Tevatron collider. One specific analysis, the D0 search of a Higgs boson decaying to a pair of W bosons with a signature including dilepton pairs and missing transverse energy, is described in detail. The potential reach of future searches for a SM Higgs using data from the Tevatron collider is also discussed.

1 Introduction

During the last four decades, the standard model (SM) of particle physics has provided extremely accurate predictions covering a wide range of physics phenomena, and it has been tested extensively. Despite its predictive power, there are reasons to believe that the SM is incomplete. In particular, particles could not have a mass within the SM without spoiling the consistency of the theory. The Higgs mechanism—proposed in 1964 by Higgs [1], Brout, Englert [2], and later developed by Guralnik, Hagen, and Kibble [3]—restores the SM consistency by introducing a doublet of scalar bosons with a symmetry-breaking vacuum state. It also postulates the existence of a scalar particle, the Higgs boson, that has not been observed yet and whose mass is not predicted by the theory.

Particle physics experimentalists have searched for the Higgs boson for more than two decades, narrowing down the range of the mass values (m_H) within which the Higgs boson could still be found. Searches at the Large Electron Positron collider provided a direct lower limit of $m_H > 114.4$ GeV at a confidence level (CL) of 95% [4]. Precision measurements of electroweak data provided an indirect 95% CL upper constrain $m_H < 157$ GeV [5]. Direct searches for a Higgs boson with mass in the range $115 - 200$ GeV became possible only recently with the dataset being collected at the Tevatron collider. The Higgs boson branching fractions depend on m_H: low-mass searches are usually characterized by a final state with a pair of b-tagged jets (for $m_H \lesssim 135$ GeV $H \to b\bar{b}$ is the dominant mode), while high-mass searches are usually characterized by a final state with a pair of W bosons (for $m_H \gtrsim 135$ GeV $H \to WW^*$ is the dominant decay mode).

In this study we present a search for a heavy Higgs boson decaying to a pair of W bosons ($H \to WW^*$), which are identified through their leptonic decay ($W \to \ell\nu$). In spite of the relatively small leptonic branching fraction of the W (corresponding to $\sim 30\%$), the $H \to WW^* \to \ell\nu\ell\nu$ channel is the one providing the best sensitivity for the search of a high-mass Higgs boson: the clear signature of its final state, with two oppositely charged leptons and the missing transverse energy, allows for the identification of rare events (such as the production of a Higgs boson) among the large number of background ones, which are due to electroweak and quantum-electrodynamics processes. For example, the cross section for the production of a Higgs boson is expected to be $\sigma(p\bar{p} \to H) < 1$ pb at $\sqrt{s} = 1.96$ TeV; after accounting for branching ratios, lepton identification efficiencies, and selection requirements, 42 signal events are expected in the dataset for $m_H = 165$ GeV. The cross section for the electroweak dilepton production (through an intermediate Z/γ^*) is larger than the signal one by more than two orders of magnitude.

In this search, we select events by applying few simple kinematic requirements (as described in Sec.2). Using a multivariate technique (described in Sec.3), we then separate the large background contribution from a potential Higgs boson signal. In Sec.4, we compute limits on the cross section for SM Higgs boson production in the mass range $115 < m_H < 200$ GeV. The results discussed here constitute an update to the high-mass Higgs boson searches presented elsewhere [6].

2 Samples and Event Selection

The data sample used in this study was recorded at the D0 experiment [7] between April 2002 and June 2009, and corresponds to an integrated luminosity of 5.4 fb^{-1}. We consider three signal production mechanisms: gluon-gluon fusion ($gg \to H$), associated production (WH/ZH), and vector boson fusion ($gg \to q\bar{q}H$). We consider the following background processes: diboson production, W(+jets), Z/γ^*(+jets), $t\bar{t}$, and multijet production. Predictions for the signals yields are computed with PYTHIA [8] simulations, and normalized to NNLO calculations. Background predictions for W(+jets) and Z/γ^*(+jets) are simulated with ALPGEN [9] and normalized to NNLO predictions [10]. The other backgrounds are simulated with PYTHIA; the $t\bar{t}$ background is normalized to NNLO calculations [11] and the diboson backgrounds are normalized to NLO calculations [12]. The multijet background is estimated from data, using an orthogonal sample with two leptons having the same charge.

We perform the event selection in two steps, described in the following two paragraphs. In the first step ("preselection"), we select events that have the main features expected for a candidate signal event. In the second step ("final selection"), we apply tighter criteria, aimed at rejecting those events that are likely to be due to background processes.

At the preselection stage we require each event to have two high-p_T leptons reconstructed; the two leptons can be two electrons (e^+e^-), an electron and a muon ($e^\pm\mu^\mp$), or two muons ($\mu^+\mu^-$). The minimum transverse momentum depends on the lepton type: electrons must have $p_T^e > 15$ GeV, while for muons $p_T^\mu > 10$ GeV (but in a dimuon event, the muon with highest p_T must have $p_T^\mu > 20$ GeV). Further details on the algorithms used for electron and muon reconstruction can be found in the full paper [13] and the references therein.

The criteria used to reject background events are based on the kinematic variables of the event. In particular, they are aimed at suppressing the most significant background process, namely Z/γ^*(+jets). We require the azimuthal opening angle between the two leptons, $\Delta\phi(\ell\ell)$, to be either $\Delta\phi(\ell\ell) < 2.0$ rad (for e^+e^- and $e^\pm\mu^\mp$) or $\Delta\phi(\mu\mu) < 2.5$ rad (for $\mu^+\mu^-$). The missing transverse energy $\not{E}_T$, is required to be $\not{E}_T > 20$ GeV (for e^+e^- and $e^\pm\mu^\mp$), or $\not{E}_T > 25$ GeV (for $\mu^+\mu^-$). In e^+e^- and $e^\pm\mu^\mp$ events, the scaled missing transverse energy [14], $\not{E}_T^{Sc}$, is required to be $\not{E}_T^{Sc} > 6$. The minimum transverse mass, M_{min}^T $(l, \not{E}_T)$, defined as the smaller of $M_T(\not{E}_T, \ell_1)$ and $M_T(\not{E}_T, \ell_2)$, is required to be M_{min}^T $(l, \not{E}_T) > 20$ GeV ($> 30 GeV$ for e^+e^-).

After reducing the total number of events from more than 700 thousands (at preselection) to less than three thousands (at final selection), we train a multivariate discriminant to improve the separation between signal and background.

3 Multivariate Discriminant

By exploiting the correlations between the quantities characterizing the reconstructed events, multivariate techniques can provide an improved signal-background discrimination compared to simple cut-based analysis.

We use an Artificial Neural Network (NN) multivariate discriminant, as implemented in the TMultiLayerPerceptron class within the ROOT package [15]. Twelve variables are used as inputs to the NN, after checking that all are properly modeled by the MC simulation so as not to induce a bias in the NN output. Variables providing the best discrimination between signal and background include: transverse momenta (of each lepton and of the dilepton system), the invariant mass $M_{\ell\ell}$, and the missing transverse energy $\not{E}_T$. A detailed description of the NN inputs is provided in the full paper [13]. The artificial neural network is organized in one input layer, one hidden layer, and one output layer; for each one of the nodes in a layer, an output value is computed from the node's inputs, using a sigmoid activation function.

We train and optimize the final discriminant for each of the mass values that we consider, namely from 115 GeV to 200 GeV in 5 GeV steps. That is, the parameters of the activation functions are optimized so as to provide the best separation between signal and background events; this optimization is performed using a steepest-descent algorithm provided within the ROOT package [15]. In Fig. 1(a) we show the distribution (at final selection) of the NN output trained for $m_H = 165$ GeV, the mass region where this analysis has the greatest expected sensitivity. Fig.1(b) shows the distribution of the output of the same NN, after subtracting the expected background. The systematic uncertainty, represented by an error band corresponding to ± 1 standard deviation, is obtained after fitting the data to the background-only template [16].

For all values of m_H that we consider, we do not observe any excess of data that could correspond to Higgs boson events. We therefore proceed to compute upper limits on the SM cross section for Higgs boson production.

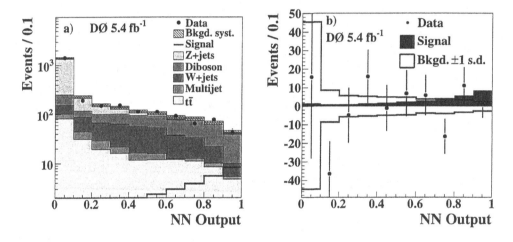

Figure 1: (color online) (a) The neural network output after final selection, with a signal corresponding to m_H=165 GeV. (b) Data after subtracting the fitted background (points) and SM signal expectation (filled histogram) as a function of the NN output for m_H=165 GeV. Also shown is the ± 1 standard deviation (s.d.) band on the total background after fitting.

4 Limits

From the NN distributions, a profile likelihood is computed; a detailed description of this test statistic can be found in Ref. [17]. Limits on the SM cross section for Higgs production are determined using a modified frequentist (CL_s) approach [18], as implemented in the COLLIE package [19].

When calculating the limits, we take into account the statistical uncertainty associated with each bin from the NN distributions, as well as two types of systematic uncertainties: uncertainties that affect only the number of expected events

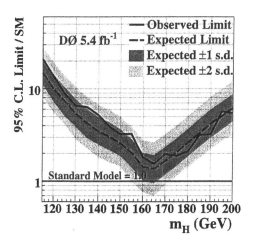

Figure 2: (color online) Upper limit on Higgs boson production cross section obtained with the $H \to WW^* \to \ell\nu\ell\nu$ channel.

(normalizations), and uncertainties that also affect the shape of the NN output distribution (shapes). Each one of the systematic uncertainties is propagated through the data-analysis steps, thus providing a shape or normalization uncertainty for the expected NN distribution. A detailed description of the systematic uncertainties can be found in the full paper [13].

Fig. 2 presents, as a function of m_H, the upper limit on the cross section for the SM Higgs boson production. The limit is reported at 95% confidence level as a ratio to the predicted SM cross section. This study reaches its maximum sensitivity for $m_H = 165$ GeV where, with 5.4 fb^{-1}, the observed upper limit is 1.55 times the SM prediction. The inclusion of more data in the $H \to WW^* \to \ell\nu\ell\nu$ search, as well as improvements to the analysis techniques, allowed the D0 experiment to produce a preliminary update of their previous result [13]. With this updated limit, shown in Fig. 3, the D0 combined sensitivity is close to the SM expectation for $m_H = 165$ GeV.

The latest combination of the CDF and D0 searches, reported in Ref. [21] and shown in Fig. 3, excludes at 95% CL the existence of a SM Higgs boson with mass $158 < m_H < 175$ GeV. At low mass, the Tevatron combination confirms up to $m_H < 110$ GeV the LEP exclusion of a light SM Higgs. According to the most recent sensitivity projections from the ATLAS [22] and CMS [23] experiments, the LHC should be able to confirm (or refute) the high-mass Tevatron exclusion already with 1 fb^{-1} of data collected at $\sqrt{s} = 7$ TeV. In the low-mass region, searches at the LHC will require more than 1 fb^{-1} of data in order to match the Tevatron sensitivity. This is due to the fact that for low values of m_H the $H \to b\bar{b}$ decay mode is the predominant one, and the LHC environment makes this search more difficult.

294

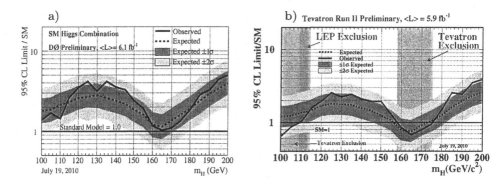

Figure 3: (color online) (a) Combined D0 limit on Higgs boson production cross section, from Ref. [20]. (b) Combined Tevatron limit on Higgs boson production cross section, from Ref. [21].

5 Conclusions

We report the results of a search for the SM High boson in final states with two leptons (e^+e^-, $e^\pm\mu^\mp$, or $\mu^+\mu^-$) and missing transverse energy. The dataset used in this search was recorded with the D0 detector at the Fermilab Tevatron collider, and corresponds to 5.4 fb^{-1} of proton-antiproton collisions. After applying kinematic selection criteria, a multivariate technique (Neural Network) is used to increase the separation of signal candidate events from the background. No significant excess of signal candidates is observed, and limits on the SM cross section for the production of a Higgs boson are set. The combination of the $H \to WW^* \to \ell\nu\ell\nu$ limit with the ones from other channels analyzed at D0, reaches a sensitivity close to the SM prediction for a Higgs boson with a mass $m_H = 165$ GeV. A SM Higgs boson with mass in the range $158 - 175$ GeV is excluded by the latest Tevatron combined result. The ATLAS and CMS experiments should be able to confirm this exclusion region with 1 fb^{-1} that the LHC is expected to deliver during its 2011 run.

6 Acknowledgments

We thank the staffs at Fermilab and collaborating institutions, and acknowledge support from the DOE and NSF (USA); CEA and CNRS/IN2P3 (France); FASI, Rosatom and RFBR (Russia); CNPq, FAPERJ, FAPESP and FUNDUNESP (Brazil); DAE and DST (India); Colciencias (Colombia); CONACyT (Mexico); KRF and KOSEF (Korea); CONICET and UBACyT (Argentina); FOM (The Netherlands); STFC and the Royal Society (United Kingdom); MSMT and GACR (Czech Republic); CRC Program, CFI, NSERC and WestGrid Project (Canada); BMBF and DFG (Germany); SFI (Ireland); The Swedish Research Council (Sweden); and CAS and CNSF (China).

References

[1] P. Higgs, *Phys. Rev. Lett.* **12**, 132 (1964), *Phys. Rev. Lett.* **13**, 508 (1964)

[2] F. Englert, R. Brout, *Phys. Rev. Lett.* **13**, 321 (1964)

[3] G.S. Guralnik, C.R. Hagen, T.W.B. Kibble, *Phys. Rev. Lett.* **13**, 585 (1964)

[4] ALEPH, DELPHI, L3, OPAL Collaborations and The LEP Working Group for Higgs Boson Searches, *Phys. Lett.* B **565**, 61 (2003)

[5] LEP Electroweak Working Group, Tevatron Electroweak Working Group, SLD Electroweak and Heavy Flavour Groups, arXiv:0911.2604 (2009)

[6] D. Gerbaudo for the D0 collaboration, FERMILAB-CONF-10-178-E (2010)

[7] B. Abbott *et al.* (D0 Collaboration), *Nucl. Instrum. Methods* A **565**, 463 (2006)

[8] T. Sjöstrand *et al.*, *Comput. Phys. Commun.* **135**, 238 (2001)

[9] M.L. Mangano *et al.*, *JHEP* **0307**, 001 (2003)

[10] R. Hamberg, W.L. van Neerven, and T. Matsuura, *Nucl. Phys.* B **359**, 343 (1991)

[11] S. Moch and P. Uwer, *Phys. Rev.* D **78**, 034003 (2008)

[12] J.M. Campbell and R.K. Ellis, *Phys. Rev.* D **60**, 113006 (1999)

[13] V.M. Abazov *et al.* (D0 Collaboration), *Phys. Rev. Lett.* **104**, 061804 (2010)

[14] V.M. Abazov *et al.* (D0 Collaboration), *Phys. Rev. Lett.* **96**, 011801 (2006)

[15] I. Antcheva *et al.*, *Comput. Phys. Commun.* **180**, 2499 (2009)

[16] W. Fisher, FERMILAB-TM-2386-E (2006)

[17] W.Fisher, D0 Note 5309 (2006)

[18] A. Read, *J. Phys.* G **28**, 2693 (2002)

[19] W. Fisher, Collie: A Confidence Level Limit Evaluator, D0 Note 5595f (2008)

[20] The D0 Collaboration, D0 Conference Note 6094 (2010)

[21] The CDF and D0 Collaborations, FERMILAB-CONF-10-257-E, (2010)

[22] T. Masubuchi for the ATLAS Collaboration, "ATLAS Higgs Sensitivity for 1/fb of data at the LHC running at 7 TeV," presented at the 35th international conference on High Energy Physics, Paris (2010)
The ATLAS Collaboration, ATL-PHYS-PUB-2010-009 (2010)

[23] M. Gataullin for the CMS Collaboration, "Prospects for Higgs boson searches with CMS," presented at the 35th international conference on High Energy Physics, Paris (2010)

Twisted strings in Extended Abelian Higgs Models*

Árpád Lukács

MTA RMKI, H-1525 Budapest 114, P.O.Box 49, Hungary

June 19, 2012

Abstract

We study string solutions in two-component Abelian Higgs models with globally $U(1) \times U(1)$ symmetric potentials. Two component Abelian Higgs models with spontaneous symmetry breaking are divided into two classes, according to the number of Higgs field components obtaining a nonzero vacuum expectation value. In one class, where one field component is nonzero in the vacuum, new, presumably stable vortices are found.

Introduction

Recently, there has been an upsurge of interest in string and string-like solutions in a range of field theoretic and solid state models with multi-component scalar fields (or order parameters). Particular examples include superconducting cosmic strings in the Witten-model [1], the $SU(2)_{\text{global}} \times U(1)$ semilocal model and the electroweak theory [3–5], and vortices for the nonlinear Schrödinger equation in nonlinear optics [12], and in the Ginzburg–Landau effective theory of the non-conventional superconductor MgB_2 [6–11].

In the present paper, we examine vortex solutions in two component Abelian Higgs theories carrying a magnetic flux. Depending on the parameters of the self-interaction potential of the scalar fields, either a single component (1VEV) or both components of the scalar field (2VEV) obtain a nonzero vacuum expectation value.

In the 1VEV case, we examine the existence of twisted string solutions, i.e. strings in which there is a relative phase (twist) between the two complex scalar components varying linearly along the direction of the straight string [14]. We show that such strings can only exist in the 1VEV case. It is important to note that the Abrikosov-Nielsen-Olesen (ANO) vortex solutions of the Abelian Higgs model with a single charged scalar field can be embedded in 1VEV two-component Abelian Higgs models. Therefore in such 1VEV models the ANO vortices coexist with the family of twisted vortex solutions with the same amount of magnetic flux, and importantly twisted vortices have lower energy. In the generic 1VEV case, we examine the existence of a lowest energy limiting case of twisted strings, which is a presumably stable solution, existing simultaneously with the embedded ANO strings and the family of twisted strings.

We also examine the existence of string solutions in the 2VEV case. These strings are not twisted, and similar to those known previously in models of two-band superconductivity ([6–11]) and nonlinear optics [12].

---

*Joint work with P. Forgács.

1   Two-component extended Abelian Higgs models

After some rescalings of the fields and the space-time coordinates, the Lagrangian of two-component extended Abelian Higgs models can be written as

$$\mathcal{L} = \frac{1}{e^2}\left\{-\frac{1}{4}F_{\mu\nu}F^{\mu\nu} + (D_\mu\phi)^\dagger(D^\mu\phi) - V(\phi,\phi^\dagger)\right\}, \tag{1}$$

with $D_\mu = \partial_\mu - iA_\mu$ and

$$V = \frac{\beta_1}{2}(|\phi_1|^2 - 1)^2 + \frac{\beta_2}{2}|\phi_2|^4 + \beta'|\phi_1|^2|\phi_2|^2 - \mu_2^2|\phi_2|^2, \tag{2}$$

where e is the charge of the fields, β_1, β_2, β' and μ_2^2 are tunable parameters (in the case of solid state models, material constants), $\phi = (\phi_1, \phi_2)^T$ and $\phi^\dagger = (\phi_1^*, \phi_2^*)$. The potential V in eq. (2) is the most general $U(1) \times U(1)$ invariant potential. In the theory given by eq. (1), one of the $U(1)$ groups is gauged, the gauge symmetry acts on the fields as $\phi \to \exp(i\chi)\phi$, $A_\mu \to A_\mu + \partial_\mu\chi$, where $\chi = \chi(x)$ is the gauge function. The remaining $U(1)$ symmetry is global, and it acts on the fields as $\phi_1 \to \exp(-i\alpha)\phi_1$, $\phi_2 \to \exp(i\alpha)\phi_2$, where α is a constant. (In the Witten model [1], the potential is the same, however, both $U(1)$ groups are gauged, as a result, there are two vector fields.)

The theory defined in eq. (1) is a member of the family of semilocal models, i.e. gauge theories with additional global symmetries [4]. A well-known example of these is the $SU(2)_{\text{global}} \times U(1)_{\text{local}}$ semilocal model, given by

$$\beta_1 = \beta_1 = \beta' = \mu_2^2 =: \beta, \tag{3}$$

i.e. in which case the potential is $V = \beta(\phi^\dagger\phi - 1)^2/2$. This model can be thought of as a limit of the Salam-Weinberg model, where the coupling of the non-Abelian field is set to zero [4].

Two-component Abelian Higgs models can be divided into two classes, according to whether only one component of the scalar field obtains a nonzero vacuum expectation value (1VEV case), or both components do (2VEV case). The number of components with nonzero VEV depends on the parameters of the potential V in eq. (2).

There is one condition the parameters β_1, β_2, β' have to satisfy in both classes, ensuring that the theory be stable ($V \to \infty$ for large field values), namely

$$\beta_{1,2} > 0, \quad \beta' > -\sqrt{\beta_1\beta_2}. \tag{4}$$

The potential has been constructed in such a way that $(1, 0)$ is a minimum of the potential in the $(\eta_1, 0)$ direction. If the condition

$$\left.\frac{\partial^2 V}{\partial\phi_2^*\partial\phi_2}\right|_{\phi=(1,0)^T} = \beta' - \mu_2^2 > 0. \tag{5}$$

holds, it is also a local minimum, therefore in the 1VEV case, the condition (5) has to be satisfied.

In order that the global minimum of the potential be at the field values $(1, 0)$,

$$V(\phi_1 = 0, \phi_2 = \eta_2) = \frac{\beta_1}{2} - \frac{\mu_2^4}{2\beta_2} > 0 \tag{6}$$

has to hold with $\eta_2^2 = \mu_2^2/\beta_2$. Condition (6) corresponds to that out of the two possible local minima $(1, 0)$ and $(0, \eta_2)$ the first one be the global minimum. This can be assumed without loss of generality (because otherwise the second component would be the one obtaining a VEV, and the two components could be interchanged). In the 2VEV class, the VEVs can be calculated by minimising V,

$$\eta_1^2 = \frac{\beta_1\beta_2 - \mu_2^2\beta'}{\beta_1\beta_2 - (\beta')^2}, \qquad \eta_2^2 = \frac{\beta_1(\mu_2^2 - \beta')}{\beta_1\beta_2 - (\beta')^2}. \tag{7}$$

The field equations obtained from the Lagrangian (1) read

$$\partial^\rho F_{\rho\mu} = i\{(D_\mu\phi)^\dagger\phi - \phi^\dagger D_\mu\phi\}, \\ D_\rho D^\rho\phi = -\partial V(\phi^\dagger, \phi)/\partial\phi^\dagger. \tag{8}$$

2 Twisted strings

2.1 Twisted string Ansatz and vortex profile equations

Twisted strings are cylindrically symmetric solutions of the field equations (8). Here, a field configuration is considered cylindrically symmetric, if rotations around an axis or translations along its direction can be compensated with suitable gauge transformations [13]. Let us use cylindrical coordinates, with the z axis being the axis of rotational symmetry. With a suitable Lorentz boost, these configurations can be rendered static, and thus be brought to the form [14]

$$\phi_1(r, \vartheta, z) = f_1(r)e^{in\vartheta}, \qquad A_\vartheta(r, \vartheta, z) = na(r), \\ \phi_2(r, \vartheta, z) = f_2(r)e^{im\vartheta}e^{i\omega z}, \qquad A_3(r, \vartheta, z) = \omega a_3(r), \tag{9}$$

with $A_0 = 0$, $A_r = 0$. Substituting (9) into the field equations (8) yields the following vortex profile equations:

$$\frac{1}{r}(ra_3')' = 2a_3|f|^2 - 2f_2^2, \\ r\left(\frac{a'}{r}\right)' = 2f_1^2(a-1) + 2f_2^2(a-m/n), \\ \frac{1}{r}(rf_1')' = f_1\left[\frac{(1-a)^2n^2}{r^2} + \omega^2a_3^2 + \beta_1(f_1^2 - 1) + \beta'f_2^2\right], \\ \frac{1}{r}(rf_2')' = f_2\left[\frac{(na-m)^2}{r^2} + \omega^2(1-a_3)^2 + \beta_2f_2^2 - \mu_2^2 + \beta'f_1^2\right]. \tag{10}$$

Boundary conditions for eq. (10) at the origin are determined by demanding regularity of the functions, while the boundary conditions at infinity originate in the requirement that the energy density decays radially, giving $f_1, a \to 1$ and $f_2, a_3 \to 0$ in the 1VEV case. In the 2VEV case, when f_2 also assumes a VEV, $f_{1,2} \to \eta_{1,2}$, where $\phi = (\eta_1, \eta_2)$ is a minimum of the potential. However, note, that in the latter case, no twisted vortices are possible, since a_3 in the covariant derivative cannot cancel the z-derivative of ϕ_2 and leave $D_3\phi_1 = 0$ intact at the same time. Therefore, in what follows, we can focus on the $f_2 \to 0$ as $(r \to \infty)$ case.

It is important to note here, that in the 1VEV case, all solutions $(\tilde{\phi}, \tilde{A}_\mu)$ of the Abelian Higgs model can be embedded in the two-component theory as $\phi_1 = \tilde{\phi}$, $A_\mu = \tilde{A}_\mu$, most importantly, Abrikosov-Nielsen-Olesen (ANO) strings can be embedded. Their profile equations can be obtained from eq. (10) by setting $f_2 = a_3 = 0$.

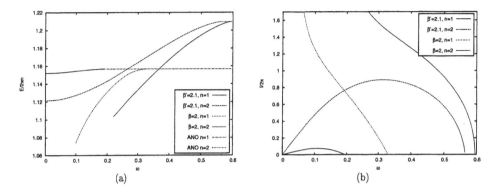

Figure 1: The energy E and the current, $\mathcal{I}$ as a function of the twist ω

2.2 Energy

The energy density calculated from the Lagrangian (1) for time independent configurations is

$$
\begin{aligned}
\mathcal{E} =& \frac{1}{2}\left[\frac{n^2(a')^2}{r^2} + \omega^2(a_3')^2\right] + (f_1')^2 + (f_2')^2 \\
&+ \frac{n^2(1-a)^2}{r^2}f_1^2 + \frac{(na-m)^2}{r^2}f_2^2 + \omega^2(a_3^2f_1^2 + (1-a_3)^2 f_2^2) + V(f_1, f_2)\,.
\end{aligned}
\tag{11}
$$

with $V(f_1, f_2) = \frac{\beta_1}{2}(f_1^2 - 1)^2 + \frac{\beta_2}{2}f_2^4 + \beta' f_1^2 f_2^2 - \mu_2^2 f_2^2$. The total energy (per unit length) of the vortex is given as $E = 2\pi \int_0^\infty r dr \mathcal{E}$. It can shown that E is a monotonously growing function of β_1, β_2, β', ω and monotonously decreasing in μ_2^2. Some energy vs. twist curves can be seen in figure 1(a).

2.3 Current

The conserved current corresponding to the global $U(1)$ symmetry of the theory (1) is given by

$$
j_\mu^3 = -i(\phi_1^* D_\mu \phi_1 - \phi_2^* D_\mu \phi_2 - \phi_1(D_\mu \phi_1)^* + \phi_2(D_\mu \phi_2)^*)\,,
\tag{12}
$$

which agrees with the third isospin component of the global $SU(2)$ current of the semilocal theory [14]. Due to the z-dependent relative phase between the two field components, $j_3^3 \neq 0$, thus a global current flows along the string. The total current flowing in the string is given as $\mathcal{I} = \int d^2x j_3^3$. Plugging the Ansatz (9) into (12), the relevant current component is

$$
j_3^3 = 2\omega a_3(f_1^2 - f_2^2) + 2\omega f_2^2\,.
\tag{13}
$$

The dependence of the global current $\mathcal{I}$ on the twist ω is depicted in figure 1(b). There the current for $\beta_{1,2} = \mu_2^2 = 2$, $\beta' = 2.1$ and for the $SU(2)$ symmetric case $\beta_{1,2} = \beta' = \mu_2^2 = 2$ is compared. As we shall see, in the general case there exists an $\omega = 0$ limiting configuration. In these cases, the total current of the string $\mathcal{I} \to 0$ as $\omega \to 0$. This behaviour agrees with what is expected from the $r \to \infty$ asymptotic properties of the solutions. In the $SU(2)$ symmetric case, the current is a monotonous function of the twist, and it diverges for $\omega \to 0$ [14]. In those cases where there is no $\omega = 0$ solution, the current – twist relation is very similar to that of the $SU(2)$ symmetric case.

2.4　Range of solutions

Numerically we have found that twisted string solutions exist for $0 < \omega < \omega_b$, where the upper limit is a function of the parameters β_1, β_2, β' and μ_2^2 of the potential and n, m of the vortex, similarly to the $SU(2)$ symmetric case [14].

At the upper limit, twisted strings bifurcate with embedded ANO strings. It is by now well known that embedded ANO vortices are unstable to small perturbations of the f_2 variable [3], and that this instability corresponds to the aforementioned bifurcation [14]. Close to the bifurcation, a systematic expansion of the solution in a bifurcation parameter ϵ has been carried out in ref. [15] in the $SU(2)$ symmetric case. The analysis of ref. [15] can be repeated in the present case with minimal modifications. The systematic expansion of a twisted vortex near the bifurcation point can be then written as :

$$f_1 = f_1^{(0)} + \epsilon^2 f_1^{(2)} + \dots \qquad \begin{aligned} a &= a^{(0)} + \epsilon^2 a^{(2)} + \dots \\ a_3 &= \epsilon^2 a_3^{(2)} + \dots \\ \omega &= \omega_b + \epsilon \omega_1 + \epsilon^2 \omega_2 + \dots \end{aligned} \qquad (14)$$

where $a^{(0)}, f_1^{(0)}$ denotes the ANO vortex, whose equations can be read off from equations (10) by putting $f_2 = a_3 = 0$. In the above expansion, we have omitted some terms linear in ϵ, that turned out to be zero. For details, and the Taylor expanded equations, see [15, 16].

For brevity's sake, here we only show the leading order equation,

$$(D_2^{(0)} + \omega_b^2) f_2^{(1)} := -\frac{1}{r} \left(r f_2^{(1)\prime} \right)' + \left[\frac{(na^{(0)})^2}{r^2} + \omega_b^2 - \mu_2^2 + \beta'(f_1^{(0)})^2 \right] f_2^{(1)} = 0. \qquad (15)$$

The expansion coefficients ω_i are dictated by the conditions for the cancellation of resonance terms. The procedure yields

$$\epsilon = \sqrt{\frac{1}{\omega_2}(\omega - \omega_b)} + \dots. \qquad (16)$$

The lower endpoint of the interval $0 < \omega < \omega_b$ is even more important. As the energy of the vortices gets smaller for decreasing values of the twist, the $\omega \to 0$ behaviour of the vortices becomes an interesting problem. In those cases, when there is a vortex solution with zero twist, it is expected to be stable.

Let us consider now the $\omega \to 0$ limit. The asymptotic behaviour of f_2 for large r is $\sim \exp(-r\sqrt{\beta' - \mu_2^2 + \omega^2})/\sqrt{r}$. Therefore, if $\beta' > \mu_2^2$, there exists a solution with $\omega = 0$ and $A_3 = 0$. In this case, the profile equations are obtained from eq. (10) by setting $\omega = 0$. such a solution for $\beta_1 = \beta_2 = \mu_2^2 = 2$ and $\beta' = 2.1$ can be seen in Figure 2(a). Note again that these vortices have significantly smaller energy than the embedded ANO vortex, and lower energy than all twisted ones, and therefore we expect them to be stable.

We have calculated the $\omega = 0$ limiting configurations numerically, for a number of values of the parameters β_1, β_2, β' and μ_2^2, and found that the limiting solutions exist for a wide range of the parameters. Fixing two out of the three parameters β_1, β' and μ_2^2, there is a range of the third parameter, for which the $\omega = 0$ vortices exist, e.g. above a critical value of μ_2^2 if β_1 and β' is fixed. (The other endpoint of the interval is given by the stability conditions (5) and (6).) Close to the critical value of the varied parameter, we have performed a similar analysis as close to the bifurcation.

If the exponent $\sqrt{\beta' - \mu_2^2 + \omega^2}$ becomes zero as $\omega \to 0$, i.e. $\beta' = \mu_2^2$, a different type of solution, with a $1/r$ type radial decay of f_2 may exist. In this case, the asymptotic form of f_2 for $r \to \infty$ is $f_2 \sim F_2 r^{-1}(\log r)^{-1/2}$, where $F_2 = 1/\sqrt{\beta_2}$ is a constant. For such a solution, see figure 2(b).

302

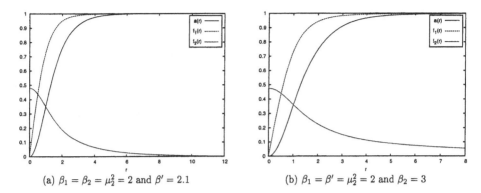

(a) $\beta_1 = \beta_2 = \mu_2^2 = 2$ and $\beta' = 2.1$ (b) $\beta_1 = \beta' = \mu_2^2 = 2$ and $\beta_2 = 3$

Figure 2: Zero twist solutions

In the $\beta_1\beta_2 = \beta'^2$, $\mu_2^2 = \beta'$ case, there seems to be no limiting solution. In these cases, as the twist ω decreases, the profile functions reach their asymptotic values farther from the origin. This way, the string expands and its energy density becomes more dilute. In ref. [14], this behaviour has been described with a scaling argument in the $SU(2)$ symmetric case, which can be generalized to the non $SU(2)$ symmetric case without major changes.

2.5 $n = 2$ vortices and the vortex-vortex interaction

In the Abelian Higgs model and its extensions, besides single vortex solutions, multivortex solutions have been found. If the vortices are well separated, it is possible to describe these configurations as vortices moving under the influence of forces acting among them. Forces between a pair of vortices can be calculated as the derivative of the energy of a two-vortex configuration with respect to their separation [12, 17]. The nature of the intervortex forces plays a fundamental role in the behaviour of superconductors. In Type I superconductors, the force is attractive, whereas in Type II ones, repulsive. In the case of conventional superconductors, which can be described with a one-component Ginzburg–Landau free energy, Type I corresponds to $\beta < 1$ and Type II to $\beta > 1$.

The physics of multicomponent theories is richer. It is therefore of high importance to obtain information on the nature of forces acting among vortices different than the embedded ANO solutions. The simplest way to obtain information on the repulsive or attractive nature of the inter-vortex forces is to calculate the energy of two $n = 1$ vortices, $2E(n = 1)$ and compare this to the energy of an $n = 2$ vortex, $E(n = 2)$. If $2E(n = 1) < E(n = 2)$ then two separated one flux quantum vortices are energetically preferable over an $n = 2$ vortex; this is the case in Type II superconductors. In Type I superconductors, the opposite is true, $2E(n = 1) > E(n = 2)$. In the present paper we shall apply this method to twisted vortices. The separation dependence of the intervortex forces; the distinction between scenarios analogous to Type I/II and Type-1.5 superconductivity requires the calculation of two-vortex solutions, and thus falls beyond our scope. The separation dependence of intervortex forces has been studied in the important case of the two-gap superconductor MgB$_2$ (2VEV case) in ref. [8], where the force was found to be attractive at long distances and repulsive at short distances. See also [7].

For potentials where the coefficients in the potential (2) are of similar magnitude, we have found that the generic behaviour is that even for $\beta_i > 1$, if the twist ω is not too close to the bifurcation, $n = 2$ twisted vortices are energetically preferred over two $n = 1$ ones (Type-I behaviour, see figure 1(a)). However, there are cases, where Type-II behaviour occurs, e.g. for

$\beta_1 = \beta' = \mu_2^2 = 2$, $\beta_2 = 3$, $E(n = 2) > 2E(n = 1)$ Type-II behaviour persists for all values of the twist.

In Ref. [14] the $SU(2)$ symmetric case has been examined. There it has been shown that although for $\beta > 1$ both (embedded) ANO vortices and twisted ones close to the bifurcation exhibit Type II behaviour, farther from the bifurcation, $n = 2$ vortices with small enough twist have lower energy than two $n = 1$ vortices.

In the $\omega = 0$ limit, we have found that the typical behaviour is that if both $n = 1$ and $n = 2$ strings with nonzero f_2 exist, then one $n = 2$ string has less energy than two $n = 1$ strings (similarly to Type I superconductivity). However, for $\beta_1 > 1$, for embedded ANO strings show the opposite behaviour (Type II superconductivity). Close to the bifurcation, we have found that true two-component strings also behave like embedded ANO ones.

As shown in section 2.2., the energy increases with growing β_1, β_2, β' and decreases for increasing μ_2^2. In the energy density (11) the coefficient of μ_2^2 is larger than that of β' and β_2, therefore one can expect vortices to have the smallest energy on the boundary of the region allowed by the stability condition (6), in the $\beta_2 \to \infty$ limit. In this case, stability

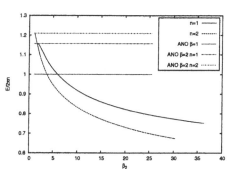

Figure 3: Energy as a function of β_2 on the stability boundary, $\beta_1 = 1$. The energy of a $\beta = 1$ ANO string is also shown for comparison.

condition eq. (5) requires β' also grow with β_2. Also, the lowest energy is achieved in the $\omega \to 0$ limit. As expected, *we have found vortices with very low energy values in this limit.* Their properties can be qualitatively explained by rescaling the field f_2 as $\tilde{f}_2 = \sqrt{\beta'} f_2$. In this way in the equation of the component f_1, β_2, β' and μ_2^2 does not appear, and the equation of $\tilde{f}_2$ assumes the form

$$\frac{1}{r}(r\tilde{f}_2')' = \tilde{f}_2 \left[\frac{(na-m)^2}{r^2} + \frac{\beta_2}{\beta'}\tilde{f}_2^2 - \mu_2^2 + \beta' f_1^2 \right]. \tag{17}$$

In the large β_2 limit, $\mu_2^2 \sim \sqrt{\beta_2\beta_2}$, $\beta' \sim \sqrt{\beta_1\beta_2}$, therefore, in order that the right hand side remains finite,

$$\tilde{f}_2^2 \sim \beta_1(1 - f_1^2) \quad (\beta_2 \to \infty) \tag{18}$$

has to hold. However, plugging this into the equation of the first component f_1, we can see that this limit is similar to the $\beta \to 0$ limit of ANO vortices. We have obtained ANO vortex profiles for very low values of β_1, and found that their energy can be much lower than 2π (e.g. for $\beta = 10^{-6}$, $E/2\pi = 0.17103$). Let us also note here, that the in the $\beta \to 0$ limit, the asymptotic behaviour of the energy is $2/\log\beta$ [19], which is in very good agreement with our numerical data. For numerical data, see Figure 3.

3 Vortices with both scalar fields having nonzero VEV in the vacuum

Let us now examine the case when condition (5) is not satisfied. In this case, at the minimum of the potential, both scalar field components are nonzero. In order that a string has finite energy per unit length, the fields have to lie asymptotically on the vacuum manifold and simultaneously all covariant derivatives have to tend to zero. In this case, the Ansatz (9) is only compatible

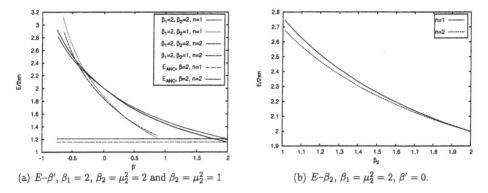

(a) E–β', $\beta_1 = 2$, $\beta_2 = \mu_2^2 = 2$ and $\beta_2 = \mu_2^2 = 1$ (b) E–β_2, $\beta_1 = \mu_1^2 = 2$, $\beta' = 0$.

Figure 4: Energy of 2VEV vortices as a function of β' and β, respectively

with these requirements if $\omega = 0$ (otherwise $a_3(r \to \infty) = 0$ would be required for $D_3\phi_1$ to vanish at $r \to \infty$ and $a_3(r \to \infty) = 1$ for $D_3\phi_1$) and $m = n$ (similarly, for $a(r \to \infty) = 1$ to compensate the angle derivatives of both fields).

Note, that it is also possible to obtain vortex solutions with $n \neq m$ (fractional flux vortices in non-conventional superconductors, see [9]), however, the energy of these is logarithmically divergent. The regularisation of such solutions (e.g. excluding a certain region around the origin, or a finite radius) would require further physical input (e.g. concerning the validity of the effective theory on small scales), and thus is beyond the scope of the present paper.

In the $n = m$, $\omega = 0$ case, the profile equations can be obtained from (10) by setting $a_3 = \omega = 0$ and $n = m$. In the fully symmetric $\beta_1 = \beta_2 =: \beta$, $\mu_2^2 = \beta$ case, the equations of f_1 and f_2 are identical. By setting $f := f_1 = f_2$ the profile equations (10) become equivalent to those of the ANO strings, without the standard rescalings of the fields and coordinates. In this case, a solution of the field equation can be obtained as an embedding of the well known ANO vortex (although with different parameters and a nonstandard scaling of the fields). The energy of such vortices can be calculated with the energy density (11) by replacing V with $V - V_0$ where V is the potential and $V_0 = V(\phi_{1,\text{vac}}, \phi_{2,\text{vac}})$. Unfortunately, due to the last term, V_0, one cannot obtain analytical results on the parameter dependence of string energy. For numerical data, see figures 4(a)–4(b). In figure 4(a), we have also drawn a dashed line marking the energy of the ANO string for $\beta = 2$, to show that the energy of the 2VEV strings tends smoothly to that of the ANO string when reaching the transition predicted by the stability condition (5).

In the 2VEV case one can include in the potential the $U(1) \times U(1)$ non-invariant terms

$$\frac{\gamma_1}{2} \left[(\phi_1^* \phi_2)^2 + (\phi_1 \phi_2^*)^2 \right] + \gamma_2 \left[\phi_1^* \phi_2 + \phi_1 \phi_2^* \right] . \tag{19}$$

Such terms appear in solid-state models (two-band superconductors [6, 7, 9–11]) and nonlinear optics [12]. In this case however, the phase of the fields cannot be changed separately, therefore the Ansatz has to be generalized as

$$\phi_1(r, \vartheta, z) = f_1(r) e^{in\vartheta}, \quad \phi_2(r, \vartheta, z) = e^{i\delta} f_2(r) e^{im\vartheta} . \tag{20}$$

By plugging in the above Ansatz into the equations of motion, it is clear that only $\delta = 0, \pi$ and $\pi/2$ are consistent with them. It is clear from the boundary conditions emerging from the condition that at infinity Φ has to take a value on the vacuum manifold, such cylindrically symmetric solutions exist if there is a vacuum solution $(\eta_1, \eta_2 e^{i\delta})$. If either $\gamma_1 = 0$ or $\gamma_2 = 0$, this condition is fulfilled. In those cases, one can proceed as shown above in the $U(1) \times U(1)$ symmetric cases. For details, see [16].

4 Conclusion

In the present paper we have presented twisted string solutions in two component Abelian Higgs models with $U(1)$ symmetric potentials. We have obtained the vortex solutions numerically, by integrating their radial (profile) equations.

We have shown that for the existence of the twisted string solutions, the potential has to be $U(1) \times U(1)$ symmetric, and at its minimum, exactly one of fields components has to be nonzero (1VEV case). In this case, the Abrikosov-Nielsen-Olesen strings can be embedded in the two component theory, and the twisted string solutions coexist with the ANO ones. We have also examined the bifurcation of twisted strings with embedded ones.

The energy per unit length of twisted strings decreases monotonously with decreasing twist, with the embedded string having the largest energy. We have shown that for a wide range of parameters, there is a zero twist limiting solution. As these are lowest energy solutions, they are presumably stable. We have also obtained the zero twist solutions numerically.

In those models where both scalar field components have a nonzero VEV (2VEV case), another type of strings can be found. In the most symmetric of these cases, it can be shown that the solutions of the Abelian Higgs model (with different parameters) can be embedded in the extended model, by taking the two scalar fields equal. Similar solutions to these have been found in theories with non-symmetric potentials as well.

Acknowledgements The author would like to thank Péter Forgács for permission to present results of joint work in the talk and for his help in preparing the manuscript for the proceedings.

The author would also like to thank the school organisers for the opportunity to participate at the school, and for the financial support. The work presented in the talk has been supported by OTKA grants No. NI68228 and K61636.

References

[1] E. Witten, *Superconducting string*, Nucl. Phys. **B 249** (1985) 557–592.

[2] T. Vachaspati and A. Achúcarro, *Semilocal cosmic strings*, Phys. Rev. **D 44** (1991) 3067.

[3] M. Hindmarsh, *Existence and stability of semilocal strings*, Phys. Rev. Lett. **68** (1992) 1263; *Semilocal topological defects*, Nucl. Phys. **B 392** (1993) 461-492.

[4] A. Achúcarro and T. Vachaspati, *Semilocal and electroweak strings*, Phys. Rept. **327** (2000) 427.

[5] D. Haws, M. Hindmarsh, N. Turok, *Superconducting strings or springs?*, Phys. Lett. **B 209** (1988) 255.

[6] M.E. Zhitomirsky and V.-H. Dao, *Ginzburg-Landau theory of vortices in a multigap superconductor*, Phys. Rev. **B 69** (2004) 054508.

[7] Victor Moshchalkov, Mariela Menghini, T. Nishio, Q.H. Chen, A.V. Silhanek, V.H. Dao, L.F. Chibotaru, N.D. Zhigadlo, and J. Karpinski, *Type-1.5 Superconductivity*, Phys. Rev. Lett. **102** (2009) 117001.

[8] Egor Babaev and Martin Speight, *Semi-Meissner state and neither type-I nor type-II superconductivity in multicomponent systems*, Phys. Rev. **B 72** (2005) 180502.

[9] Egor Babaev, Juha Jäykkä and Martin Speight, *Magnetic Field delocalization and Flux Inversion in Fractional Vortices in Two-Component Superconductors*, Phys. Rev. Lett. **103** (2009) 237002.

[10] J. Geyer, R.M. Fernandes, V.G. Kogan and J. Schmalian, *Interface energy of two band superconductors*, Phys. Rev. **B 82** (2010) 104521.

[11] E.H. Brandt and M.P. Das, *Attractive vortex interaction and the intermediate-mixed state of superconductors*, J. Supercond. Nov. Magn. (2010) 1-11.

[12] L.M. Pismen, *Vortices in Nonlinear Fields*, Oxford University Press, 1999.

[13] P. Forgács and N.S. Manton, *Space-time symmetries in gauge theories*, Commun. Math. Phys. **72** 15 (1980).

[14] Forgács, P., Reuillon, S. and Volkov, M.S., *Twisted superconducting semilocal strings*, Nucl. Phys. **B 751** (2006) 390–418.

[15] P. Forgács and Á. Lukács, *Instabilities of twisted strings*, JHEP **12** (2009) 064.

[16] P. Forgács and Á. Lukács, *String solutions in two component Abelian Higgs models*, in preparation.

[17] N.S. Manton and P.M. Sutcliffe, *Topological Solitons*, Cambridge University Press, 2004.

[18] E. B. Bogomol'nyi, *Stability of classical solutions*, Sov. J. Nucl. Phys. **24** (1976) 449; H. J. de Vega and F. A. Schaposnik, *A classical vortex solution of the Abelian Higgs model*, Phys. Rev. **D 14** (1976) 1100.

[19] A. Yung, *Vortices on the Higgs Branch of the Seiberg–Witten Theory*, Nucl. Phys. **B 562** (1999) 191-209.

Positronium Hyperfine Splitting

Akira Miyazaki,

Department of Physics, Graduate School of Science

and International Center for Elementary Particle Physics (ICEPP),

The University of Tokyo, 7-3-1 Hongo, Bunkyo-ku, Tokyo, 133-0033, Japan

miyazaki@icepp.s.u-tokyo.ac.jp

Abstract

Positronium is an ideal system for the research of QED in the bound state. The hyperfine splitting of positronium (Ps-HFS: about 203 GHz) is a good tool to test QED and also sensitive to new physics beyond the Standard Model. Previous experimental results show $3.9\,\sigma$ (15 ppm) discrepancy from the QED O $\left(\alpha^3 \ln 1/\alpha\right)$ prediction. We point out probable common systematic errors in all previous experiments. I measure the Ps-HFS in two different ways. (1) A prototype run without RF system is described first. (2) I explain a new direct Ps-HFS measurement without static magnetic field. The present status of the optimization studies and current design of the experiment are described. We are now taking data of a test experiment for the observation of the direct transition.

1   Introduction

Positronium (Ps), the electron-positron bound state, is a purely leptonic system. The energy difference between *ortho*-positronium (*o*-Ps, $^3S_1$ state) and *para*-positronium (*p*-Ps, $^1S_0$ state) [1] is called hyperfine splitting of positronium (Ps-HFS). It is a good target to study bound state QED precisely. The Ps-HFS value is approximately 203 GHz (0.84 meV), which is significantly larger than hydrogen HFS (1.4 GHz). About one third of this large value is contributed by a quantum oscillation as shown in Fig. 1: $o\text{-Ps} \to \gamma^* \to o\text{-Ps}$ [2] . Since some hypothetical particles, such as a milicharged particle, can participate in the quantum oscillation to shift Ps-HFS value, its precise measurement provides a probe into new physics beyond the Standard Model.

Measurements of the Ps-HFS have been performed in 70's and 80's [1, 2]. The results were consistent with each other, and combined precision of 3.3 ppm is obtained. They were consistent with O $\left(\alpha^2\right)$ calculation of the QED available at that time. The corrections of O $\left(\alpha^3 \ln 1/\alpha\right)$ have been calculated using NonRelativistic QED (NRQED) in 2000 [3]. The new prediction is $\Delta_{\text{HFS}}^{\text{th}} =$ 203.391 69(41) GHz, where the uncertainty is the unknown nonlogarithmic O $\left(\alpha^3\right)$ term estimated in an analogous way to the HFS of muonium. This

---

[1] Although *p*-Ps decays mainly into two photons with lifetime of 125 ps, it takes 142 ns for *o*-Ps to decay. This is because *o*-Ps can only decay into three photons which is strongly suppressed by invariant matrix and kinematics. Two photon decay of *o*-Ps is forbidden by C conservation.

[2] *Ortho*-Ps has the same quantum number as a photon.

calculated value differs from the measured value of $\Delta_{\mathrm{HFS}}^{\mathrm{exp}}$ =203.388 65(67) GHz by 3.9σ as shown in Fig. 2. This discrepancy may be due to common systematic errors in the previous measurements or to new physics beyond the Standard Model.

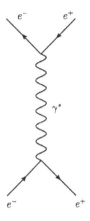

Figure 1: o-Ps contribution to Ps-HFS

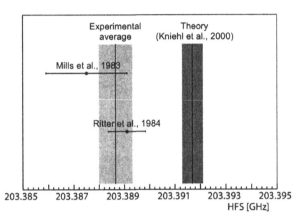

Figure 2: The discrepancy of Ps-HFS value

In all previous measurements, the Ps-HFS value was not directly measured, since 203 GHz was too high to produce and control. Zeeman splitting of o-Ps has been measured instead. A static magnetic field makes Zeeman mixing between $m_z = 0$ spin state of o-Ps and p-Ps. As a result, the energy level of $m_z = 0$ state of o-Ps becomes higher than $m_z = \pm 1$ state. This Zeeman splitting, which is approximately proportional to Ps-HFS, is a few GHz frequency under about 1 Tesla magnetic field. Static magnetic field is applied in RF cavities where positronium is produced. Zeeman transition from o-Ps of $m_z = \pm 1$ to o-Ps of $m_z = 0$ has been observed.

We point out the following three possibilities as the common systematic errors in these indirect measurements.

1. They may underestimate the non-uniformity of the magnetic field.

2. The unthermalized o-Ps contribution can result in an underestimation of the material effect.

3. RF systems to cause the transition might not be sufficiently stable.

Direct measurement of Ps-HFS without any magnetic fields is a main topic. I show the status of a prototype experiment in section 3. It is completely free from the systematic errors of the magnetic field. A proposal about the second possibility is summarized by A. Ishida, et $al.$ [4]. The third point is also promising. The experiment with quantum oscillation instead of the RF source is performed. I report the result of this experiment in section 2.

2 Measurement with quantum oscillation

We used a completely different method from previous one (i.e. without a RF source). This method is theoretically proposed by V.G. Baryshevsky et al.[5], in which a quantum oscillation between two Zeeman energy levels of o-Ps is measured in a static magnetic field. Positrons emitted from a β^+ source are polarized in the direction of their momentum due to parity violation in the weak interaction. Consequently, the resulting o-Ps is also highly polarized. This o-

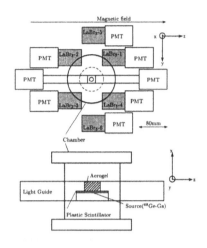

Figure 3: Schematic diagram of the setup

Ps is a superimposed state of above two energy eigenstates of o-Ps in a static magnetic field. When a perpendicular magnetic field of about 0.1 T is induced to the polarization axis of this o-Ps, this quantum oscillation can be detected as an observable *oscillation* in timing spectra of o-Ps decay.

This approach is quite different from the previous all experiments. It is free from systematic errors originating from the high power light source and the RF cavity with high Q-value. Instead of them, precise measuring technique of decay curve is crucial. Especially, time-to-digital converter (TDC) with high performance is essential for precise measurement of the timing spectrum. Therefore, these two different approaches are complementary. Both experimental approaches are necessary to understand the discrepancy.

Figures 3 show a schematic diagram of the experimental setup of quantum oscillation. The upper figure shows the entire experimental setup. The magnetic field direction is along the z-axis. The $LaBr_3(Ce)$ scintillators are placed in the yz-plane. They detect γ-rays with high energy resolution of 4.0 % (FWHM) at 511 keV and high timing resolution of 200 ps (FWHM). The direction of the β^+ emitted from the ^{68}Ge-Ga source is along the x-axis. The bold circle is a vacuum chamber. The coordinate system is also shown. The lower figure is a magnified view of the vacuum chamber, in which the ^{68}Ge-Ga source, the thin plastic scintillator and the silica aerogel are located. This plastic scintillator tags positrons, which go into the silica aerogel. Positronium is formed in this aerogel to decay into γ-rays. The signal from plastic scintillator comes at approximately the formation time of positronium. The time of the signal from $LaBr_3$ is the decay time. We took a delayed coinsidence between them to make a decay curve of positornium. The time is measured with direct clock TDCs (5 GHz: timing resolution of 200 psec). These TDCs have excellent integral and differential linearities.

Figures 4 show the measured time spectra at a magnetic value of 100 mT (left figure) and 135 mT (right figure). In both figures, the data points are plotted with error bars while the solid lines show the best fit results. A result of a prototype experiment is obtained as Δ_{HFS}^{exp} =203.324±0.039(stat.)±0.015(sys.) GHz. The accuracy is 200 ppm, which is an improvement by a factor of 90 over the previous experiment which used the similar method [6]. This result is consistent with both theoretical calculations and previous precision measurements of transition. However, we showed that we can improve this result to compete the most precise measurement with some simple improvements of our detection system. (in this paper [7])

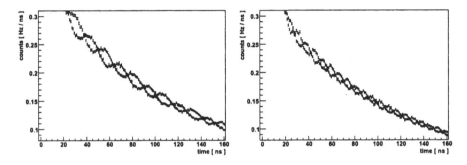

Figure 4: Timng spectra of the decay curve of o-Ps

3 Direct measurement of Ps-HFS

We are planing to directly measure the HFS transition, which does not need a static magnetic field. It is thus free from the systematic error from the field mentioned in section 1. However, the direct transition from o-Ps to p-Ps has very small probability of 3×10^{-8} s^{-1}, since this transition is $M1$ transition, and the Ps-HFs is extremely large. Therefore, a powerful radiational field of 203 GHz is essential so as to stimulate the direct transition.

The frequency of 203 GHz is just intermediate between optical light and radiowave. There was no high power light source for spectroscopy in sub-THz region. We are developing a new light system in this region. A frequency tunable radiational source is necessary to measure a whole shape of the resonance curve [3] . Our first target is to just observe the direct transition from o-Ps to p-Ps [4] . In order to accomplish this goal, we are developing following three new *optical* devices.

1. Sub-THz to THz light source called *gyrotron*,

2. Efficient transportation system of *mode converter*,

3. Parallel etalon with high-finesse called *Fabry-Pérot cavity*.

These are explained from the next subsection.

---

[3]RF system in previous precise indirect measurements of 3 GHz was not tunable, neither. They changed the strength of a magnetic field to shift the Zeeman splitting to tune effectively the resonance frequency of Zeeman transition.

[4]This can be detected as an increase of two photon-decay ratio.

3.1 Gyrotron

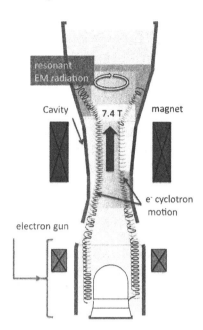

Figure 6: Picture of the gyrotron for the HFS experiment.

Figure 5: Schematic of gyrotron

The gyrotron[8] is a novel high power light source for sub-THz to THz frequency region. It has been developed in the field of nuclear fusion. The structure of gyrotrons is shown in Fig. 5. The electrons are emitted from the DC electron gun, concentrated and rotated as cyclotron motion in the superconducting magnet. The cyclotron frequency f_c is given by

$$f_c = \frac{eB}{2\pi m_0 \gamma}, \tag{1}$$

where B is the magnetic field strength, m_0 is the electron rest mass, and γ is the relativistic factor of the electron. A cavity is placed at the maximum magnetic field in which resonant frequency is given by

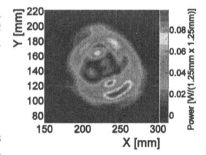

Figure 7: Power profile

$$f = \frac{1}{2\pi} \sqrt{\left(\frac{\chi_{mn}}{R}\right)^2 + \left(\frac{l\pi}{L}\right)^2}, \tag{2}$$

where R, L are radius and length of the cavity, respectively. l is an index of longitudinal mode. m, n are indices of transverse mode. χ_{mn} is a root of differential bessel function. This cavity frequency is tuned just to the cyclotron frequency to enhance the monochromatic light. The electrons stimulate cyclotron resonance maser in the cavity. The produced coherent photons are guided to the output port through the window, while electrons are dumped at a collector.

We developed a gyrotron operating at $f_c = 203$ GHz with $B = 7.364$ Tesla, $\gamma \sim 1.02$, which is shown in Fig. 6. The stable power of 300 W is obtained at the output window of gyrotron. The frequency width, which is determined by B uniformity and γ spread by thermal distribution of electrons, is expected to be less than 1 MHz. It is narrow enough to control the resonance at the Fabry-Pérot cavity. Measured result with a similar gyrotron shows the frequency width is less than 10 kHz [9]. Although the frequency can be tuned by changing the γ factor with different acceleration of electrons, the tuning range is limited by the resonant width of the cavity to several hundreds of megahertz.

Figure 7 shows the power profile of the radiation at taken with an infrared (IR) camera. The profile has a circular polarization called TE03 mode. Unfortunately, the mode inside Fabry-Pérot cavity is a linearly polarized *gaussian mode*. Therefore, the original gyrotron output cannot couple with Fabry-Pérot cavity. That's why a mode conversion is necessary to use gyrotron power efficiently.

3.2 Mode converter

Transportation system is composed of three parabolic mirrors called M0, M1 and M2 as shown in Fig. 8. The first parabolic mirror M0 converts polarization from circular to linear. M1 and M2 simply change the shape of power distribution from bi-gaussian to gaussian. Then, plain mirror M3 reflects radiaion and introduces it into Fabry-Pérot cavity. The power distribution is successfully converted into gaussian-like mode. Coupling between input light and Fabry-Pérot cavity is now about 60%. However, the transformation efficiency is 30%, because the output of the current gyrotron is not optimized. As a result, about 20% of the original radiation from gyrotron can resonate in the cavity.

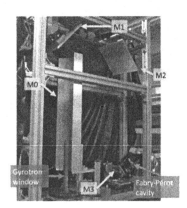

Figure 8: Transportation

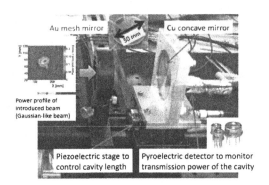

Figure 9: Fabry-Pérot cavity

3.3 Fabry-Pérot cavity

Photons produced at the gyrotron are transported and accumulated in a cavity. Since 203 GHz photons can be treated optically at the centimeter or larger size scale, we use a Fabry-Pérot cavity, which consists of two opposing mirrors to confine photons between them. Unlike RF cavities, the confinement in the Fabry-Pérot cavity is 1-dimensional while the other four sides are open as shown in Fig.9. A golden mesh mirror is used on the input side of the cavity to introduce photons from gyrotron. A copper concave mirror is used on the other side.

The two most important characteristics of a cavity are *finesse* $\mathcal{F}$ and *input coupling* C. With the reflectivity of mesh mirror R_f and concave mirror R_e, finesse is defined as

$$\mathcal{F} = \frac{\pi \left(R_f R_e (1 - A)\right)^{1/4}}{1 - \left(R_f R_e (1 - A)\right)^{1/2}}, \tag{3}$$

where A is the medium loss inside the cavity. Round-trip times N of photon in the cavity is given by

$$N = \mathcal{F}/2\pi, \tag{4}$$

Therefore finesse characterizes the capability of the cavity to store photons inside. To maximize the $\mathcal{F}$, power losses must be minimized. There are 3 types of loss, diffraction loss, medium loss and ohmic loss. With the confinement of photons by the concave mirror, diffraction loss is negligible in our cavity. Medium loss in gas[5] is measured as about 0.1%. Ohmic loss occurs at the mirrors, which is around 0.15% at the copper mirror and less than 1.0% at the mesh mirror.

Input coupling is the fraction of input power matched to the cavity mode. It is an important parameter to efficiently introduce photons into the cavity. In our cavity, the input coupling is determined by transmittance of the input mesh mirror.

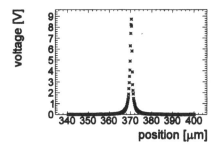
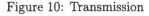
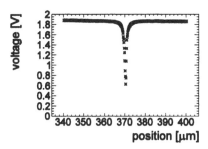

Figure 10: Transmission   Figure 11: Reflection

Figure 9 shows the test setup, which is comprised of a mesh mirror on the mirror mount and a concave mirror on a piezo stage. Transmitted power is measured through a small hole on the concave mirror. On the other hand, reflected power is measured outside of the mesh mirror. When we shifted the cavity length precisely by the piezo stage, Breit-Wigner resonance was observed in the transmitted power monitor as shown in Fig. 10. Here, the horizontal axis is a position of the piezo stage, and the vertical axis is the output of power monitor.

---

[5]Mixture of nitrogen 0.9 atm and isobutane 0.1 atm.

Finesse is obtained from the width of the resonance Γ. Finesse is calculated by

$$\mathcal{F} = \frac{\lambda/2}{\Gamma}, \tag{5}$$

where $\lambda = 1.47$ mm is wavelegth of 203 GHz. We got finesse of about 650, which is equivalent to 100 times round-trip according to Eq.4

Figure 11 shows the measured reflection power. Input coupling is given by

$$C = 1 - \frac{V_{peak}}{V_{baseline}}. \tag{6}$$

Here, V_{peak} is a voltage at peak decreasing from $V_{baseline}$, the voltage of the baseline of the reflected power. We achieved input coupling of 67%. This large value is mainly due to a good mode conversion explained in the last subsection. The current status of the power is summarized in Table 1. The power of 10 kW is accumulated in the cavity.

Table 1: The summary of radiation power with our devices

Device	Efficiency	power (W)
Gyrotron	1	300
Mode converter	0.30	90
Fabry-Pérot cavity	$0.60 \times 100 \times 2$	about 10,000

3.4 Detection System

Figure 12 shows a photograph of positronium production and signal detection system. Gyrotron power is introduced to the cavity via the mesh mirror, and accumulated inside Fabry-Pérot cavity. This cavity is placed inside a gas chamber filled with mixture gas of 0.9 atm nitrogen and 0.1 atm isobutane. The positron is emitted from β^+-source. The ^{22}Na β^+-source is located 20 mm above the cavity. In order to generate start timing, the emitted positrons pass through a β-tag scintillator, with thickness of 100μm. The signal from plastic scintillator transits through light guide made of acryl, to reach photomultiplier (PMT). Such kind of β-tagging system was also used in the experiment with quantum oscillation explained in section 2.

A lead collimator, with thickness of 10 mm, is placed under the plastic scintillator so as to select the positrons which go into the cavity. It also works as a shield to protect LaBr$_3$(Ce) scintillators from accidental photons [6].

Positron forms positronium with an electron in the gas. *Para*-positronium annihilates into two 511 keV photons immediately, while o-Ps remains with lifetime of 142 ns to decay into three photons, whose energy are continuous and less than 511 keV. A signal of the transition from o-Ps to p-Ps under 203 GHz is a *delayed-two-photon* event. Four LaBr$_3$ scintillators surround the chamber to detect photons. Two photon-decay can be easily separated from three photon-decay with this energy information. The LaBr$_3$ scintillators also have good

---

[6]They are mainly 1275 keV and 511 keV photons emitted around the source.

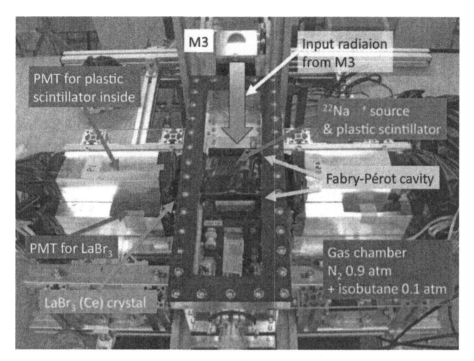

Figure 12: A photograph of gas chamber and detection system

timing resolution to separate delayed events (i.e. signal of transition) from prompt events [7] . to improve signal to background ratio significantly [8] .

The signal collection efficiency and background rates were estimated using Monte Carlo simulation (GEANT4). There are three major background processes. The first one is an three γ contamination from o-Ps. The second one is a pick-off background. A positron in o-Ps interacts with a electron in a matter [9] only to annihilate into two photons. This process is called pick-off annihilation, and becomes background in our measurement. The last one is an accidental pileup process. In order to eliminate these backgrounds, we selected back-to-back signal in LaBr$_3$ scintillator and imposed condition that smeared energy deposit is 511keV$\pm 3\sigma$.

The obtained power of 10 kW is used for the simulation. Figure 13 shows an expected spectra for one month of data taking. The estimated rate is also summarized in Table 2. In this table, "ON" means the signal under 203 GHz radiation while "OFF" means that without radiation [10] . The main background is three γ contamination. And the other two backgrounds are the same size of the signal. We can clearly see the transition within one month.

---

[7]Almost all the prompt events are two photon-decay.

[8]In case of the power of 10 kW in the cavity, S/N is estimated to be improved 16 times, when a timing window is imposed from 50 to 250 ns in decay curve.

[9]In this case nitrogen and isobutane.

[10]In Table 2, total (ON) - total (OFF) is not equal to signal. Because, the background events associated with o-Ps decreases under high power resonance radiation (i.e. ON). A Part of o-Ps transits into p-Ps to decay earlier than the timing window. As a result, total (ON) - total (OFF) becomes less than expected signal rate.

Table 2: The summary of signal estimation

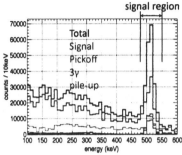

Figure 13: Signal estimation

process	rate (mHz)
signal	63
pickoff	93
3γ	300
pileup	100
total (ON)	560
total (OFF)	530

4   Summary

There is a large discrepancy between theory and experiment in Ps-HFS value. We suspect some common systematic errors in previous experiments. The prototype experiments without RF system was already performed. It was a complementary method against previous experiments, and the accuracy of 200 ppm was obtained. We are now tackling the direct measurement without any magnetic fields. It is the first trial for sub-THz spectroscopy with $M1$ transition. We have developed a high power 203 GHz radiation source called gyrotron, mode converter and Fabry-Pérot cavity. Monte Carlo simulation of the detection system shows that the observation of Ps-HFS is feasible. We are now taking data of a test experiment for the direct transition from o-Ps to p-Ps. The signal is expected to be observed in a month.

These experiments are collaborated with Y. Sasaki, A. Ishida, T. Yamazaki, T. Suehara, T. Namba, S. Asai, T. kobayashi, H. Saito, M. Yoshida, K. Tanaka, M. Ikeno, A. Yamamoto, T. Idehara, I. Ogawa, Y. rushizaki and S. Sabchevski.

References

[1] A. P. Mills, Jr. *et al.* Phys. Rev. Lett. **34** (1975) 246; A. P. Mills, Jr. *et al.* Phys. Rev. A **27** (1983) 262.

[2] M. W. Ritter, *et al.* Phys. Rev. A **30** (1984) 1331.

[3] B. A. Kniehl and A. A. Penin, Phys. Rev. Lett. **85** (2000) 5094.

[4] A. Ishida, G. Akimoto, Y. Sasaki, *et al.* arXiv:1004.5555 (2010).

[5] V. G. Baryshevsky *et al.* J. Phys. B **22** (1989) 2835.

[6] S. Fan, *et al.* Phys. Lett. A **216** (1996) 129.

[7] Y. Sasaki, A. Miyazaki, *et al.* arXiv:1002.4567 (2010), *accepted by Phys. Lett. B.*

[8] T. Idehara, et al. IEEE Trans. Plasma Sci. **27** (1999) 340.

[9] T. Idehara, et al. *Appl. Magn. Reson.* **34** (2008) 265.

How I Failed to Find
Any New Fundamental Particles

Michael Mulhearn*

December 15, 2010

Abstract

We consider two attempts to discover new fundamental particles: a combination of searches for Standard Model Higgs boson production at DØ and an earlier search for direct production of Dirac magnetic monopoles at CDF. No new fundamental particles are discovered and in each case 95% CL upper limits on the cross-section are presented.

1   Introduction

In the standard model (SM), the spontaneous breakdown of the electroweak symmetry generates masses for the W and Z bosons and produces a scalar massive particle, the Higgs boson, which has so far eluded detection. The discovery of a Higgs boson would top a remarkable list of experimentally confirmed SM predictions.

The existence of magnetic monopoles would add symmetry to Maxwell's equations and would make charge quantization a consequence of angular momentum quantization, as first shown by Dirac [1]. While not predicted to exist at energies accessible to modern colliders, the existence of such magnetic monopoles would not violate any known physical law. Despite numerous experimental searches, Magnetic monopoles have also not been discovered [2].

2   Searches for the SM Higgs

The mass of the Higgs m_H is unknown, but indirect constraints from electroweak measurements and direct searches for the Higgs boson [3]

---

*On Behalf of the CDF and DØ Collaborations

suggest that the value is most likely between 100 and 200 GeV, the region currently probed at the Fermilab Tevatron. In this proceeding, we combine the results of direct searches for SM Higgs bosons in $p\bar{p}$ collisions at $\sqrt{s} = 1.96$ TeV recorded by the DØ experiment [4]. The combination [5] includes searches for Higgs bosons produced through gluon-gluon fusion (GGF, $gg \to H$), vector boson fusion (VBF, $q\bar{q} \to q'\bar{q}'H$), and in association with vector bosons ($q\bar{q} \to VH$). The analyses use data corresponding to integrated luminosities ranging from 2.1 to 6.7 fb^{-1}, collected from 2002 to 2010. The primary Higgs boson decay modes examined are $H \to b\bar{b}$, $H \to W^+W^-$, $H \to \tau^+\tau^-$ and $H \to \gamma\gamma$. The searches are organized into analysis subsets comprising different Higgs production and decay modes, each designed to isolate a particular final state. The analyses were designed to be mutually exclusive after the event selections.

For $m_H < 135$ GeV the primary Higgs decay is $H \to b\bar{b}$. The GGF and VBF production channels are not considered for this decay mode, as the fully hadronic final state suffers from prodigious background at the Tevatron. But the associated production channels $WH \to \ell\nu b\bar{b}$, $ZH \to \nu\bar{\nu}b\bar{b}$ and $ZH \to \ell\ell b\bar{b}$, where $\ell = e, \mu$, have reconstructed leptons or missing transverse energy in the final state and therefore smaller backgrounds. To distinguish the decay $H \to b\bar{b}$ from background processes involving light quarks or gluons, jets are identified as likely containing b-quarks (b-tagged) if they pass loose or tight requirements on the output of a multivariate analysis treated to separate b-jets from light-jets [6]. The candidate events are classified as double-tagged (DT) if at least two jets are b-tagged and single-tagged (ST) if only one jet is b-tagged. For these analyses, each lepton flavor of the V boson decay ($\ell = e, \mu$) is treated as an independent channel. Additional sensitivity is obtained from the $t\bar{t}H \to t\bar{t}b\bar{b}$ channel which examines events with up to three b-tagged jets.

For $m_H > 135$ GeV the primary Higgs decay is $H \to W^+W^-$. In this decay mode, the GGF and VBF production mechanism have reasonable background when at least one of the W bosons decays leptonically. We consider three final states of opposite-signed leptons: $WW \to e^+\nu e^-\nu$, $e^\pm\nu\mu^\mp\nu$, and $\mu^+\nu\mu^-\nu$. The $H \to e^\pm\nu\mu^\mp\nu$ analysis further separates events in three final states with 0 jets, 1 jet, and two or more jets. A separate analysis considers the $H \to W^+W^- \to \ell\nu q\bar{q}$ process. In all $H \to W^+W^-$ decays with $m_H < 2M_W$, at least one of the W bosons will be off mass shell. There is a small contribution from $H \to ZZ$ decays, particularly in the $H \to W^+W^- \to ee/\mu\mu\nu\nu$ searches. In all cases, lepton selections include both electrons and muons ($\ell = e, \mu$), while τ leptons are included in the simulation and the selections have acceptance for secondary leptons from $\tau \to e/\mu$ decays. For

Table 1: List of analysis channels and corresponding integrated luminosity L. In cases where sub-channels use different datasets, a range of integrated luminosities is presented.

Channel	L (fb^{-1})	Channel	L (fb^{-1})
$WH{\rightarrow}\ell\nu b\bar{b}$,	5.3	$H{\rightarrow}W^+W^-{\rightarrow}e\nu\mu\nu$,	6.7
$ZH{\rightarrow}\nu\bar{\nu}b\bar{b}$,	5.2-6.4	$H{\rightarrow}W^+W^-{\rightarrow}ee/\mu\mu\nu\nu$	5.4
$ZH{\rightarrow}\ell\ell b\bar{b}$,	4.2-6.2	$H{\rightarrow}W^+W^-{\rightarrow}\ell\nu q\bar{q}$	5.4
$H{\rightarrow}\gamma\gamma$	4.2	$VH{\rightarrow}VW^+W^-$	5.3
$t\bar{t}H{\rightarrow}t\bar{t}b\bar{b}$	2.1	$X+H{\rightarrow}\tau\tau b\bar{b}/q\bar{q}\tau\tau$	4.9

$VH{\rightarrow}VW^+W^-$ production, we search for leptonic V boson decays with three final states of same-signed leptons: $VWW \rightarrow e^\pm e^\pm + X$, $e^\pm\mu^\pm + X$, and $\mu^\pm\mu^\pm + X$.

The combined senstivity is improved by including searches optimized for additional Higgs boson decay modes. The $X+H{\rightarrow}\tau\tau b\bar{b}/q\bar{q}\tau\tau$ analysis selects the $\tau\tau$ plus dijet final state with one τ decaying to μ and the other decaying hadronically. This analysis is sensitive to $ZH \rightarrow \tau\tau b\bar{b}$, $VH \rightarrow q\bar{q}\tau\tau$, GGF and VBF. The $H \rightarrow \gamma\gamma$ analysis includes signal contributions from GGF, VBF, and associated production.

The analyses used in this combination are listed in Table 1. Since the previous DØ SM Higgs combination [7], we have updated the $WH{\rightarrow}\ell\nu b\bar{b}$, $ZH{\rightarrow}\nu\nu bb$, $ZH{\rightarrow}\ell\ell b\bar{b}$, $VH{\rightarrow}VW^+W^-$, and $H{\rightarrow}W^+W^-{\rightarrow}e\nu\mu\nu$ analyses. The $H{\rightarrow}W^+W^-{\rightarrow}\ell\nu q\bar{q}$ channel is a new addition to the combination.

3   Final Discriminants and Systematic Uncertainties

The $H \rightarrow \gamma\gamma$ channel uses the diphoton mass as the final discriminant used to search for a Higgs signal, while the ttH channel uses the scalar sum of the transverse momentum from the leading jets in the event. For all other channels, multivariate analyses are trained to discriminate signal from background, with the output distributions taken as the final discriminant. Systematic uncertainties are quantified by their effect on the final discriminant.

The systematic uncertainties differ between analyses for both the signals and backgrounds. Here we summarize only the largest contributions. Most analyses carry an uncertainty on the integrated luminosity of 6.1% [8], but when possible the overall normalization is instead determined from the NNLO Z/γ^* cross section using data events near the peak of $Z \to \ell\ell$ decays. The $H \to b\bar{b}$ analyses have an uncertainty on the b-tagging rate of 1-9%. These analyses also have an uncertainty on the jet measurement and acceptances of $\sim 7\%$. All analyses include uncertainties associated with lepton measurement and acceptances, which range from 1-5% per lepton in the final state. The largest contribution for all analyses is the uncertainty on the background cross sections at 6-30% depending on the analysis channel and specific background. These values include both the uncertainties on the theoretical cross section calculations and the uncertainties on the higher order correction factors. The uncertainty on the expected multijet background is dominated by the statistics of the data sample from which it is estimated, and is considered separately from the other cross section uncertainties. Several analyses incorporate shape-dependent uncertainties on the kinematics of the dominant backgrounds, derived from the potential deformations of the final variables due to generator and background modeling uncertainties. Further details on the systematic uncertainties are provided in Ref. [5].

4   The SM Higgs Combination

The outcome of an experiment is more consistent with either the signal-plus-background (S+B) hypothesis or the background-only (B) hypothesis as quantified by the ratio of the Poisson likelihoods for each hypothesis. Distributions for the log likelihood ratio (LLR) are obtained by generating Poisson fluctuations of the B and S+B hypotheses. By integratring these LLR distributions up to the LLR for the experiment, we determine the confidence intervals CL_B and CL_{S+B}. Limits are set by adjusting the signal cross section until $CL_S = CL_{S+B}/CL_B = 1 - \alpha$, with a 95% CL limit obtained for $\alpha = 0.95$ [9].

All systematic uncertainties originating from a common source are treated as correlated in the combination. For the background rate, they are generally comparable to the signal expectation, so the treatment of systematics is an important part of the limit calculation. The systematic uncertainties are included as nuisance parameters which adjust the B and S+B predictions. To minimize their impact on the sensitivity, the likelihood of the B and S+B hypotheses used to calculate each LLR value are first maximized by independent fits of the nuisance parameters.

As no significant signal-like excess is observed, we derive limits on the SM Higgs boson production $\sigma \times B(H \to X)$. To accomodate diverse contributions from processes with different cross-sections and sensitivities, we present our results in terms of the ratio of 95% CL upper cross section limits to the SM predicted cross section as a function of Higgs boson mass. The SM prediction for Higgs boson production would be considered excluded at 95% CL when this limit ratio falls below unity. Figure 1 shows the expected and observed 95% CL cross section limit as a ratio to the SM cross section in the probed mass region ($100 \leq m_H \leq 200$ GeV), with all analyses combined. These results are also summarized in Table 2. The expected and observed LLRs are shown in Figs. 1 and 2.

Table 2: Combined 95% CL upper limits on $\sigma \times B(H \to X)$ for SM Higgs boson production. The limits are reported in units of the SM production cross section times branching fraction.

m_H (GeV)	100	105	110	115	120	125	130	135	140	145	150
Expected:	1.80	1.86	2.13	2.31	2.60	2.67	2.82	2.59	2.40	2.19	1.87
Observed:	1.40	1.69	1.44	2.65	3.50	4.16	3.16	4.17	3.53	3.29	2.43

m_H (GeV)	155	160	165	170	175	180	185	190	195	200
Expected:	1.62	1.21	1.14	1.36	1.60	1.92	2.40	2.93	3.40	3.96
Observed:	1.93	1.17	1.03	1.10	1.35	1.86	2.86	3.27	4.44	4.97

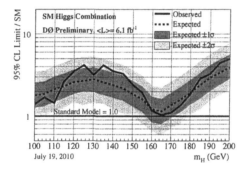

Figure 1: Expected (median) and observed 95% CL cross section upper limit ratios for the combined analyses as a function of m_H. The bands indicate the 68% and 95% probability regions where the limits can fluctuate in the absence of signal.

322

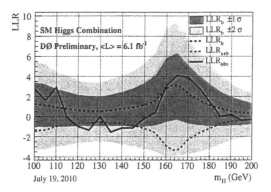

Figure 2: Log-likelihood ratio distribution for the combined analyses as a function of m_H. The bands indicate the 68% and 95% probability regions where the limits can fluctuate in the absence of signal.

5 Search for Magnetic Monopoles

Magnetic monopoles have magnetic charge g satisfying the Dirac quantization condition:

$$\frac{ge}{\hbar c} = \frac{n}{2} \iff \frac{g}{e} = \frac{n}{2\alpha} \approx 68.5 \cdot n$$

where n is an integer and α is the fine structure constant. We consider an $n = 1$ monopole with mass less than 1 TeV/c^2, spin $\frac{1}{2}$, and no hadronic interactions. Monopoles are accelerated by a magnetic field and are highly ionizing due to the large value of g/e.

The monopole search [10] uses a 35.7 pb^{-1} sample of $p\bar{p}$ collisions at $\sqrt{s} = 1.96$ TeV produced by the Fermilab Tevatron and collected by the CDF II detector during 2003 using a special trigger. The detector consists of a magnetic spectrometer including silicon strip and drift-chamber tracking detectors and a scintillator time-of-flight system, surrounded by electromagnetic and hadronic calorimeters and muon detectors [11]. CDF uses a superconducting solenoid to produce a 1.4 T magnetic field. The field is parallel to the beam direction, which is taken as the z direction, with ϕ the azimuthal angle, and r the radial distance in the transverse plane.

The important detector components for this search are the central outer tracker (COT) [12] and the time-of-flight (TOF) detector [13], both positioned inside the solenoid. The coverage of the cylindrical COT extends from a radius of 40 cm to 137 cm and to pseudo-rapidity $|\eta| \sim 1$. The COT consists of eight superlayers, each containing 12 layers of sense wires. The COT makes position measurements for track reconstruction as well as integrated charge measurements for determining a particle's ionization energy loss dE/dx. The COT is surrounded by 216 TOF scintillator bars, which run parallel to the beam line and form a cylinder of radius 140 cm. Each TOF bar is instrumented with a photomultiplier tube (PMT) on each end. The TOF measures both the time and height of PMT pulses; the pulse height is typically used to correct for discriminator-threshold time slewing. Due to their large ionization and massive production of delta rays, monopoles in scintillator with velocity $\beta > 0.2$ are expected to produce more than 500 times the light from a minimum-ionizing particle (MIP) [2].

We have built and commissioned a highly ionizing particle trigger [14] that requires large light pulses at both ends of a TOF scintillator bar. The trigger was designed to detect monopoles efficiently while consuming less than 1 Hz of the CDF data acquisition bandwidth. The trigger thresholds of about 30 MIPs are well below the expected response to a monopole and have a negligible effect on the trigger efficiency.

In the CDF detector, a monopole is accelerated along the uniform solenoidal magnetic field in a parabola slightly distorted by relativistic effects. The TOF acceptance is estimated from Monte Carlo simulation, by extending the GEANT simulation [15, 16, 17, 18] to handle magnetic monopoles, including the acceleration from the magnetic field, energy loss and multiple scattering. When calculating the TOF acceptance, we simulate the heuristic pair production mechanism proposed by Ref. [19]. The TOF acceptance for monopole pairs simulated with GEANT is shown in Figure 3.

Monopoles curve in the rz plane, in sharp contrast to electrically charged particles, which curve in the $r\phi$ plane. A specialized reconstruction program isolates monopole candidates using data from the COT. Candidates consist of coincident track segments composed entirely of hits with large ionization, consistent with a straight line in the plane perpendicular to the magnetic field.

The COT electronics encodes the integrated charge as the width of a hit, which is the ionization measurement used for monopole candidate selection. A typical MIP produces hit widths of about 20 ns. An extrapolation of the non-linear COT response for ordinary particles predicts that monopoles would produce hit widths of about 230 ns (1000 MIPs), still within the dynamic range of the COT. We do not use this extrapolation. Instead we cut in the tail of the width distribution from ordinary tracks, found to be at 140 ns (50 MIPs) in minimum-bias data collected with an open trigger highly efficient for inelastic $p\bar{p}$ collisions. Hits with charge below this amount are not

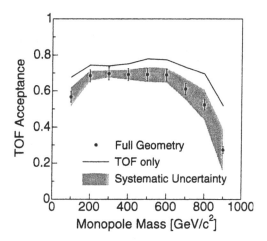

Figure 3: The acceptance of the TOF for monopole pairs, as a function of monopole mass. The band indicates the total systematic uncertainty.

considered by the monopole reconstruction.

The default COT tracking algorithm first reconstructs track segments in each of eight superlayers. It checks for hits loosely consistent with a straight-line, using a tolerance of 20 ns. The identified hits in each segment are then fit to a circular trajectory. In the monopole algorithm, the segments are required to be composed entirely of high-ionization hits. Also, because a monopole can be as slow as $\beta \sim 0.1$ with changing transverse velocity, the usual timing assumption ($t_{\text{flight}} = r/c$) cannot be used. Instead, the time of flight to each superlayer is varied between r/c and $10r/c$ in 5 ns increments.

A monopole candidate consists of several ϕ-coincident, low-curvature segments. From Monte Carlo simulation, we choose a loose cut on the segment curvature $\rho < 0.001$ cm^{-1}, which for an electron would correspond to $p_T > 4$ GeV/c. Likewise, the ϕ tolerance is a loose 0.2 radians. The remaining cuts are on the minimum number of hits needed in a segment and on the total number of ϕ-coincident segments required for a monopole candidate. By ignoring the width cut, the segment-finding algorithm efficiency is measured in an independent data sample using high-p_T tracks. In this manner, we choose a highly efficient cut requiring seven coincident superlayers with at least eight hits in each segment. This has a 94% efficiency with a 1% statistical uncertainty. For these cuts, the efficiency for finding high-mass monopole pairs calculated with the Monte Carlo simulation is

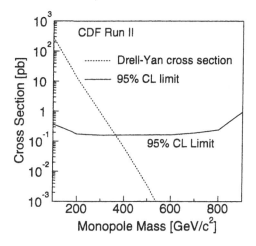

Figure 4: The 95% CL cross-section upper limit versus magnetic monopole mass. The theory curve for Drell-Yan monopole pair production intersects at the mass limit $m > 360$ GeV/c^2.

nearly 100%. The efficiency for high-p_T electrons in simulation, after removing the width cut, is also nearly 100%.

None of the $130,000$ events from the monopole trigger sample passes the candidate requirements, and we report a limit [18]. Monopole production limits are typically reported by the cross-section upper limit as a function of monopole mass to minimize the dependence on a particular production model. The cross-section exclusion limit is shown in Figure 4. The limit excludes monopole pair production for cross sections greater than 0.2 pb at the 95% confidence level for monopole masses between 200 and 700 GeV/c^2. For the Drell-Yan mechanism, this implies a mass limit of $m > 360$ GeV/c^2 at the 95% confidence level. In the nearly five years since its publication [10], this result has remained the most stringent direct experimental limit on magnetic monopole production.

6   Acknowledgments

The author thanks the CDF and DØ Collaborations for providing the results presented here.

References

[1] P. A. M. Dirac, Proc. Roy. Soc. Lond. **A133**, 60 (1931).

[2] G. Giacomelli and L. Patrizii (2003), hep-ex/0302011.

[3] R. Barate et al., Phys. Lett. B **565**, 61 (2003);
The LEP Electroweak Working Group,
http://lepewwg.web.cern.ch/LEPEWWG/.

[4] V. Abazov et. al., Nucl. Instrum. Meth. A **565**, 463 (2006).

[5] DØ Collaboration, DØ Note 6094-CONF.

[6] T. Scanlon, FERMILAB-THESIS-2006-43.

[7] DØ Collaboration, DØ Note 6008-CONF.

[8] T. Andeen et al., FERMILAB-TM-2365.

[9] W. Fisher, FERMILAB-TM-2386-E.

[10] A. Abulencia et al., Phys. Rev. Lett. **96** 201801 (2006);

[11] R. Blair et al., FERMILAB-PUB-96-390-E.

[12] T. Affolder et al., Nucl. Instrum. Meth. **A526**, 249 (2004).

[13] C. Grozis et al., Int. J. Mod. Phys. **A16S1C**, 1119 (2001).

[14] G. Bauer et al., FERMILAB-PUB-05-331-E.

[15] R. Brun, F. Bruyant, M. Maire, A. C. McPherson, and P. Zanarini, *GEANT3*, CERN-DD/EE/84-1.

[16] G. Bauer et al., Nucl. Instrum. Meth. **A545**, 503 (2005).

[17] P. Schieferdecker, Undergraduate thesis, Ludwig Maximillian University (2001).

[18] M. J. Mulhearn, Ph.D. thesis, Massachusetts Institute of Technology (2004).

[19] G. R. Kalbfleisch, W. Luo, K. A. Milton, E. H. Smith, and M. G. Strauss, Phys. Rev. **D69**, 052002 (2004), hep-ex/0306045.

DIPLOMAS

Twenty five Diplomas were open for competition among the participants. They have been awarded as follows:

- John S. BELL Diploma to:

  Leron BORSTEN
  Imperial College, London. UK

- Patrick M.S. BLACKETT Diploma to:

  Mohd Danish AZMI
  Aligarh Muslim University, India

- Nicola CABIBBO Diploma to:

  Nicola AMBROSETTI
  Bern University, Switzerland

- James CHADWICK Diploma to:

  Orest HRYCYNA
  Jagellonian University, KRAKÓW, Poland

- Sidney COLEMAN Diploma to:

  Lasma ALBERTE
  LMU, MÜNCHEN, Germany

- Richard H. DALITZ Diploma to:

  Petr DUNIN-BARKOWSKI
  ITEP, MOSCOW, Russia

- Paul A.M. DIRAC Diploma to:

  Philipp BURDA
  ITEP, MOSCOW, Russia

- Bruno FERRETTI Diploma to:

  Dmitriy GALAKHOV
  ITEP, MOSCOW, Russia

- <u>Vladimir N. GRIBOV</u> Diploma to:

  Sergey MIRONOV
  ITEP, MOSCOW, Russia

- <u>Robert HOFSTADTER</u> Diploma to:

  Michael MULHEARN
  Fermi Lab, BATAVIA, IL, USA

- <u>Gunnar KÄLLEN</u> Diploma to:

  Vasyl ALBA
  ITEP, MOSCOW, Russia

- <u>Seymour J. LINDENBAUM</u> Diploma to:

  Gianluca INGUGLIA
  University of Groningen. The Netherlands

- <u>Yuval NE'EMAN</u> Diploma to:

  Tristan DENNEN
  University of California at Los Angeles – UCLA, USA

- <u>Giuseppe P.S. OCCHIALINI</u> Diploma to:

  Roberto PREGHENELLA
  INFN and University of Bologna, Italy

- <u>Oreste PICCIONI</u> Diploma to:

  Abram KRISLOCK
  Texas A&M University, TX, USA

- <u>Bruno PONTECORVO</u> Diploma to:

  Regina RESCIGNO
  University of Salerno, Italy

- <u>Giampietro PUPPI</u> Diploma to:

  Chris PAGNUTTI
  University of Western Ontario, London, ON, Canada

- <u>Isidor I. RABI</u> Diploma to:

  Akira MIYAZAKI
  The University of Tokyo, Japan

- Giulio RACAH Diploma to:

  Bianca Letizia CERCHIAI
  University of Milano, Italy

- Bruno ROSSI Diploma to:

  Alessandro GRELLI
  Utrecht University, The Netherlands

- Julian S. SCHWINGER Diploma to:

  Maximilian M. SCHMIDT-SOMMERFELD
  CERN, Geneva, Switzerland

- Victor F. WEISSKOPF Diploma to:

  Andrey KHMELNITSKIY
  CERN, Geneva, Switzerland

- Eugene P. WIGNER Diploma to:

  Arpad Laszlo LUKACS
  Hungarian Academy of Sciences, Budapest, Hungary

- Bjorn H. WIIK Diploma to:

  Davide GERBAUDO
  Princeton University, Princeton, NJ, US

- Chien Shiung WU Diploma to:

  Yaxian MAO
  CERN, Geneva, Switzerland

AWARDS

The Awards for the best 'New Talent – Speakers' have been attributed to:

i) Leron BORSTEN
Imperial College London, London, UK
for his original presentation in
MATHEMATICAL PHYSICS

ii) Akira MIYAZAKI
The University of Tokyo, Japan
for his original presentation in
DETECTOR PHYSICS

iii) Philipp BURDA
ITEP, Moscow, Russia
for his original presentation in
THEORETICAL PHYSICS

iv) Michael MULHEARN
Fermi Lab, Batavia, IL, US
for his original presentation in
EXPERIMENTAL PHYSICS

Award for <u>Best Scientific Secretary</u> given to:
Maximilian M. SCHMIDT-SOMMERFELD
CERN, Geneva, Switzerland

Award for <u>Best Question</u> given to:
Yaxian MAO
CERN, Geneva, Switzerland

Award for <u>Best Student</u> given to:
Philipp BURDA
ITEP, Moscow, Russia

PARTICIPANTS

Vasyl ALBA
Ukraine

ITEP
MOSCOW, Russia

Lasma ALBERTE
Latvia

LMU
MÜNCHEN, Germany

Nicola AMBROSETTI
Italy

Bern University
BERN, Switzerland

Richard ARNOWITT
USA

Texas A&M University
COLLEGE STATION, TX, USA

Alexey ARTSUKEVICH
Russia

P.N. Lebedev Institute of Physics – FIAN
MOSCOW, Russia

Mohd Danish AZMI
India

Aligarh Muslim University
ALIGARH, India

Zvi BERN
USA

University of California at Los Angeles – UCLA
LOS ANGELES, CA, US

Alessandro BETTINI
Italy

INFN and University of Padova
PADOVA, Italy *and*
Canfranc Underground Laboratory
CANFRANC, Spain

Leron BORSTEN
The Netherlands

Imperial College London
LONDON, UK

Raphael BOUSSO
USA

Lawrence Berkeley National Laboratory
BERKELEY, CA, US

Philipp BURDA
Russia

ITEP
MOSCOW, Russia

Bianca Letizia CERCHIAI
Italy

University of Milano
MILANO, Italy

Luisa CIFARELLI
Italy

University of Bologna
BOLOGNA, Italy

Fabrizio COCCETTI
Italy

Enrico Fermi Centre
ROME, Italy

Yves DÉCLAIS
France

LAPP/IN2P3/CNRS
ANNECY LE VIEUX Cedex, France

Tristan DENNEN
USA

University of California at Los Angeles – UCLA
LOS ANGELES, CA, US

Ievgen DUBOVYK
Ukraine

National Academy of Sciences of Ukraine
KHARKOV, Ukraine

Michael James DUFF
UK

Imperial College London
LONDON, UK

Petr DUNIN-BARKOWSKI
Russia

ITEP
MOSCOW, Russia

Sergio FERRARA
Italy

CERN, GENEVA, Switzerland
and
LNF–INFN
FRASCATI, Italy

Pavlo FROLOV
Ukraine

National Academy of Sciences of Ukraine
KHARKOV, Ukraine

Dmitriy GALAKHOV
Russia

ITEP
MOSCOW, Russia

Davide GERBAUDO
Italy

Princeton University
PRINCETON, NJ, US

Steve GIDDINGS
USA

University of California
SANTA BARBARA, CA, US

Alessandro GRELLI
Italy

Utrecht University
UTRECHT, The Netherlands

Barbara GUERZONI
Italy

INFN and University of Bologna
BOLOGNA, Italy

Dieter HAIDT
Germany

EPJC–DESY
HAMBURG, Germany

Rolf-Dieter HEUER
Germany

CERN
GENEVA, Switzerland

Orest HRYCYNA
Poland

Jagellonian University
KRAKÓW, Poland

Gianluca INGUGLIA
Italy

University of Groningen☐
GRONINGEN, The Netherlands

Juansher JEJELAVA
Georgia

F. Andronikashvili Institute of Physics
TBILISI, Georgia

Andrey KHMELNITSKIY
Russia

CERN
GENEVE, Switzerland

Christian P. KORTHALS-ALTES
The Netherlands

CNRS–Luminy
MARSEILLE, France

Abram KRISLOCK
Canada

Texas A&M University
COLLEGE STATION, TX, US

Nadiia KRUPINA
Ukraine

Kharkov Physics and Technology Institute
KHARKOV, Ukraine

Jakub KUCZMARSKI
Poland

Warsaw University
WARSZAWA, Poland

Maxim KURKOV
Russia

St. Petersburg State University
ST. PETERSBURG, Russia

Shih-Chang LEE
Taiwan

Institute of Physics of Academia Sinica
TAIPEI, Taiwan

Arpad Laszlo LUKACS
Hungary

H.A.S. C.R.I.P. Hungarian Academy of Scie
BUDAPEST, Hungary

Sharareh MAJIDI

University of Helsinki

Iran

Yaxian MAO
China

Alessio MARRANI
Italy

Sergey MIRONOV
Russia

Akira MIYAZAKI
Japan

Michael MULHEARN
USA

Francesco NOFERINI
Italy

Piermaria J. ODDONE
USA

Antonio ORTIZ VELASQUEZ
Mexico

Chris PAGNUTTI
Canada

Vladyslav P. PAUK
Ukraine

Dmitriy PONOMAREV
Russia

Alexandr POPOLITOV
Russia

Roberto PREGHENELLA
Italy

HELSINKI, Finland

CERN
GENEVA, Switzerland

Stanford University
STANFORD, CA, US

ITEP
MOSCOW, Russia

The University of Tokyo
TOKYO, Japan

Fermi Lab
BATAVIA, IL, US

Enrico Fermi Centre
ROME, Italy

Fermi Lab
CHICAGO, IL, US

Instituto de Ciencias Nucleares – UNAM
MEXICO City, Mexico

University of Western Ontario
LONDON, ON, Canada

Taras Shevchenko National Kiev University
KIEV, Ukraine

P.N. Lebedev Institute of Physics – FIAN
MOSCOW, Russia

ITEP
MOSCOW, Russia

Enrico Fermi Centre
ROME, Italy
and
INFN and University of Bologna

BOLOGNA, Italy

Regina RESCIGNO
Italy

Università degli Studi di Salerno
SALERNO, Italy

Mikhail SALYKIN
Russia

St. Petersburg State University
ST. PETERSBURG, Russia

Maximilian M. SCHMIDT-SOMMERFELD
Germany

CERN
GENEVE, Switzerland

Horst STOECKER
Germany

GSI
DARMSTADT, Germany

Gerardus 't HOOFT
The Netherlands

Utrecht University
UTRECHT, The Netherlands

Samuel C.C. TING
USA

MIT
CAMBRIDGE, MA, US

Steve VIGDOR
USA

BNL
UPTON, NY, US

Andrey VINOGRADOV
Russia

I.V. Kurchatov Institute of Atomic Energy
MOSCOW, Russia

Lucia VOTANO
Italy

LNGS
L'AQUILA, Italy

Horst WENNINGER
Germany

CERN
GENEVA, Switzerland

Ali ZAHABI
Iran

University of Helsinki
HELSINKI, Finland

Farhana ZAIDI
India

Aligarh Muslim University
ALIGARH, India

Antonino ZICHICHI
Italy

CERN, GENEVA, Switzerland
and
INFN - University of Bologna, BOLOGNA
and
Enrico Fermi Centre, ROME, Italy